Hermann Haken: From the Laser to Synergetics

Bernd Kröger

Hermann Haken: From the Laser to Synergetics

A Scientific Biography of the Early Years

Springer

Bernd Kröger
Tübingen
Germany

ISBN 978-3-319-11688-4 ISBN 978-3-319-11689-1 (eBook)
DOI 10.1007/978-3-319-11689-1

Library of Congress Control Number: 2014949170

Springer Cham Heidelberg New York Dordrecht London

Printed on acid-free paper

Springer International Publishing AG Switzerland is part of Springer Science+Business Media (www.springer.com)

Contents

Tables and Figures

Figures

Chapter 1
Introduction

Our conceptions of reality changed dramatically during the 20th century. The theories of quantum mechanics and relativity have altered our view of the composition of matter and of the structure of the universe. During the 1950s fundamental new insights into evolution were provided by the disclosure of the genetic code. This was followed in the 1960s by the development of a new paradigm, when the dynamic interactions of open systems that dominate nature became the focus of scientific research interest. The concept of self-organizing processes caused by "chance and necessity" (Jacques Monod) and the insight that deterministic processes can lead to non-predictable final states aroused wide interest, not only in the scientific community. A highly emotional debate took place about whether mankind would enter into a new historical phase, creating a "new dialogue with nature" (Prigogine and Stengers). Belonging to these theories about self-organization is "synergetics", the new "science of working together" that Hermann Haken formulated in the 1970s.[1]

While quantum-mechanics,[2] relativity theory[3] and the genetic code[4] have been studied in detail by historians, this has yet to be done for the topic of self-organization.[5] The first beginnings can be found in the 1980s[6] but a comprehensive analysis is not yet available. One of the reasons might be that the term "self-organization" is not defined in a way that is unambiguous, and it comprises a broad range of topics. Its broad spectrum is indicated by keywords like "general systems theory", "cybernetics", "autopoiesis", "hyper-cycle theory", "dissipative Systems" and "synergetics".

[1] He used the word *Synergetics* already in 1971 in the title of his first publication on the subject. (H. Haken & Graham, 1971).

[2] There exist numerous publications on the history of quantum-mechanics. See for instance: (Cassidy, 1992), (Darrigol, 1992), (Hermann, 1969), (Hund, 1984), (Jammer, 1989), (Kragh, 1994), (Mehra & Rechenberg, 1982ff.) An indispensable survey of the current literature is given by the annual review Isis Current Bibliography.

[3] On the history of relativity theory see: (Holton, 1973), (Pais, 1986), (Mehra, 1974), (Miller, 1981), as well as (Hentschel, 1990).

[4] See (Crick, 1981), (Watson, 2003), (Judson, 1996), (Olby, 1974), (Wilkins, 2003).

[5] See chapter 9 of this book.

[6] See especially (Krohn, Küppers, & Paslack, 1987).

B. Kröger, *Hermann Haken: From the Laser to Synergetics*,
DOI: 10.1007/978-3-319-11689-1_1

The initiative for this work came from Klaus Hentschel, who informed the author about the theoretical physicist Hermann Haken, one of the originators of quantum-mechanical laser theory and the founder of synergetics. Hermann Haken has been generous in his willingness to let me inspect and use his private archive[7] and the documents from his personal file at the University of Stuttgart, which tipped the balance in favour of writing this biography. These primary sources give new insights into the life and work of Hermann Haken, especially into the early scientific developments of his career.

Some words about the scientific career of Hermann Haken might be helpful to give a better understanding of why I have concentrated on the early development of synergetics. In 2012 Hermann Haken celebrated his 85th birthday. He studied mathematics and physics at the Universities of Halle and Erlangen where, in 1951, after receiving his doctoral degree in mathematics, he started his scientific career. His theoretical research on excitons[8] – a topic of solid state physics – gained him an international reputation, and he was invited for research stays in Great Britain and the United States of America. It was a fortunate coincidence that Haken was staying at the famous Bell Laboratories in New York when the discovery of the laser was announced in 1960. While in the USA, Haken received and accepted an offer to become Chair of Theoretical Physics at the *Technische Hochschule Stuttgart*. He immediately started a research programme to develop a quantum-mechanical theory of the laser. In a race against time with two other US-American research groups, Haken and his co-workers solved the problem. In 1970 he finally published his comprehensive English text "Laser-Theory" in the series "Handbook of Physics". While thoroughly investigating the theoretical features of the laser in 1969, Hermann Haken and his doctoral student Robert Graham (born 1942) noticed that the creation of coherent radiation at the laser threshold resembles a phase transition; but, different from phase transition in closed systems (for instance, when helium is cooled to near absolute zero and the phenomenon of super-conductivity occurs), the laser is an open system, which is continuously exchanging energy with the environment. Haken now turned to the question of what conditions must be fulfilled to allow such dynamic systems to develop stable and ordered states. Looking for analogies in different scientific areas, he observed that other phenomena in physics, chemistry and even biology can be described with the mathematical tools he had developed for describing the laser. In the early 1970s, this led him to postulate synergetics, the theory of the self-organizing interaction of parts in open systems. He could demonstrate that many open systems far from thermal equilibrium can be described by few "order parameters" that determine the behaviour of the overall system. At the same time, it was

[7] In the following "Archive Haken".

[8] An exciton is the name of a "quasi"-particle within a solid (semi-conductor or insulator), composed of an electron and a "hole" (missing electron). The exciton can move like a normal particle within the solid. In his early years, Hermann Haken had been intensely occupied with the mathematical theory of excitons and became a renowned scientist in this subject.

realized that such open or "dissipative systems" (Prigogine) are abundant in nature, while closed systems, until then mostly dealt with in physics, represent special cases. Haken started an interdisciplinary and multidisciplinary research programme to analyse and demonstrate the analogies and regularities in as many scientific fields as possible. He initiated the later named ELMAU Conferences to which he invited leading scientists from all over the world in an annual or bi-annual interval. This was made possible by a generous grant from the *Volkswagenwerk Foundation*, which supported synergetics as one of its priority programmes over a period of ten years, starting in 1980.

The structure and breakdown of synergetics is shown by a chart (Fig. 1) by Hermann Haken:[9]

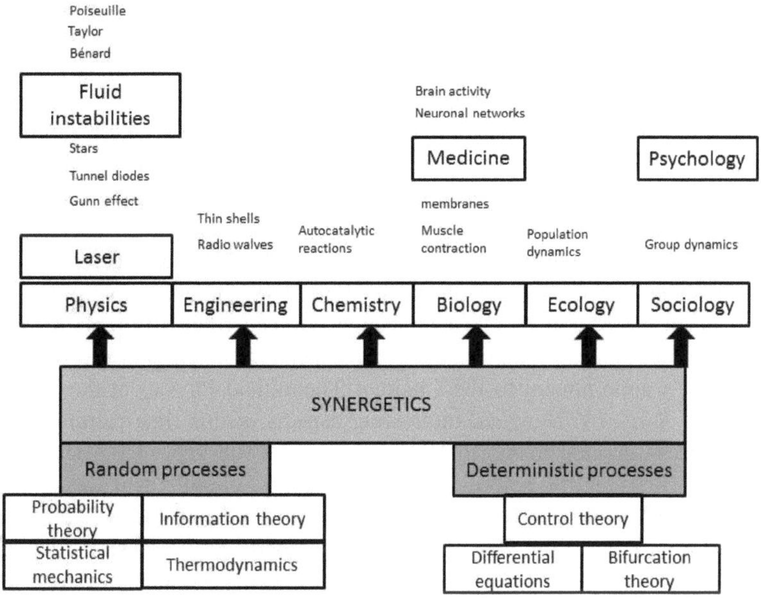

Fig. 1 Theories that determine the structure of *Synergetics* and the resulting applications of synergetics in different scientific fields

In the 1980s and 1990s, the phenomena of pattern recognition and pattern formation in the visual system, as well as questions of movement coordination, directed Haken's research interest to the medical and physiological fields. It was inevitable that he finally became interested in human brain research, the human brain being the most complex synergetic system in existence. He was the first to propose that thoughts might be the order parameters of the neuron network of the brain. Lastly, during the 1990s synergetics brought fresh ideas to the field of

[9] Re-drawn from a sketch of Hermann Haken, dating from the early 1980s. Supplemented by the fields of medicine and psychology (Archive Haken). See also page 197.

psychology, introducing to it the notions of phase transition, order parameters and the harmonic cooperation of sub-systems in the sense of self-organization.

Hermann Haken's research activities spanned more than sixty years. During the whole period he stayed at the University of Stuttgart, rejecting several invitations to take up highly esteemed chairs at other institutions. He published more than 600 articles and wrote 23 textbooks, many of which were translated into foreign languages. His extraordinary accomplishments in solid state physics, laser theory and *Synergetics* were acknowledged throughout the world. He received seven honorary doctoral degrees, became a member of thirteen scientific academies and, in 1984, was elected member of the exclusive *Order pour le Mérite*.[10]

Against this background, an exploration of the special aspects of Hermann Haken's activities seems necessary. We focus on the early years of synergetics up to the middle of the 1980s and emphasize the role that laser theory played in the development of this scientific endeavour.

Chapter two presents some unknown biographical details on Hermann Haken's youth that can be deduced from various sources and from his interviews. He moved from Halle in Saxony to Nuremberg-Erlangen where he first was student and later on assistant at the University of Erlangen, as described in **Chapter three**. During this period he got in contact with the renowned solid state physicist Walter Schottky, who was working at the research laboratory of the Siemens Company at Pretzfeld. At the end of this period Haken received an invitation from the theoretical physicist Herbert Froehlich, who invited him to participate in a research stay in Cambridge (Great Britain). Froehlich became a longtime friend and important discussion partner for Haken. **Chapter four** deals with Haken's appointment to the Chair of Theoretical Physics at the *Technische Hochschule Stuttgart*. New and interesting aspects of this time resulted from the right of access granted to me to Haken's personal file kept at the University of Stuttgart.[11] The scientific activities of the "Stuttgart School" (as it was called later on) with regard to the development of the quantum-mechanical laser theory is the central topic of **Chapter five**. It came as surprise that no detailed historical description concerning the development of laser theory existed. Because the laser played a central role in Haken's thinking, this topic is elaborated on at some length and the race between the research groups is explained: the US American Willis Lamb Jr. (including Marvin Scully), the US American Melvin Lax (including William Louisell) and Hermann Haken (in cooperation with Wolfgang Weidlich, Hannes Risken, Herwig Sauermann and Robert Graham, as well as other members of the "Stuttgart School").

The development leading towards synergetics and its genesis are portrayed in **Chapter six**. Central to this was the insight that the laser effect corresponds to a phase transition in an energetically open system. Phase transition theory was a highly discussed topic in many physical fields and a continuing subject of

---

[10] The number of living German members is restricted to 40 persons that should be distributed equally to the three different sections of the order.

[11] Once again I would like to thank Hermann Haken for this generous permission.

scientific conferences at that time. Open systems, named "dissipative systems" by the Russian-Belgian chemist Ilya Prigogine, had been examined in chemistry, especially the Bénard phenomenon (the self-organized creation of convection cells in fluids) and the Belousov-Zhabotinsky reaction (a periodically oscillating change in colour). At the same time, at several interdisciplinary conferences in Versailles running under the title of "From theoretical Physics to Biology", the question of whether life can be explained solely by the laws of the natural sciences was discussed. Biophysicist Manfred Eigen from Goettingen answered this question in the affirmative and in 1971 presented his "hyper-cycle" model, an auto-catalytic explanatory approach to prebiotic genesis. Haken had got to know Eigen and Prigogine in the context of the Versailles Conferences and he was an active participant at these meetings. At the invitation of Manfred Eigen, Haken was a constant guest at the annual Winter Seminars organized by Eigen at the Alpine ski resort of Klosters, where questions of biogenesis and systems theory were discussed intensely.

During this time, Hermann Haken gained the conviction that in nature many open systems can be mathematically described by the methods he had developed for laser theory (an open system as well). In 1971, he published (jointly with his student Robert Graham) an article entitled *Synergetik – die Lehre vom Zusammenwirken*[12] in the popular science journal *Umschau.* In the following year, 1972, he initiated the first of eleven ELMAU Conferences where he discussed the applications of the new concept with other scientists. The basic insight was that in an open system, at a critical point where the system experiences a phase transition, a breaking of symmetry occurs as well and the system takes on a stable state, driven by fluctuations. A certain state (in scientific speech "modes") dominates the development of the system via self-feedback, leading to the coherent behaviour of the other parts of the system: they "order" the system, hence the name "order parameter". The other parts of the system that follow the order parameters are going to be "slaved", to use one of Haken's terms.[13] This mechanism leads to a drastic reduction in the parameters necessary for the mathematical description of the system. This is why highly complex systems, for instance in chaos theory, can be described by only three variables (or order parameters). In 1975, Haken's confidence in his synergetic approach increased when he succeeded in calculating the Bénard phenomenon, the Belousov-Zhabotinsky reaction and the Lorenz equations of hydrodynamics with the help of the mathematical tools he had invented for the laser problem. This research period was concluded with the publication of his fundamental textbook "Synergetics: An Introduction" and by the organization of the second ELMAU Conference.

It was only from 1977-78 that Haken actively started to promote the word "synergetics" as a new "science of working together". **Chapter seven** describes in detail the three strategies used by Haken. For one thing, he gave as many talks and lectures as possible at different conferences and other meetings, relating the

---

[12] Synergetics – the science of working together.

[13] A phrase of Haken without any anthropomorphic connotations.

synergetics method to the conference topic in question. In many cases that led to new insights into the subject. When synergetics was chosen as priority programme for the *Volkswagenwerk Foundation*, the grant allowed him to organize annual resp. bi-annual meetings on the applications of synergetics at Schloss Elmau in the Bavarian Alps. The third ingredient of his promotion strategy was the book series, "Springer Series in Synergetics" with the Springer scientific publishing company, which, with Haken as its senior editor, grew to more than seventy-five volumes. In addition to these many activities, he and his co-worker Arne Wunderlin extended the mathematical tools of synergetics during this time. In particular, the generalized Ginzburg-Landau equations and the method of the "slaving principle" were put on a sound basis. In addition, the different research activities at Haken's institute are described in this chapter.

From the middle of the 1980s the Haken's interest shifted towards the fields of medicine, physiology and psychology. Even though this time is not the main period investigated in this book, in **Chapter eight** we present the most important lines of development and the key actors in Haken's work for the sake of completeness.

Chapter nine describes the role of synergetics in comparison to other approaches to self-organization theories, but we want to emphasize that the history of modern theories of self-organization is highly complex and has only been investigated in a rudimentary way. The definitive history of this subject needs still to be written. Therefore, this chapter tries to collect different references from the work of Hermann Haken where he expressed his views on other theoretical system *Ansätze* ("approaches").

A short summary of the scientific part of Hermann Haken's life and work is given in **Chapter ten**, which is followed by five appendixes, including a comprehensive bibliography of Haken's publications.

This book is the first biography of the life and work of Hermann Haken. However, a nearly unknown first-hand biographical source was unexpectedly found: the book by Hermann Haken entitled *"Nel senso della sinergetica*", published in Italian, turned out to be a popular-scientific autobiography.[14] The German manuscript of this work published in 2005 was found in Haken's archive and this autobiography will be cited in this biography.

The history of laser theory and of synergetics have not been written by other authors. Haken himself has published some articles on this subject, for instance *"Geschichte der Synergetik"*[15] and especially *"Entwicklungslinien der Synergetik 1 und 2"*.[16] But these works are not historical narrations;[17] instead, they describe

[14] (Hermann Haken, 2005).

[15] (Hermann Haken, 1988c). "History of Synergetics".

[16] (Hermann Haken, 1988a), (Hermann Haken, 1988b). "Lines of development of Synergetics 1 and 2".

[17] Regarding the problematic nature of the writing of history by scientists involved in the subject, see (Brush, 1995) and (Heilbron, 1987).

the logical developments of Haken's research activities and thoughts and do not take much account of parallel developments by other scientists.[18]

Due to the missing literature on the history of synergetics, Haken's scientific publications form the primary source of this book. A rough classification gives the overview shown in Table 1:

Table 1 Quantity of Haken's publications during the years from 1950 to 1990, sorted by subject (10-year periods).

Period	Solid State	Laser	Synergetics	Mathematics	Other
1950–1960	30	-	-	1	-
1961–1970	7	33	-	6	4
1971–1980	31	10	31	19	9
1981–1990	4	19	124	25	35

The interviews performed with Hermann Haken and his colleagues are another important source of information. Between September 2010 and October 2012 five detailed interviews, lasting several hours in total, were carried out. These interviews have been recorded, transcribed and subsequently, after revision by Haken, released and deposited at the University of Stuttgart Archive. Additional interviews were conducted: with Haken's longtime colleague and friend Wolfgang Weidlich and with his former students Robert Graham, Fritz Haake, Rudolf Friedrich and Herbert Ohno. Other interview partners were the biophysicist Manfred Eigen and Ruth Winkler-Oswatitsch.[19]

Hermann Haken's personal archive at the University of Stuttgart allows many additional insights into his work. But it has to be mentioned that its content is highly fragmentary. Many documents were lost in the removal of the physics institutes from central Stuttgart to the new campus in Stuttgart-Vaihingen. In addition, Haken did not file his letters and only kept material he needed for lecturing and talks. Nevertheless, some important documents could be recovered, for instance, a list of participants and speakers at the important conference on "optical pumping" at Heidelberg in 1962. Here, Haken met with Willis Lamb Jr and discussed the position of semi-classical laser theory. No published conference proceedings and no other written records exist of this event. Also found was the handwritten plan for the 1969-70 lecture on systems far from thermodynamic equilibrium, where Haken coined the word "synergetics".

---

[18] See Chapter nine.

[19] The transcriptions of the interviews are deposited at the *University Archive Stuttgart* (Archive Haken) as well.

Chapter 2
Hermann Haken: Youth and University Education

Hermann Paul Josef Haken was born 12 July, 1927, in the German town of Leipzig. He was the eldest son of Karl and Magdalena, who were Roman Catholics. The maiden name of his mother was Vollath, explaining the dedications in some of his later books.[20] However, the place of birth of the young Hermann was not where his family lived. Their home was in Halle, some 40 kilometres from Leipzig, where his father worked as a librarian at the university. In the late 1920s, life was difficult for the largest part of the population due to the world economic crisis. As life in the countryside was in some respects easier than in the towns, his parents decided that he should spend a large part of his youth in lower Bavaria, the homeland of his mother.[21] Many years later, Hermann Haken remembered that the profession of his father meant that early on he came into contact with literature, although, as he remarked, "more or less belles lettres and philosophical books"[22] and not often scientific works.

On the occasion of his induction into the *Heidelberger Akademie der Wissenschaften*, Haken provided unique insight into his family history:

> My paternal ancestors originally came from Scandinavia and then moved to Pomerania. My father, who had been Librarian (*Bibliotheksrat*), from his early youth was interested in for genealogical research. He found traces dating back to the fifteenth century. Our ancestors were quite often Protestant pastors or lawyers. After my grandfather's parents died very young, he and his brother grew up in the house of his uncle, who was the mayor of Stettin. By the way, the *Haken-Terrace* in Stettin is named after

---

[20] See his famous book of 1977, *Synergetics.* Hermann Haken, *Synergetics - An Introduction. Nonequilibrium Phase Transitions and Self-Organization in Physics, Chemistry and Biology* (1977). Dedicated "To the Memory of Maria and Anton Vollath".

[21] See CV Hermann Haken (typewritten) for his application as "Privatdozent" at the University of Erlangen, 1956. (Personal file [Personalakte]) Haken, Blatt 51 (Archive Haken; Archive of the University of Stuttgart).

[22] Interview, Hermann Haken from 21.09.2010, p. 5.

B. Kröger, *Hermann Haken: From the Laser to Synergetics*,
DOI: 10.1007/978-3-319-11689-1_2

> him. One of my ancestors, who had been born exactly two hundred years before me, also engaged in the natural sciences. The family on my mother's side came from Bavaria and Austria. They had been civil servants, businessmen and farmers. My parents got to know each other in Leipzig, half way so to speak. They got married during the years of Depression after World War I, and in 1927 I came into the world."[23]

After attending primary school, Hermann Haken, from 1938 until the end of World War II, was a student of the Oberschule der Franckeschen Stiftungen[24] at Halle. On 16 August, 1946, he took his high school diploma.[25] He got the second best grades in nearly all subjects, having not attended school sports.[26]

A chronic disease prevented him from military service. Even many years later he was grateful to the doctor who gave him a medical certificate:

> 1937/1938, I had a chronic intestinal disease. Something like clinical dysentery that had been extremely painful and displeasing. The treating physician was Professor Nitschke,[27] whose son I met later on at the University of Stuttgart where he was a professor of history.[28] Professor Nitschke signed me off military service over the whole war-period. By this, one might say, he saved my life.[29]

In the certificate of eligibility for university entrance,[30] it is noted that "Herrmann Haken wants to become a teacher at secondary school." An affinity for teaching seemed to show up very early.

Already during his boyhood, Hermann Haken showed a great interest in natural sciences and technology. He was especially interested in airplanes, due to the fact that the house of his parents was situated in the entry lane of the military airport of Halle. Also, the book written by Hermann Oberth on planetary rocket travel influenced him strongly. His dream was to become an aircraft engineer, but this was not possible after the war, because in Germany it was forbidden to run an airplane industry.

---

23 Inaugural address at the *Heidelberger Akademie der Wissenschaften 1990* (Archive Haken; Archive of the University of Stuttgart).

24 Academic high school.

25 Personal file (Personalakte) Haken (Archive Haken; Archive of the University of Stuttgart).

26 Copy of the certificate of eligibility for university entrance; 16 August, 1946 (Personal file Haken, pp. 19 and 20; Archive of the University of Stuttgart).

27 Professor Alfred Nitschke (1898–1960), director of the children's hospital at the *University of Halle* during the years 1938–1946.

28 August Nitschke (born in 1926 at Hamburg), from 1960 professor of medieval history at the *Historisches Institut der Universität Stuttgart*, of which he was a co-founder in 1967.

29 Interview with Hermann Haken from 21.09.2010, p. 5 (Archive Haken).

30 See footnote 40.

So, finally, his field of study was decided by his mother. Haken remembered:

> The question which field of study I should take arose at the end of World War II. I wanted to study physics. My parents had given me, as a present, a popular book about physics. I do not remember the title – something like "You and Physics". That had had a certain influence on me. So I told my parents, but my mother said that she disliked it. This is the first time that I talk about it: at that time a rumour had spread that the Russian army – we lived in the Soviet occupied zone – had displaced physicists to Russia. Later on, we were told that they had gone voluntarily. In short, my mother feared I might become a gifted physicist [laughs] and be displaced. She proposed that I should study mathematics. Therefore, as a well-behaved son, I studied mathematics at the University of Halle."[31]

He did his basic studies in mathematics at the University of Halle, where he acquired an education that he described as "tangible classical mathematics given by very good professors."[32] His academic masters were, he remembered,[33] the mathematicians Harry Schmidt,[34] Heinrich Brandt[35] and Heinrich Jung.[36] An assistant professor called Weber[37] urged him to take the required intermediate examination before leaving for the University of Erlangen, which he did in 1948. As mentioned earlier, the city of Halle belonged to the Soviet occupation zone and his parents decided that Hermann should go to his relatives in Nuremberg to finish his studies. However, there was a small problem. At that time, there was no university in Nuremberg where one could study mathematics. Therefore, he had to join the nearby *University of Erlangen* that was easily reached by means of public transport from Nuremberg. Hermann Haken brought with him a letter of recommendation from his academic teacher Brandt from Halle addressed to the

---

[31] Interview with Hermann Haken from 21.09.2010, p. 6 (Archive Haken).

[32] Interview with Hermann Haken from 21.09.2010, p. 6 (Archive Haken). For the history of the *University of Halle* see Hermann-J. Rupieper, "Beiträge zur Geschichte der Martin-Luther-Universität 1502–2002" (2002).

[33] Cited from Haken, Hermann: " Selbstorganisation in Naturwissenschaft, Technik und Gesellschaft" (Festvortrag anlässlich der Tagung der Deutschen Mathematikervereinigung in Halle 2002; preprint (Archive Haken, p. 1).

[34] Harry Schmidt (1894–1951), German mathematician. Professor of applied mathematics at the *University of Halle* from 1945.

[35] Heinrich Brandt (1886–1954), German mathematician. Professor of mathematics at the *University of Halle* from 1930.

[36] Heinrich Jung (1876–1953), German mathematician, disciple of the physicist Walter Schottky. From 1920, he was professor at the *University of Halle*, where he retired in 1948.

[37] Probably Otto Weber, disciple of August Gutzmer.

mathematician Otto Haupt[38] in Erlangen. He was received well and even got a small job as a student assistant.

Because the first years of Hermann Haken's life in Erlangen have not been documented, we give a longer excerpt from the interview:

> *Hermann Haken:* As soon as I was there [Erlangen], I immediately visited one of the professors called Specht[39]:
>
> "I would like to study for a doctoral degree".
> "Any idea about what topic"?
> "Infinite groups".
>
> Then, he gave me a topic that was quite different from what I had in mind with "infinite groups", namely the so-called identity problem with groups. He suggested a book by Reidemeister.[40]
>
> At that time the identity problem was unsolved. Generally speaking, these groups consist of non-commutative finite elements, but can also be of commutative character. These elements are then grouped into arithmetic expressions called "words". Of these "words" some will be arbitrarily set to 1. Then, the problem is that due to these defined relations, one should decide whether two "words" are equivalent or not. I racked my mind about this question and could solve part of the problem. Only years later, I learned that the general problem in question is unsolvable, according to the theorem of Gödel.[41] But this I did not know at that time. I discovered a group of classes where it is possible. But, on the other hand, I noticed also that sometimes it is not possible, because you will get into an indefinite recursion, without an end.

---

[38] Otto Haupt (1887–1988), German mathematician. Full professor of mathematics at the *University of Erlangen* from 1921.

[39] Wilhelm Otto Specht (1907–1985), professor of mathematics from 1948 until 1972 at the *University of Erlangen.*

[40] Kurt Reidemeister (1893–1971), professor of mathematics at the *University of Königsberg*.

[41] Kurt Gödel (1906–1978), German mathematician and one of the most prominent logicians of the twentieth century. The first incompleteness theorem states that no consistent system of axioms is capable of proving all truths about the relations of the natural numbers. For any such system, there will always be statements about the natural numbers that are true, but that are unprovable within the system (i.e. cannot be solved). See also Wolfgang Stegmüller, *Unvollständigkeit und Unentscheidbarkeit. Die metamathematischen Resultate von Goedel, Church, Kleene, Rosser und ihre erkenntnistheoretische Bedeutung* (1973). For biographical data about Gödel see, for instance, John W. Dawson jr, *Das logische Dilemma. Leben und Werk von Kurt Gödel* (2007).

Interviewer: Your article appeared in the journal *Mathematischen Zeitschrift*[42] that is renowned among mathematicians. The topic you dealt with is very close to physics because group theory is very important for theoretical physics.

Hermann Haken: That is an important point you make. At that time, I had already read a book by Weyl[43] – group theory in quantum mechanics. Even before that time in Halle, there was Professor Brandt.[44] Later, I also learned group theory with Professor Specht – the representation theory of groups by means of matrices. [...]
At the end of 1950, I finished my work. But then I read in the journal *Zentralblatt der Mathematik,* that a Russian mathematician – the name is still in my mind – Tartakowski,[45] had solved the general problem. I informed my professor who then said "if that is the case, then you cannot complete your doctorate with your results. So just take your diploma." Originally, I intended to obtain my doctorate on this topic. So I took my diploma. But, in the meantime, I got Tartakowski's article. Thankfully, I was able to read it in its original Russian language. After World War I, my father had a Russian friend whom he taught German. From that time on, my father tried to learn Russian, which he started to do in the mid 1930s. I had to control his learning progress and so, according to my youth, I learned the language even faster than him. We also got to know a family from Kiev, ethnic Germans living abroad. During the time, when the husband had to go to war, his wife, who spoke Russian perfectly, taught me.

Interviewer: So you were able to read the original article of Tartakowskis?

Hermann Haken: Yes. And I found to my satisfaction that Tartakowski did not solve the general problem; only a part of it, a special aspect. So again, I went to my professor and told him. Subsequently he said:

---

[42] " Zum Identitätsproblem bei Gruppen," Mathematische Zeitschrift 56 (1952), 335–362.

[43] Hermann Weyl (1885–1955), important mathematician and physicist. The work in question is probably "Gruppentheorie und Quantenmechanik," Leipzig, Hirsel Verlag. 1928).

[44] Heinrich Brand (1886–1954), professor of mathematics at the *University of Halle*.

[45] Wladimir A. Tartakowski, Russian mathematician, professor at the *University of Leningrad* (St. Petersburg).

> "If that is the case, then you work out some more examples and complete your doctorate". So I finished my diploma work, supplemented by some amendments, and also my thesis. I took my doctoral degree in 1951 in Erlangen in July. I remember it very well because it was the day before my birthday."[46]

During the last year of his studies, Haken got a job as assistant lecturer (*wissenschaftliche Hilfskraft*) at the mathematical institute of the university from summer 1949 till the following summer. Finally, on 14 November, 1950, he took his exam in mathematics and was awarded an A grade ("sehr gut"). His minor subject was theoretical physics. The exam was taken by Helmut Volz.[47] After his examination Haken switched from mathematics to theoretical physics. This was possible because he got the assistant lecturer position from a young lady, "Fräulein Stud. Assessor Edith Bosch", who later was to become his wife.[48] Thus, in the end, he could realise his stated aim and start a career in physics.

After having overcome the difficulties with the work of Tartakowski, as described by Haken, in July 1951, he took his doctoral degree, "Dr. philosophiae naturalis," only seven months later.[49] At that time, the only assistant professor position at the institute was occupied, so Hermann Haken continued as assistant lecturer. Fortunately, the job holder Helmut Tietze abandoned his position and left university only a month later. On the 1 September, 1952, Hermann Haken was finally appointed assistant professor at the *University of Erlangen.*[50]

In the following four years, Haken came to be one of the leading German solid state theorists (see next chapter) and habilitated in the summer of 1956. His professorial dissertation was entitled "Zur Quantentheorie des Mehrelektrons im

---

[46] Interview with Hermann Haken from 21.09.2010, pp. 7–8 (Archive Haken). In his inauguration speech for the *Heidelberger Akademie der Wissenschaften*, Hermann Haken disclosed a peculiar detail concerning his doctoral exam:
"[…] only three months before the examination, I discovered that besides mathematics and physics, according to Erlangen doctoral regulation, I needed a third subject. My future wife, who also studied physics and I got to know at the institute, had also studied mineralogy. So I went to the mineralogy professor and asked him. But he said that in only three months I could not learn enough about mineralogy for the examination. That is why I chose philosophy and had to prepare for Immanuel Kant's work *Kritik der reinen Vernunft.* Here, I got a grade A ("sehr gut"), whereas in physics I only reached a grade C."

[47] Copy of the certificate on the diploma-examination in mathematics at the *University of Erlangen.* (Personal File Haken page 20; University Archive Stuttgart).

[48] Copy of the application for scientific lecturer. (Personal file Hermann Haken pp. 3/4; University Archive Stuttgart). The renumeration was 200 Deutsche Mark a month.

[49] With the best possible German grade: „summa cum laude".

[50] Letter of appointment dated 8th august 1952. (Personal file Haken page 24; University Archive Stuttgart). The renumeration was 3.400 DM per annum plus 606 DM housing allowance.

schwingenden Gitter."[51] Referees of his work were Helmut Volz and the famous physicist Friedrich Hund, at that time teaching at the University of Frankfort/M. Both experts rated the work as "excellent". Volz wrote[52]:

> The present work of Mr. Haken is connected to a range of publications on the same complex of problems. It is a synoptic view of the subject and an essential extension of these works and shows how the obtained results affect different aspects of solid state physics that are in the focus of actual scientific interest. [....]
> The mathematical derivations make use of the most modern methods of field theory. Their mastery, presented in the appendix, shows that Haken has outstanding command of these methods.

Friedrich Hund[53] especially appreciated the scientific methodology of Haken's work[54]:

> For me the principal value of the scientific work of Mr. Haken seems to be that he tackles the problems in question at the beginning in a very wide and general manner. Then, he takes the best obtainable general methods, uses a proper tool that allows a deep insight into the structure of the problem and only then gets the solution of the special problem. This approach preserves the connection to vividness.

In these remarks by Friedrich Hund we have *in nuce* the most concise description of Haken's scientific methodology. He used this approach also in the following decades when working on the problems of laser theory and synergetics.

On 24 September, 1956, Hermann Haken obtained his *venia legendi*, the licence to teach physics at the university and was appointed as associate professor. Volz tried to get a senior assistant position for Haken because several other universities wooed him. In a very short time he received offers for the position as

---

[51] The work was published in an abbreviated form in the journal Zeitschrift für Physik 146 (1956), 527 – 554.

[52] Copy of the expert opinion Helmut Volz, concerning the professional dissertation of Dr. Hermann Haken (Personal file Haken pp. 62/63; University Archive Stuttgart).

[53] Friedrich Hund (1896– 1997), German physicist. He was a PhD student of Max Born in Göttingen, one of the founders of quantum mechanics. For many years Hund, together with Werner Heisenberg, was full professor of theoretical physics in Leipzig. From 1957 onwards, Hund was professor of theoretical physics in Göttingen. See also Manfred Schröder, "Hundert Jahre Friedrich Hund: Ein Rückblick auf das Wirken eines bedeutenden Physikers" (1996).

[54] Copy of the "second opinion concerning the habilitation work Hermann Haken" (Personal file Haken p. 63; University Archive Stuttgart).

associate professor or head of department from the *Bergakademie Clausthal*, the *University of Saarbrücken* and the *Technical University of Darmstadt*.[55] The Natural Science faculty finally succeeded, by switching positions with the Philosophy faculty, in getting a paid lectureship[56] for Haken and thus retaining his services in Erlangen.

In parallel to the first steps in the scientific career of the young Hermann Haken, there were also important developments in his private life. His mother died as early as 1953. Subsequently, his father, who had retired, moved from Halle to Erlangen in the same year.[57]

As we have seen, Hermann Haken got to know his future wife Edith Bosch at the physics institute in Erlangen. She did her doctorate in experimental solid state physics supervised by Rudolf Hilsch.[58] Edith Bosch[59] was four years Hermann Haken's senior and a native of Fürth, a nearby town of Erlangen. They married 5 September, 1953, one year after his appointment as scientific lecturer. It was not long before the family grew: their eldest daughter Maria came into the world on 7 March, 1954, a son Karl-Ludwig three years later on 4 January, 1957, and, finally, the youngest daughter Karin was born in 1962.[60]

---

[55] Application of the faculty of Natural Sciences of the *University of Erlangen* directed to the *Bayerische Staatsministerium für Unterricht und Kultur*, 18.12.1956 (Personal file Haken; University Archive Stuttgart).

[56] *Diätendozentur* means a position that was partially paid for by the number of students attending and in part by a basic salary.

[57] Private communication with Hermann Haken.

[58] Rudolf Hilsch (1903–1972), German physicist. From 1941, he was professor of experimental physics in Erlangen. In 1953, he was chair in physics at Göttingen.

[59] Born 28.2.1923 (Personal file Haken).

[60] Personal file Hermann Haken (Archive Haken, University Archive Stuttgart).

Chapter 3
The Erlangen Years: Solid State Physics 1950 – 1960

The *Friedrich-Alexander University of Erlangen* was founded as early as 1743. Physics, as a separate discipline, existed since 1857. One can track the history of natural sciences in Erlangen according to the biographies of the institution's respective scientists.[61] The first physics professor was Friedrich Kohlrausch, who died only one year later. Skipping this interesting story, we concentrate on the developments during the first decades in the twentieth century, when research in Erlangen started to focus on solid state physics. During these years, the physics chair was in control of members of the *School of Göttingen*, or the students of Robert W. Pohl.[62] In the years 1926 until 1939 Bernhard Gudden[63] was the departmental chair. After 1941, he was succeeded by Rudolf Hilsch,[64] another senior assistant who studied with Pohl. Hilsch, in collaboration with Pohl, constructed a semiconductor amplifier in 1938 and thus can be called one of the first experimental solid state physicists in Germany.

The history of the early years of solid state physics has been analyzed in great detail. To appraise the theoretical works of Hermann Haken during his time in

---

[61] (Wachter, 2009).

[62] Robert W. Pohl (1884 - 1976) was an influential German experimental physicist and from 1919, a professor at the University of Göttingen. His research focused on crystal optics (Farbzentren) and he was one of the pioneering researchers in semiconductor physics (crystal detectors). Pohl was very successful in placing his students on many experimental physics chairs at German universities, as Arnold Sommerfeld did too in theoretical physics at that time. For biographical details see (Teichmann, 2001).

[63] Bernhard Gudden (1892 - 1945) was a German physicist. He completed his PhD in 1919 with R.W. Pohl in Göttingen and then was his assistant. During the period from 1926-1940 Gudden was Professor of Physics in Erlangen and then at the Karls-University in Prag. (Cited after: Mollwo, Erich, „Gudden, Bernhard Friedrich Adolf", in: Neue Deutsche Biographie 7 (1966), pp. 249. [Online]; URL: `http://www.deutsche-biographie.de/pnd117576018.html`).

[64] Rudolf Hilsch (1903 - 1972), was a German physicist. From 1941, he was Professor of Physics in Erlangen. He was physics chair in Göttingen in 1953 until his retirement.

B. Kröger, *Hermann Haken: From the Laser to Synergetics*,
DOI: 10.1007/978-3-319-11689-1_3

Erlangen, we sketch the most important developments in solid state physics until the early 1950s, taking most of the material from secondary sources.[65]

The electrical and magnetic properties of materials had been intensely investigated by physicists during the 19th century. Many empirical values and relations were found and often presented in graphical form; for instance the Wiedemann-Franz law that describes the ratio of the thermal conductivity to the electrical conductivity with respect to the temperature. Due to its practical application, the electrical conductivity of metals was the focus of interest. The advent of atomic theory in the early twentieth century showed that conductivity must have something to do with the mobility of the electrons in the atomic lattice. But why some materials would be good conductors and others nearly not at all could not be explained by the early theories. During the years 1927 to 1935, Arnold Sommerfeld, together with his scholars Wolfgang Pauli, Werner Heisenberg, Felix Bloch, Rudolf Peierls and others, developed a quantum-mechanical theory of metals based on the movement of electrons called "Elektronentheorie der Metalle." Heisenberg, who in 1927 had become a professor of physics in Leipzig, was very interested in solid state physics at that time.[66] His PhD student Felix Bloch, joining him after studying with Erwin Schrödinger in Zürich, considered the wave function of an electron in a three-dimensional lattice and solved the corresponding Schrödinger-equation in his dissertation. For a perfect periodical lattice, the result was a plane wave that was modulated by a periodic function according to the lattice period.[67] The different conductivities of the materials thus were the result of "impurities" within previous understandings of the ideal lattice. These "impurities" had been the central research topics of Robert W. Pohl in Göttingen. But Bloch also found another result. He calculated that, taking into account the Pauli exclusion-principle, the energy levels of the electrons would not be discrete (like in a free atom) but would form so called energy-bands. At about the same time Hans Bethe and Rudolf Peierls demonstrated that the conductivity of a material does not only depend on the electron mobility. The electrons must also be able to find free space in the energy band to move around. If there are no free spots, the material behaves as an insulator.[68] The quantum-mechanical treatment of the crystal-lattice thus explained the difference between conductivity and material being an insulator for the first time. Only shortly afterwards, in 1931, Heisenberg showed that in nearly fully occupied energy bands "missing" electrons ("holes") behave as if they were positive electrons. The combination of electrons and "holes" in solids, called "excitons," became the main research topic of Hermann Haken twenty years later.

---

[65] (Lilian Hoddeson, Braun, Teichmann, & Weart, 1992). (Serchinger, 2008). (Handel, 1996). (Handel, 1994). (Eckert, 1990). (Eckert, 1993). (L. Brown, Pais, & Pippard, 1995), especially the chapter "Superfluids and Superconductors" by A.J. Leggett p. 913 – 966, and (Teichmann, 1988).

[66] (Handel, 1999), p. 13.

[67] (Bloch, 1928).

[68] (Kragh, 1994), p. 368.

In 1931, Heisenberg had another young guest researcher in Leipzig. Due to a grant by the *Rockefeller Foundation*, the British physicist Alan H. Wilson came over from Cambridge. Wilson simplified the Bloch-Peierls theory by assuming that the quasi free electrons in a metal formed open resp. closed shells (electronic band structure; conductivity by valence bands), in analogy to the valence electrons of the atoms.[69] Wilson gave two talks about this subject in Leipzig. The second colloquium was attended by a group of solid state physicists from Erlangen under the direction of B. Gudden. In his obituary of Alan Wilson, Ernst Sondheimer[70] described the situation at that time:[71]

> The experimental situation about metals, semiconductors and insulators was still very confused at that time; thus it was thought that the peculiar resistance curves of germanium and silicon might be due to oxide layers and that these substances, when sufficiently pure, would probably be metals. Wilson gave his theory of the difference between metals and insulators [...] but he left open the question of whether true ('intrinsic') semiconductor exist. In Gudden's view, no pure substance was ever a semiconductor; he believed that their conductivity was always due to impurities acting either as donors or acceptors of electrons."

The knowledge about solid state physics in the early thirties was summarized in a comprehensive review article that in 1933 appeared in the renowned *Handbuch der Physik*, written by Bethe and Sommerfeld (though mostly by Bethe).

Even with this progress in the knowledge and understanding of electric conductivity and certain magnetic properties, one basic phenomenon still waited understanding: superconductivity. Early in 1911, the Dutch physicist Heike Kamerlingh-Onnes found that, if one approaches very low temperatures (several degrees above absolute zero), some materials lose their electrical resistance completely. An applied current (i.e. electrons) will continue to flow without energy loss. In addition, in 1933, Walter Meißner and Robert Ochsenfeld, working at the *Physikalisch-Technischen Reichsanstalt* in Berlin, could demonstrate that at those low temperatures an applied magnetic field from the outside was forced out of the metal. This could not be explained in terms of classical physics. To say it plainly: superconductivity was a great mystery. According to the status of the quantum-mechanical theory at that time, the electron waves should interfere with the impurities of the ideal lattice or with the fluctuating atomic cores. In this way, they would lose some energy leading to a warming of the conductor. But due to some unknown reason the electrons escaped this effect at the very low temperatures.

---

[69] (Wilson, 1931).

[70] Ernst H. Sondheimer (born 1923) was a British theoretical physicist and a disciple of A. H. Wilson.

[71] (Sondheimer, 1999), p. 552.

Like many other young German scientists of the post-Second War period, Hermann Haken was not very fluent in the English language. Before the war, Germany had been the leading language in physics publications. Many foreign scientists, British as well as American, studied theoretical physics at German universities, for instance in Munich, Leipzig and Zürich, and published in German. That was the reason why after the war most students tended to learn modern physics from textbooks written in German.[72] Hermann Haken recalled:

> To study the problems of solid state physics the textbook of Herbert Froehlich "Elektronentheorie der Metalle"[73] was like a bible for us students – also for me. [...] Somehow it had to be the case that the scattering effect of the lattice in the superconducting state had been shut off or could not come into operation. To understand this deactivation one thought that the electrons were coupled together and thus would gain an element of "rigidity" in their movement. [...] Many theoretical models for this interaction between electrons that should have accounted for this rigidity effect had been tried. [...] In our seminars we analyzed all different theories, not only the one from Heisenberg; but even we, as students, remarked that they all came up against a difficulty. Indeed, all these theories could not be used."

From the very beginning, Herbert Froehlich, exercised an important influence over the scientific career of Hermann Haken through his textbook. This even intensified at the end of the fifties. Froehlich was born in 1905 in Rexingen (Baden-Württemberg) and completed his PhD in 1930 with Sommerfeld, the topic having been some problems of the photoelectric effect. After a short period working as an assistant to Sommerfeld, he went to Freiburg where in December 1932 he became an adjunct professor. Only months later, he emigrated to Great Britain where for many years he worked in close collaboration with Walter Heitler and Heinz London. Together they accomplished important contributions to the theory of di-electricity in solids. Finally, in 1948, he was elected professor of theoretical physics at the University of Liverpool, a position he hold until his retirement in 1973.[74]

In respect to superconductivity Haken remarked

> Herbert Froehlich from the University of Liverpool made a sensational proposal which he substantiated with an extensive article.[75] The idea was that the lattice oscillations themselves would produce the necessary interactions for a kind of attraction between the electrons. That would have resulted in the fact that the interactions that produced the resistance on one hand; on

---

[72] (Haken, 2005), pp. 6/7.
[73] (Fröhlich, 1936).
[74] A detailed appreciation of the life and work of Herbert Fröhlich provides (Hyland, 2006).
[75] (Fröhlich, 1950b).

> the other hand would be responsible for the fact that the resistance did not occur at all. Really a sensational concept![76]

Froehlich, however, could not provide a thorough mathematical theory on this subject. But his theory explained the so called "isotope effect" of super-conductivity,[77] stimulating the research in super-conductivity in the early 1950s. Haken was also inspired:

> My own work in solid state physics [...] was strongly inspired by Fröhlich's work. Thus I started by treating models of the interaction between an electron and a single highly excited lattice vibration. Later, I dealt with Fröhlich's model of the one-dimensional superconductor, [...] this, later on, became part of my Habilitations-schrift.[78]

Simultaneously with these developments in super-conductivity theory, in Erlangen there had been developments that had been important for Hermann Haken as well: his affiliation to the research laboratory of the *Siemens Corporation* at Pretzfeld and its most important scientists Walter Schottky and Eberhard Spenke.

In 1948, the *University of Erlangen* established a second physics chair (applied physics) that was awarded to Erich Mollwo.[79] Mollwo as well as Hilsch were pupils of Pohl and originated from the *Göttingen School*. It was not by chance that two solid state physicists were working at the *University of Erlangen* because there was a close collaboration with the laboratories of the *Siemens Corporation*. Later on, in 1953, Hilsch got a call from the University of Göttingen and was elected as the successor of his former teacher Pohl. In Erlangen he was succeeded by Rudolf Fleischmann who turned away from solid state physics and addressed himself to the highly topical fields of nuclear and elementary particle physics.

In 1948, when Hermann Haken came to Erlangen to finish his studies, there had been no chair for theoretical physics. Since 1946, theoretical physics was taught by Helmut Volz,[80] who was appointed to the rank of extraordinary professor.[81]

---

76 (Haken, 2005), p. 7.

77 (Fröhlich, 1950a).

78 (Haken, 2006), p. 3.

79 Erich Mollwo (1909 – 1993) was a German physicist and pupil of R. W. Pohl. From 1948, he was a professor of applied physics at the *University of Erlangen*. (Rechenberg, Helmut, "Mollwo, Erich", in: Neue Deutsche Biographie 18 (1997), p. 7; (online): URL:`http://www.deutsche-biographie.de/pnd137949650.html`.)

80 (Hentschel & Hentschel, 1996). Appendix F: Biographical Profiles. Also short biographical notes of the *University Erlangen-Nürnberg*, `uni-erlangen.de/universitaet/ehrenpersonen/helmut-volz.shtml` (recalled 27.6.2011).

81 Upgraded to Full Professor in 1958.

Helmut Volz was not an educated theoretician. Born in 1911 in Göppingen, he studied mathematics and physics at the *University of Tübingen*. He wanted to become a secondary-school teacher[82] and took the necessary first and second state examinations for this job. But then he decided to go on and complete a doctoral degree with the famous experimental physicist Hans Geiger,[83] the inventor of the Geiger-Müller-counter, finishing his studies in 1935. With the help of a grant from the *University of Tübingen*, he held a postdoc position in Leipzig at the institute of Werner Heisenberg for two years. We can call this his theoretical physics initiation. In 1937, Geiger, who in the meantime had taken a chair at the *Technische Hochschule Berlin-Charlottenburg,* offered Volz a position as assistant professor. He stayed in Berlin until 1944. Volz was exempt from military service from 1940 onwards, presumably through the influence and request of Geiger, and the German "Heereswaffenamt" (army ordnance office) called him back to Berlin. His integration into the activities of the "Uranverein" (uranium society) had never really transpired. We only can state that he worked experimentally with Geiger and presented his professional dissertation in 1943, which was titled "Wirkungsquerschnitte für die Absorption langsamer Neutronen,"[84] a topic that was important for the uranium society.[85]

We can see that Volz, with the exception of the two years he spent as an assistant to Heisenberg, was not a genuine theoretical physicist whose specialty would have been theoretical quantum mechanics. Accordingly, Haken had to teach himself quantum theory[86]:

> It was important to me to understand quantum theory. Because I did not know about an affirmative textbook I had to be reliant on myself. Some years later I got knowledge of a translation of a Russian textbook written by Blochinzew[87] that had taken a didactical approach resembling the one I had chosen".

It is one of the lasting merits of Helmut Volz to have recognized the high potential of Hermann Haken and to get him into a position at the Institute of Theoretical Physics. This change of position from the mathematical institute to

---

82 See (Winnacker, 2006), p. 42.

83 Hans Geiger (1882 - 1945) was a German experimental physicist. He became famous by his collaboration with Ernest Rutherford on the research on scattering experiments with α-particles. He was the inventor of the Geiger-Müller-counter of electrical charged elementary particles.

84 Cited after (Winnacker, 2006), p. 44.

85 Translates into: "Cross-sections for the absorption of slow neutrons."

86 (Haken, 2005), p. 7.

87 Dimitri I. Blochinzew: "Die Grundlagen der Quantenmechanik". Dt. Verlag der Wissenschaften. Berlin 1953.

theoretical physics had a tremendous influence on Haken's the subsequent research. He remembered[88]:

> Why did I start in solid state physics? Hilsch and later on Mollwo had been solid state experimental physicists [in Erlangen]. Hilsch performed experiments in superconductivity. Therefore he and Volz hold a seminar where they discussed the newest articles about the theory of superconductivity. These were mostly publications how to microscopically explain superconductivity. There were articles by Welker as well.[89] He, as far as I remember, tried to explain it with magnetic currents. Now Froehlich demonstrated that superconductivity occurs the indirect way by the activity of the lattice phonons. Later the work of Cooper[90] was published. Thereafter, Bardeen[91] and Schrieffer[92] followed. I gave a talk in this seminar on one of these publications; my wife also delivered a presentation".

Through his training as a secondary-school teacher, Volz was an exceptionally good educator. Haken benefitted from it, although at first reluctantly:

> I would like to mention that Volz was an exceptionally good teacher. I believe that he got an education as a secondary-school teacher and, that's a funny story, he wanted to try this seminar education with me. When I gave talks, he always corrected me. At that time I was really angry with him, but later I noticed that I was really thankful for him. He taught me many things; for instance how to write a text clearly or how to use the blackboard efficiently"[93].

---

[88] Interview with Hermann Haken from 21.9.2010 (Archive Haken), p. 9-10.

[89] Heinrich Welker (1912 - 1981) was a German physicist. Detected the so called III-V semiconductors, especially GaAs and predicted their semiconducting properties.

[90] Leon Cooper (born 1930) was an American physicist. In 1972, he received the physics Nobel Prize for the development of the BCS-Theory of superconductivity. Quantum mechanical pairs of electrons (so called Cooper-pairs) that are the basis for superconductivity are named after him.

[91] John Bardeen (1908 - 1991) was an American physicist. He received the physics Nobel Prize twice: for his development of the transistor and for his contributions to the BCS-Theory of superconductivity. The BCS-Theory was published in 1957. See (Lilian Hoddeson & Daitch, 2002).

[92] John R. Schrieffer (born 1931), is also an American physicist. Developed the BCS-theory together with Bardeen and Cooper. Nobel Prize in 1972.

[93] Interview with Hermann Haken from 21.9.2010 (Archive Haken), p. 9.

Volz advanced the scientific career of Hermann Haken a second time: he introduced him to Eberhard Spenke[94] and Walter Schottky,[95] who worked at the solid state laboratory of the *Siemens Corporation* in Pretzfeld, near Erlangen[96]:

> „[...] Volz had a contract as external employee with Siemens and Spenke planned on writing a book on solid state physics. And he knew that there was a topic called "second quantization"; the work of Fock.[97] Volz commissioned me to translate the article written by Fock. Looking back that has been really formative, because, during finishing my own thesis, I was forced to learn the second quantization thoroughly. Later on the second quantization again and again strongly influenced my work". [...]
>
> „So I traveled to Spenke at Pretzfeld. [The laboratory] belonged to Siemens and had been relocated during the war from Berlin to Franconia due to the air raids. Spenke introduced me to Schottky".

Since 1943, Walter Schottky, at that time fifty-seven years of age, lived and worked in Pretzfeld. Being one of the pupils of the famous theoretician Max Planck, Schottky aimed at a university career. During the First World War, he shifted to the *Siemens & Halske Corporation* in Berlin, where he did electro-technical and basic physics research. After a short interplay as a professor of physics at the *University of Rostock*, in 1927 he made his choice for industrial research and returned to Siemens. Many physical phenomena are associated with his name: the Schottky barrier, the Schottky-emission, etc. He was especially interested in effects that occur at material boundaries. With his theory of the rectification at semiconductor-metal boundaries, he was an important forerunner for the detection of the transistor-effect. In 1950, when Haken got to know him,

---

[94] Eberhard Spenke (1905 - 1992) was a German industrial physicist and co-worker of Walter Schottky at the *Siemens Laboratories* in Pretzfeld. He was the first to extract the purest silicon and was therefore often named "father of the silicon semi-conductor." For life and work see especially (Handel, 1999).

[95] Walter Schottky (1886 - 1976) was an important and influential German industrial physicist who worked with the *Siemens Corporation.* He led groundbreaking research in electrical engineering and semi-conductors. For life and work of Schottky see (Serchinger, 2008).

[96] Interview with Hermann Haken from 21.9.2010 (Archive Haken), p. 11.

[97] Wladimir A. Fock (1898 - 1974)was a Russian physicist. He delivered important contributions to the development of quantum theory. In 1926 he found a generalization of the Klein-Gordon-equation and wrote important articles concerning the theory of many electrons. The publication in question is probably: Fock, Vladimir: "Konfigurationsraum und zweite Quantelung," *Zeitschrift für Physik* 75 (1932), 622 - 647.

Schottky was in his sixty-fourth year of age and a world famous physicist. But he was also known as an introverted scientist who was not easy to deal with in daily life. In science, he had developed his own terminology and thus it was hard to figure out the meaning of his arguments.[98] In fact, in 1929 Spenke had been employed expressly as his mathematical assistant to "translate" Schottky's theories into generally intelligible diction.[99] Spenke came to Pretzfeld in 1946 and established a research laboratory that has been independent from the main laboratory of Siemens in Erlangen. At his laboratory he initiated the transition from germanium to silicon as the most important semi-conductor material. The laboratory at Pretzfeld thus was at the forefront of scientific research in the early 1950s.

Hermann Haken, merely twenty-four years of age, benefited in many respects from his affinity to this top level research:

> In 1951, I too was given a small remunerated contract [of the *Siemens Corporation*], for which I had to travel to Pretzfeld on every Saturday having discussions with Spenke. He was writing a textbook on semi-conductor physics. At that occasion I was introduced to Professor Walter Schottky with whom I later published some scientific articles.[100]

The book in question was published in 1954. In the introduction to his work, Spenke expressed his thanks to the two Erlangen physicists:[101]

> Professor Helmut Volz and Dr. Hermann Haken (University of Erlangen) assisted with a detailed "translation" of the work by Heisenberg on defect electrons into the "normal" language of wave mechanics. Furthermore Dr. Hermann Haken performed a number of calculations, taken as the basis of §11[102] of chapter VII".

His contact to the scientists at the Pretzfeld laboratory led Hermann Haken to the forefront of semi-conductor research. The cooperation with Volz and Spenke nearly constrained him to the quantum theory of solid state physics. Thus, solid state physics has been the starting point of his career on many body problems.

---

[98] (Serchinger, 2008).

[99] See (Madelung, 1999).

[100] (Haken, 2005), pp. 8/9.

[101] (Spenke, 1954), p. IX.

[102] Paragraph 11 is titled "Aussagen des Bändermodells über den Leitfähigkeitscharakter eines bestimmten Kristallgitters." (Results of the band model concerning the character of conductivity of a certain crystal lattice).

> The professor of theoretical physics[103] took notice of me because he entrusted me with a special exercise in quantum theory dealing with aspects of many body quantum theory. The most elegant approach to quantum theory is via the Schroedinger equation. Here, the movement of the electron is described in a wave form, the so-called de Broglie-wave. [...] To simultaneously describe the movement of many electrons one can proceed in two, seemingly different ways. On one hand it is possible to establish a Schroedinger equation for many of those electron waves, or one uses the so called second quantization. In that case, the so called wave field of the electron is quantized for a second time. [...] In the beginning it was not at all clear, I found, how these two methods, the many wave Schroedinger equation view and the second quantization approach were interrelated. This riddle had been solved by Vladimir Fock in the Soviet Union in an elaborately detailed mathematical work. But this article had been written in a non-comprehensible way and my professor instructed me to translate it into a more common language. Luckily, I succeeded in the translation and by this way learned a great amount about the methodology of the quantum field theory. Thus I was one of the first scientists that applied this methodology to problems of solid state and semi-conductor physics.[104]

This gave rise to the first publication of Hermann Haken that appeared (with the co-authorship of Volz) in 1951 under the title "Zur Quantentheorie des Mehrkörperproblems in Festkörpern."

As a postdoc, Haken continued to occupy himself with the interaction of electrons and the atomic lattice of a solid. He developed methods that enabled him to change from the more or less static view of the problem to a dynamic treatment of the lattice. His starting point was the reflections of Herbert Fröhlich and John Bardeen for the solution of the problem of superconductivity:

> The question, whether one can solve the later phenomenon [superconductivity] by a sophisticated treatment of the electron – lattice interaction had been addressed by Froelich and Bardeen. While Froehlich took as a starting point the fact, known from modern field theory, that the quantum mechanical interactions of electrons and the movement of the atomic lattice (phonons) also induces an interaction between the electrons, Bardeen tried to explain the situation via a one

[103] Professor Volz is meant.

[104] (Haken, 2005), p. 8.

> particle model described by some kind of Hartree-Fock-Ansatz. But it appeared that in both cases the applied mathematical tools only allowed a pertubative solution. A satisfying treatment of the problem was not possible and the conclusions were fraught with great uncertainty. Therefore it seems important to look for a simpler mathematical model that can be treated in an exact mathematical way and permits to draw reliable conclusions".[105]

It is especially the last sentence that signals the basic scientific approach of Haken's methodology in problem solving: look out for a "simple" starting point that can be solved exactly and only then try to solve more and more complex questions of the problem. We will see this strategy in the later work on laser theory and Synergetics time and again.

During the following three years, Haken concentrated on the mathematical treatment of the different dynamical motions of a solid. Especially of interest to him the interactions of electrons with "positive holes" called excitons in physics jargon. These combinations were named "Wannier-Excitons" and were pictured as hydrogen-like objects moving within a semi-conductor. It seemed natural to Hermann Haken, having investigated the interaction of an electron with the atomic lattice, to look for the results treating the interaction of an exciton with the lattice:

> If an electron moves through a vibratory and easily deformed lattice it needs, so to speak, to push away atoms resulting into a higher apparent mass of the electron. An electron that moved through the lattice, surrounded by a lattice deformation, was called polaron. Froehlich made several contributions to this problem and therefore the object had been named Froehlich-Polaron. It was obvious for me to deal with the exciton as if it was made of a normal polaron and a "hole"-polaron. But, in addition, it was necessary to develop mathematical tools to take account of the stronger interaction of the polarons. Using the methods and tools that I had learned in quantum field theory I was able to show that not only there was a variation of the masses of the two particles but that there occurs also a direct interaction between the two types of polarons. Later this interaction was named in the literature "Haken-Potential."[106]

The co-operation of and acquaintance with the famous Walter Schottky was highly important for Hermann Haken. Even though Schottky had no great influence on the industrial research of Siemens, he still played an influential role

---

[105] (Haken, 1953), p. 409.
[106] (Haken, 2005), p. 10-11.

in the integration and development of the emerging solid state research community in Germany:

> After World War II, Schottky no longer had a visible influence on technological development, although he still worked for Siemens. But, being the Chairman of the "Halbleiterausschusses des Verbandes Deutscher Physikalischer Gesellschaften"[107] from 1953 onwards, he addresses himself to the new physics discipline [solid state physics] in the Federal Republic of Germany and build up a reputation as science organizer."[108]

Haken could handle the difficult traits of Schottky quite well. Their collaboration resulted in two jointly published articles that appeared in 1956 and 1958.[109] On top Schottky roped Haken in for the activities at the semi-conductor committee of the Association of the German Physical Societies. Its papers were published annually as an anthology, the editor being Schottky. For every article in the series "Semi-conductor problems" Schottky himself wrote a detailed commentary that quite often was much longer than the article itself. Often that behavior delayed publication of the book.[110] In 1957, when Schottky assigned Hermann Haken with the task of writing a review article on the "Actual status of exciton research in semiconductors,", it was seen in the physics community as a high distinction.[111]

In the postwar era, it was not often that German scientists attended foreign conferences. The rapid development of semiconductor research nevertheless made necessary an international conference that took place in Amsterdam from 29 June until 3 July, 1954. Haken was one of the few German solid state theoreticians that participated. He delivered a talk on non-radiative transitions in solids.[112] Another important conference, "The International Conference on Semiconductors and Phosphors," was held two years later in Garmisch-Partenkirchen.[113] It was here where Haken met the important Japanese theoretical solid state physicist Ryoko Kubo[114] for the first time. The two went on to form a lifelong friendship. Also attending was the Russian-French experimentalist Serge Nikitine, who was the director of the Institute of Solid State Spectroscopy at the *Université Strasbourg*.

---

[107] Translates into "Semi-Conductor Committee of the Association of the German Physical Societies."

[108] (Serchinger, 2008), p. 572.

[109] (Haken & Schottky, 1956) and (Haken & Schottky, 1958).

[110] (Madelung, 1999), p. 56.

[111] (Haken, 1957b).

[112] (Haken, 1954).

[113] Footnote on page 167.

[114] H. Haken, private communication. Interview April, 2011. There is little to be found on the biography of Kubo (1920 - 1995). See A. L. Kuzemsky, ‚Biography of Ryogo Kubo', http://theor.jinr.ru/~kuzemsky/rkubio.html, retrieved 19.03.2013 and the obituary in Physics Today: (Suzuki, 1996).

Haken had met him some months before in May 1956 at a conference on luminescence in Paris.[115] Nikitine invited Haken to give lectures at Strasbourg (1957). Later on, they continuously worked together on solid state research problems, with Haken as the theoretician in the cooperation.[116]

Table 2 Mentioning of Hermann Haken in the university calendars of the *University of Erlangen* from the winter term 1952/53 until the winter term 1959/60

Mentioning of Hermann Haken in the university calendars of the University Erlangen[117] Winter term (WS) 1952/53 till winter term 1959/60		
	Faculty	**Lecture**
WS 1952/53	Theoretical Physics Institute Directorate: H. Volz Assistant: Dr. H. Haken	Volz: Ausgewählte Kapitel aus der Quantentheorie Hilsch, Mollwo, Volz: Kolloquium über neuere physikalische Arbeiten
SS 1953	(as above)	Volz: Ausgewählte Kapitel aus der Quantentheorie II Volz mit Ass. Haken: Übungen zur Quantentheorie Hilsch, Mollwo, Volz: Kolloquium über neuere physikalische Arbeiten
WS 1953/54	(as above)	Volz mit Ass. Haken: Quantentheoretische Übungsstunde (Einführung) NN., Mollwo, Volz: Kolloquium über neuere physikalische Arbeiten
SS 1954	(as above)	Volz mit Ass. Haken: Ausgewählte Übungen zur modernen theoretischen Physik NN., Fleischmann, Mollwo, Volz: Kolloquium über neuere physikalische Arbeiten (jedes folgende Semester in dieser Besetzung)
WS 1954/55	(as above)	Volz mit Ass. Haken: Übungen aus der Atomtheorie Volz: Seminar über neuere Ergebnisse der theoretischen Physik (fast jedes Semester)
SS 1955	(as above)	Volz (ohne Haken): Quantentheorie der Absorptions-, Emissions- und Streuvorgänge
WS 1955/56	(as above)	Volz mit Ass. Haken: Quantentheoretische Übungsstunde
SS 1956	(as above)	Volz mit Ass. Haken: Quantenmechanische Übungen mit besonderer Berücksichtigung der Kernphysik

[115] „La Luminiscence des corps anorganiques", Paris. 21.May – 26.May 1956.

[116] (Haken, 1986). Together they published eight papers.

[117] *University of Erlangen*: „Personen- und Vorlesungsverzeichnis der Friedrich Alexander Universität Erlangen". Erlangen, different volumes as of semester SS 1935 till SS 1961.

Table 2 *(continued)*

WS 1956/57	(as above)	Volz mit Ass. Haken: Übungen zur Quantentheorie
SS 1957	(as above) Privatdozent Dr. H. Haken	Volz und Haken: : Seminar über neuere Ergebnisse der theoretischen Physik (fast jedes Semester) **Haken:** Einführung in die Theorie des Atomkerns
WS 1957/58	Theoretical Physics Institute Directorate: H. Volz; Assistant: Dr. W. Weidlich; Privat-dozent Dr. H. Haken	Volz und Haken: : Seminar über neuere Ergebnisse der theoretischen Physik **Haken**: Einführung in die Theorie des festen Körpers
SS 1958	(as above)	Volz und Haken: Quantenmechanische Übungsstunde **Haken:** Relativitätstheorie
WS 1958/59	(as above)	**Haken:** Quantentheorie Volz und Haken: Quantenmechanische Übungsstunde
SS 1959	(as above)	Volz mit Ass. Weidlich: Quantenmechanische Übungsstunde **Haken:** Mechanik mit Übungen
WS 1959/60	Theoretical Physics Institute Directorate: H. Volz Assistent: NN Privatdozent Dr. H. Haken	(Haken in den USA)

In addition to his research activities and the creation of an international network of colleagues working in the same field, Haken had to think on his academic career. His work on the dynamic interaction of excitons with the atomic lattice represented the content of his professional thesis: "Zur Quantentheorie des Mehrelektrons im schwingenden Gitter."[118] At the end of the summer 1956 he received the "venia legendi" for theoretical physics. By special arrangement he was able to teach as a "Privatdozent" at the university. As is custom for every young lecturer, Haken had to prepare the necessary material for his lectures. Table 2 gives a survey of Haken's activities during his years at the University of Erlangen, as they can be reconstructed from the official university calendars. His first lecture was on "Einführung in die Theorie des Atomkerns"[119] during the summer term 1957. One should bear in mind however that the inauguration of Haken as private docent only happened to be very late in September 1956. Therefore it is highly possible that the schedules of lectures had been already

[118] The article appeared in abridged form in *Zeitschrift für Physik* 146 (1956), 527 – 554. (On quantum theory of multi electrons in an oscillating lattice).

[119] Introduction into the theory of atomic nuclei.

printed; resulting in a course taught by him during the winter term that did not show up in the schedule.

Professor Dieter Fick, a former student of Haken at the *University of Erlangen*, took four terms of the "Quantum mechanical tutorial" given by Volz and Haken. He remembers that the course was only attended by three students and was nearly exclusively taught by Haken:

> During the second half of the '50s, quantum mechanics was lectured in Erlangen in a two hour course only every fourth semester. Haken has been the only lecturer to teach it [in the new manner].[120]

In 1956 and 1957, Haken's research activities concentrated on the behavior of excitons at very low temperatures as he tried to resolve the mysterious problem of superconductivity. The main effort had been working out the interaction of Wannier-excitons with the dynamics of the atomic lattice in the ground state (i.e. at low temperature). Performing the calculations one could ascribe to the excitons (bound state of an electron – hole pair) a small or a large radius, resulting in different limit values. We can keep track of Haken's approach in an article from 1957 that shows how masterly he included the newest mathematical methods in his work:[121]

> The results of the conventional variational calculus and of a method by H.J.G. Meyer are given as special cases. Also, for small radii, an improvement of former results of the author is given. [...] For the calculation, it proved comfortable to use the variation principle of Feynman--not in its original form of Feynman's path integrals, but rather in a translation into the common quantum theoretical language".

In these works, Haken came close to the solution of the problem of superconductivity. He remembered:[122]

> Retrospectively I must say that, at that time, I made a silly mistake: before Cooper's article appeared, I wrote a paper on this subject that I wanted to send to [the journal] *Nuovo Cimento*, where I also had published other works on excitons.[123] In this paper I said: if the two particles have the same charge there will be attraction [not repulsion]. In this respect, like Cooper, I was on the right track. Only I assumed that the bound particles would be very close together forming a quasi-exciton-

[120] Private communication with Dieter Fick, December, 2010.
[121] (Haken, 1957a).
[122] Interview with Hermann Haken from 21.09.2010, p. 12.
[123] (Haken, 1956).

polaron pair. There had also been a paper by Schafroth[124] that showed if there were bosons in the superconductor, then one could claim Bose-Einstein condensation and in this way explain superconductivity. Therefore, I thought: if two electrons would be close together they would form a boson and that could explain superconductivity. One mistake was of course that I thought the particles close together whereas Cooper-pairs are relatively far apart. The other point was that I was ignorant about how to include Fermi-statistics. Later I solved this problem, but then it was too late.

One point may be of interest: Bardeen was working at *General Electric*. At *General Electric* there was also a German emigrant by the name of Henry Ehrenreich.[125] I noticed later that he had ordered reprints of my exciton articles. Bardeen had knowledge of my work.

In 1957, the "BCS – Theory of Superconductivity" (**B**ardeen – **C**ooper – **S**chriefer) was published. The theory explains the condensation of electrons in solids at very low temperatures into "Cooper-pairs" that can be described by a coherent quantum mechanical wave function. The three named authors were awarded the Nobel Prize in Physics in 1972.

Haken's research was at the forefront of solid state theory and his contributions were very well recognized internationally. In May 1957 and again in April 1958, he lectured on solid state theory at the *Université de Strasbourg* at the institute of Serge Nikitine[126]. A career opportunity also arose in Germany. Due to a temporary vacancy, he was entrusted with the provisional cover as extraordinary professor of theoretical physics at the *University of Munich* in the period between May 1958 through 30 September 1958. During this time, it was reported later that he gave "a highly appreciated lecture on electrodynamics."[127] In Munich he received an important letter:[128]

[124] Max Robert Schafroth (1923 - 1959) was a Swiss physicist and a pupil and assistant of Wolfgang Pauli in Zürich.

[125] Henry Ehrenreich (1928 - 2008) was a German-American physicist. From 1955 until 1963, he was a researcher at the *General Electric Research Laboratories* in Schenectady (NY), USA). Then, he was Professor of Physics at the *Harvard School for Engineering and Applied Sciences*. Biographical details can be found in his obituary from the *Harvard University* under www.physics.harvard.edu/misc/ehrenreich.html, recalled 19.03.2013.

[126] See footnote 21 of an article that Haken presented at the 8th meeting of the *Société de Chimie Physique* in Paris 1958. (Haken, 1958).

[127] Expert assessment of Professor Fritz Bopp, on the occasion of the appointment in 1960 of Hermann Haken to the TH Stuttgart. (Personal file Haken; University Archive Stuttgart).

[128] Interview with Hermann Haken from 21.09.2010, p. 13.

> that excited me extremely. Professor Herbert Fröhlich, an idol of most German solid state physicists – including myself – wrote, if I could send him over one of my co-workers. By the form of address and the contents I concluded that Froehlich thought that I was a well-established professor. Answering the letter I pointed out that I had no co-worker because I was a young docent, but that I would like to visit him in person. Thus I had a period of residence in Liverpool, together with my wife and our two little children, where we had an interesting and inspiring time.

The necessary application by Haken[129] at the *University of Erlangen* for a three-monthly leave of absence to depart for Herbert Froehlich's Department of Physics in Liverpool was supported by Helmut Volz as well as Rudolf Fleischmann. They saw no problem with his duties in Erlangen because only the semester break after the winter term in 1958/59 and two weeks before and after would be affected. The fact also helped that, starting from the winter term, the assistant position of Haken was filled with Wolfgang Weidlich who came from Berlin to Erlangen. Later on, Weidlich became an important and close colleague of Haken's at the *Universtiy of Stuttgart.* (See chapter 4). Haken's scientific contributions were acknowledged by an invitation to the third semiconductor conference that took place in August 1958 in Rochester (USA).

> ...in 1958 I also received another invitation. This time for a conference. Apparently people had noticed my quantum field theoretical treatment of the exciton. The invitation was by John Bardeen who organized a solid state conference in New York State. John Bardeen (1908 – 1991) was a highly renowned scientist. In the early fifties of the 20th century he had invented the transistor along with William Shockley (1910-1989) and Walter Brattain (1902-1987) and later received the Nobel Prize for it together with the latter mentioned. At the end of the fifties travelling from Europe to the USA occurred by ocean liner not by airplane. I took the train to Amsterdam and, arriving at the harbor, I was stunned by the sheer size of the "New Amsterdam." The ship was much bigger and taller than a multistory house! In New York, my old friend Robert Pohl, called Bobby, took charge of me. We drove with his car, a big American cruiser – a Buick – to Ithaca, where he had a job as an experimental physicist at Cornell University".[130]

---

129 Request from Haken 7 January, 1959 (Personal file Haken; University Archive Stuttgart).

130 (Haken, 2005), p. 12.

At this third semiconductor conference, the follow up-meeting of those at Amsterdam and Garmisch-Partenkirchen, Haken took the chance to intensify his international contacts. His presentation had the title "On the theory of exitons in solids."[131] It seems remarkable that John Bardeen, in his opening address, mentioned the forthcoming talk given by Haken. The conference was attended by some 500 scientists and only about seventy, representing 14 countries, came from outside the United States. Haken met Mel Lax, Serge Nikitine, Henry Ehrenreich, K. Kobayashi and Y. Toyozawa, who later on played important roles in his scientific journey.

Back in Erlangen, duty called. During the summer term in 1959 he gave the course (and tutorial) on theoretical mechanics. But shortly afterwards his articles on exciton theory and his stay at the conference in New York bear fruit. On one side his scientific career got a great boost. He got an invitation for a probation lecture at the *Technische Hochschule Stuttgart*. On the other hand, he received honorable invitations as a visiting guest professor at *Cornell University* in Ithaca, New York, and from the solid state laboratory of the *General Electric Company* in Schenectady. Once again, he had to ask Helmut Volz and the Bavarian State Ministry of Education for a leave of absence.[132] And once again, Helmut Volz backed his former assistant:

> ...because of the fact that the many suggestions concerning the state-of-the-art research methods and the insight given into the American educational system will be of positive influence for the institute in Erlangen after the return of Dr. Haken, I would like to support the application in question warmly and confirm the official interest in his leave of absence."[133]

In September 1959 Haken travelled to the United States with his family. But this time, he traveled first class on the ocean liner Hanseatic because *General Electric* paid for the travel expenses.[134] At that time, Nobel Prize laureate Hans Bethe[135] was Professor at Cornell University. He was an idol for Haken having written the famous handbook-article "*Elektronentheorie der Metalle*" in 1933 when he was cooperating with Arnold Sommerfeld. Haken was invited to Bethe's home privately and also gave a lecture in his seminar on excitons.[136]

---

131 (Haken, 1959).

132 Application for leave of absence, winter term 1959/60. (Personal file Haken; University Archive Stuttgart).

133 Expert opinion Volz of 15.7.1959, concerning the application of Hermann Haken. (Personal file Haken; University Archive Stuttgart).

134 (Haken, 2005), p. 15.

135 Hans Bethe (1906 - 2005) was a German-American physicist. In 1962, he received the Nobel Prize in physics for the explanation of the energy production in stars. Bethe was the leading scientist of the theory of atomic nuclei and was a group leader of the theory division at Los Alamos. Biographical data see (G. Brown, 2006), also (Schweber, 2012).

136 (Haken, 2005), p. 16.

Although Haken got many inspiring impressions, the scientific production had been small during this time: "somehow I felt very tired and couldn't produce something really useful."[137] In the United States, Haken then received an invitation as guest professor to the famous *Bell Telephone Laboratory* that he conceived as a high honor.

> To be invited there was a high honor because these were the world wide leading laboratories in solid state physics. I was invited to the theory group that was internally named at Bell with the number 11 11. The group director at that time had been Phil Anderson[138], who later on was awarded the Nobel Prize for his work on metal-isolator junction. Other famous physicists had been Conyers Herring and Melvin Lax. I was especially pleased to get into contact with Gregory Wannier,[139] who had created the concept of the Wannier-Excitons. This concept had been the starting point of my scientific work and I think had been responsible for my invitation to Bell".[140]

Once again, Haken had to approach Volz and the State Ministry. After his stay at *Cornell University*, he agreed to return to Erlangen to take up his teaching duties. Shortly before Christmas, 1959, he asked for the prolongation of his leave until 31 July 1960,[141] "for the purpose of a research stay at the solid state laboratory of the *General Electric Co.* and the *Bell Telephone Co.* in the USA." He concluded his proposal by remarking:

> I rejected the proposal made by both laboratories to join their staff permanently, as well as the proposal by the solid state group of the Cornell University to extend my guest professorship and will terminate my activity in the USA definitely on 31st of July 1960 at the latest".

---

137 (Haken, 2005), p. 16.

138 Philip W. Anderson (born in 1923) is an American theoretical physicist. He won the Nobel Prize in physics in 1977. He stayed with the Bell Laboratories from 1949 till 1984, serving as theory group director from 1959 to 1961. Autobiographical data to be found at the Nobel Prize organization homepage: `http://www.nobelprize.org/nobel_prizes/physics/laureates/1977/anderson.html`, recalled 20.03.2013. For his work at Bell see (Bernstein, 1987).

139 Gregory H. Wannier (1911 - 1983) was a Swiss-American physicist. From 1961 till 1983, he was a professor at the *University of Oregon*. A short survey of his life and work is given by (Anderson, 1984).

140 (Haken, 2005), p. 18. For Haken's work at *Bell Labs* see chapter 5.

141 Application for extension of my leave of absence from 22.12.1959 (Personal file Haken; University Archive Stuttgart).

By letter dated 12 January, Volz, who in the meantime had become full professor, again supported the application of his absent colleague and the approval of the State Ministry was given on 5 February 1960.

Only four month later, on 17 May 1960, an unexpected turn of the situation occurred for the *University of Erlangen.* Wilhelm Specht, in his capacity as dean of the Natural Science faculty, informed the *Bavarian State Ministry* that Hermann Haken had received a call from the *Technical University of Stuttgart.* He had been offered the chair of theoretical physics in succession of the retired Professor Erwin Fues. The appointment-procedure took a while and on 9 November 1960, Haken informed the State Ministry about his call and asked for dismissal of the Bavarian public service retroactively to 31 August.

What can we learn from this turbulent period?

His time in Erlangen has been formative for Hermann Haken in several aspects. The co-operation with Walter Schottky and Eberhard Spenke was important, because this led him to an intense study of thc quantum field theory of solids and semi-conductors. Especially the solid comprehension of the mathematical methods of the "second quantization" was future-oriented. Haken was habilitated and, not least, Helmut Volz supported him in every aspect so that he could build up connections to an international network of solid state physicists. The person most influential to him had been Herbert Froehlich in Liverpool. All of this resulted in his successful application as full professor of theoretical physics at a renowned university.

Beyond the science world, of course, the most important events had been getting to know and marrying his wife Edith as well as the birth of his two eldest children.

Chapter 4
Appointment to the Theoretical Physics Chair at the Technical University Stuttgart

Early in 1960 Hermann Haken received a call inviting him to become full professor for the chair of theoretical physics at the Technical University Stuttgart. The tenured professorship had become vacant due to the retirement of Professor Erwin Fues who had held this chair since 1949.

The *Technische Hochschule Stuttgart* had been founded as "Real- and Gewerbeschule" in 1829. Specialisation and academisation led to its conversion into a "Technische Hochschule" (Technical University) some forty years later in 1879.[142] Finally in 1900 the school was awarded the right to award doctorates in the technical disciplines. One field of attention for the physics department was "crystal physics", a field we would today call 'solid state physics'. After World War II the experimental part of physics was represented by Hans Kneser and Heinz Pick, the latter being a pupil of Robert Wichard Pohl from Göttingen. Pick had close connections to Gudden and Hilsch in Erlangen, other disciples of Pohl. After short intermezzi as head of the department by Max Abraham and Erwin Schrödinger, theoretical physics was dominated by Peter Paul Ewald from 1921 till 1937. Ewald had studied with Arnold Sommerfeld in Munich. In 1937 he had to flee the Nazi regime, emigrating to the United States. Ewald trained assistants and graduate students who went on to be notable in the field: Fritz London, Hans Bethe, Erwin Fues and Ulrich Dehlinger, to name only a few, all excellent physicists and pioneers in the field of solid state physics. Dehlinger and Fues represented theoretical physics at the *University Stuttgart* after the Second World War.

It is fair to say that the *Technische Hochschule Stuttgart* had an excellent reputation in solid state physics and also had a longstanding tradition in the theory of this subject. Thus it is not by chance that Hermann Haken, who had established a good reputation as a forthcoming theoretician in solid state physcis during the

[142] The history of the TH Stuttgart is given by (Becker & Quarthal, 2004); (J. H. Voigt, 1979) and (J. Voigt, 1981). The situation at the physics department is presented by (Seeger, 2004). See also U. Dehlinger: "Theoretische Physik in Stuttgart 1919 – 1969." Unpublished manuscript. University Archive Stuttgart. Nachlass Dehlinger SN 33, Nr. 62.

B. Kröger, *Hermann Haken: From the Laser to Synergetics*,
DOI: 10.1007/978-3-319-11689-1_4

fifties[143], had been shortlisted for the vacant chair. However the selection process was not without problems, because the recommendation list also contained the names of Wilhelm Brenig[144] and Joseph Meixner. Brenig was even younger then Haken and worked with Heisenberg at the Max -Planck Institute in Munich. Meixner was a disciple of the Sommerfeld School, and in 1960 he was already in his 52nd year of life. Since 1948 he had been a full professor at the *RWTH Aachen* and co-editor of the new editions of the famous textbooks on theoretical physics written by Sommerfeld. Opinions about filling the post were divided among the committee: some wished to have "a colleague who was able to give a full course on the whole subject of theoretical physics".[145] This was a requirement that Haken could not yet fulfill. During the discussion Dehlinger and Pick especially took a stand for Hermann Haken. They emphasised that "a scientist should be found who had mastered the empirical material of modern quantum theory. This special field would be under consideration by the department and would be in better hands with Haken".[146] That argument turned the balance. On 12th of January 1960 the recommendation list was presented to the "Great Senate" of the *Technischen Hochschule,* with Hermann Haken at the top. Again it was emphasised:

> "Dr Haken definitely belongs to the front row of young theoretical physicists working in the field of solid state physics. He has achieved an international reputation especially with his work on excitons in non-metallic crystals. He is the only German theoretician who successfully applies general methods of quantum mechanical field theory to solid state problems. It is desirable that these two specific lines of scientific research supplement the theoretical subjects that had been treated in Stuttgart so far."[147]

The negotiations for the appointment were successfully completed during the summer and Haken took up his post on the first of October 1960. Two assistant lecturers and two auxiliary scientific posts were assigned to the chair. It was a particularity that Haken managed to get money for invited guest professors (half a post). Due to his experience in Liverpool, and in the United States at Bell Laboratories, he had come to value the intellectual stimulus of foreign experts. Part of the agreement was also permission to follow an invitation at the "Institute

---

[143] Interview W. Weidlich, 18.1.2011. Page 9.

[144] Brenig was born in 1930. In 1960, he was only thirty years of age and had not yet habilitated (normally a qualification necessary for a full professorship at that time).

[145] Minutes of the meeting of the Great Senate (Personal file Haken, University Archive Stuttgart).

[146] See footnote 110.

[147] Application of the appointments committee (succession Prof. Fues) to the Great Senate of the TH Stuttgart dated 12th January 1960. (Personal file Haken, University Archive Stuttgart).

for Fundamental Physics" in Kyoto that was led by the eminent Japanese Nobel laureate Y. Yukawa, in March and April 1961. Haken had already received and accepted the invitation in early 1960. The offices of his institute were very small and located at the Azenbergstrasse 12. Small seminars and meetings were held in Haken's office. Because the housing supply in Stuttgart was difficult, Haken also slept in his office for the first few months.

Haken was lucky in choosing his first assistant. Hannes Risken (1934 – 1994), being seven years younger than Haken, had finished his doctoral studies in Aachen with a thesis in solid state physics entitled "Zur Theorie heißer Elektronen in Many Valley Halbleitern".[148] Risken worked at the *Philips Zentrallaboratorium* in Aachen while performing his PhD work. His contact with Hermann Haken had probably come about through Dr H.J.G. Meyer, the director of the Philips Laboratory at Eindhoven, with whom Haken had been in close contact since 1954.[149] Risken was soon triggered by Haken's enthusiasm for the laser. He and Wolfgang Weidlich, who joined the group one year later, evolved into the most important members of what was later named the "Stuttgart Laser School".

We have a first-hand account from Hermann Haken on Hannes Risken:

> „Dr. Hans [sic!] Risken joined my institute 1962. [...] He was first interested in the calculation of cavity modes where he developed a new method, which allowed to find analytical expressions for the mode forms and in particular for the mode losses. His calculations are the most accurate known at present [...] He then got involved with the problem of noise of lasers by the application of Fokker-Planck techniques. He was the first who derived the photon distribution in the threshold region. [...] Dr. Risken is one of my best collaborators and I have already

---

[148] "About hot electrons in many valley semi-conductors".

[149] Hajo G. Meyer (born in 1924), was a German physicist having Jewish roots. After fleeing to the Netherlands he was deported to Auschwitz by the Nazis. He survived and studied theoretical physics. He was later responsible for the scientific research activities of the *Philips Company* in the Netherlands. The relationship between Haken and Meyer is illustrated by a personal quotation from an interview in September 2010: "There was a conference in Amsterdam. It had been my first participation at an international conference. I gave a talk on radiationless transitions. This had been a highly mathematical talk. Another lecture had been given by Hajo Meyer, also about radiationless transitions but from a more realistic point of view. After the lecture I had a discussion with him. That was really touching because Meyer was a Jew and had been detained in a German concentration camp at only 16 years of age. With the help of a security guard he was able to escape. During his time in the KZ he had taken an oath never to talk to a German again. So by chance I was the first German he talked to and that changed his mind. Later on we met quite frequently and developed a close relationship. From the scientific point of view it did not matter, but it was important from the human side". (Interview with Hermann Haken from 21.9.2010; Archive Haken; University Archive Stuttgart, p. 11).

> recommended him for professorships at other german universities. He is a very easy going fellow and it is always a great pleasure to cooperate with him."[150]

The scientific collaboration of Risken and Haken will be described in the following chapter. Risken had been awarded a research fellowship by the *Deutschen Forschungsgemeinschaft*[151] when he joined Haken as a scientific collaborator. In 1966 he became assistant lecturer and habilitated with a work titled "*Zur Statistik des Laserlichtes*"[152] in 1967. He then spent a one year post doctorate term as associate professor at the *Department of Electrical Engineering* at the *University of Minnesota* (USA). Haken tried to win him back to Stuttgart him and finally succeeded in 1969. Risken became "Wissenschaftlicher Rat" (H2) and later "Abteilungsvorsteher" (H3) at the *I. Institut für Theoretische Physik* of the (now called) *University Stuttgart.* He became an internationally renowned expert in the mathematical application of the so called Fokker Planck Equation for the laser. In consequence he had been offered a full professorship in theoretical physics at the newly founded University Ulm in 1971, which he accepted. He held this position from 1972 until his early death in 1994.[153]

Due to Risken's research grant Haken was able to appoint his former colleague from Erlangen, Wolfgang Weidlich, to the vacant assistant position. In the meantime Weidlich had habilitated in Berlin with Ludwig, his former doctoral advisor. Ludwig had left Berlin and taken on a chair at the *University Marburg* in 1963. Therefore "in theoretical physics the *Free University Berlin* was nearly deserted and I was happy to accept Haken's offer to come to Stuttgart".[154] This statement corresponds with Haken's view:

> "I knew Weidlich from our mutual time as assistants in Erlangen. Then he went back to Professor Ludwig[155] in Berlin. At the beginning of the sixties the way his professional future was developing there seemed unclear. Thankfully I had an open opportunity for an assistant, and because I personally valued him as a scientist and in his personality I offered and he took this position in [19]63.

---

150 Letter of recommendation'H. Risken from H. Haken dated 20.4.1967 (Deutsches Museum München, Archive, Nachlass Hannes Risken NL 131).

151 „German Research Foundation".

152 „On the statistic of the laser light".

153 (Schleich & Vollmer, 1994).

154 Interview Wolfgang Weidlich, 18.1.2011. (University Archive Stuttgart).

155 Günther Ludwig (1918 - 2007), was a German theoretical physicist. From 1949 till 1963 professor of theoretical physics at the *Free University Berlin*. Since then "ordinaries" at the *University Marburg*.

> In the course of retention processes I always tried to get better and better posts for my co-workers. I made a special effort to find a good position for Weidlich, especially with regard to his habilitation".[156]

Weidlich meant an important boost for Haken, because he had a thorough know-ledge of the "Second Quantization" which he had learnt in Berlin with Ludwig. Weidlich remembered the situation:

> "From 1959 till [19]63 I had been in Berlin. I used this time as efficiently as possible occupying myself with axiomatic matters and relativistic quantum theory. Haken was at home with non-relativistic theory [...] and with the second quantization. He benefitted very much of this knowledge in the development of the laser theory".[157]

There are two different approaches to tackling problems in quantum mechanics: the so called Heisenberg-Ansatz approach, using operators and the Schrödinger-Ansatz approach using a wave-equation. Both approaches finally get the same results. Haken loved to work with Heisenberg-Ansatz, whereas Weidlich was a skilled physicist working with Schrödinger-Ansatz and the corresponding master equation. In cooperation with Risken, who had been an expert in the Fokker-Planck-equation (a derivative of the master equation) the "Stuttgart School" had a comprehensive mathematical tool box at its disposal to solve the coming theoretical physics problems.

> "At the Azenbergstrasse we had a seminar in my office. Weidlich and others regularly participated. In this way our scientific collaboration started. I [Haken] pursued the subject with the quantum mechanical Langevin-equations. Risken, who has been mentioned already, interpreted the topic with the semi-classical Fokker-Planck-equation and Weidlich introduced another aspect with the density matrix-equation or, as it is otherwise known, the master equation".[158]

At the beginning of the 1960s two other developments influenced the professional situation for Haken in Stuttgart: on one hand, due to the recommendations of the "Wissenschaftsrat"[159], there was a forced expansion of the *Technischen Hochschule* into a "full-sized" university (more faculties, not only those of the technical and natural sciences). Secondly he received "calls" for tenure professorships at the universities of Bonn and Muenster.

At the end of the 1950s the German government accelerated its efforts to expand the science educational system. It seemed necessary to boost the international competitiveness of the Federal Republic of Germany. The "Wissenschaftsrat"

---

[156] Interview Haken, 16.11.2010, P. 6 (University Archive Stuttgart; Archive Haken).
[157] Interview Weidlich, 18.1.2011, P. 10 (University Archive Stuttgart; Archive Haken).
[158] Interview Haken, 16.11.2010, P. 6 (University Archive Stuttgart; Archive Haken).
[159] Science Advisory Board of the German Government.

(science advisory board), founded in 1957, was set the task "to develop a master plan for the promotion of the sciences […] and to specify and set priorities". After only three years, in 1960, the first report was issued "Empfehlungen des Wissenschaftsrates zum Ausbau der wissenschaftlichen Einrichtungen"[160]. This volume dealt with the scientific universities. The status quo and future demand for every German university and technical high school was shown meticulously .[161] The recommendations of the board had largely been implemented "one to one" in the years to come. In early 1965 Ludwig Raiser (1904 – 1980), the chairman of the advisory board, could already declare the realisation of the recommendations for the expansion of the universities "with respect to the personnel upgrading has been nearly finished. The State budgets show that from 1961 until the budgetary year 1964 (included), 1091 (1960 recommended: 1217) chairs, 2594 (1960 recommended: 2556) posts for non-professional teaching staff […] and 5145 (1960 recommended: 5557) assistant posts had been created".[162]

TH Stuttgart

Disziplin	Bestand 1960	Vom Wissenschaftsrat empfohlen
	Fakultät für Natur- und Geisteswissenschaften	
	Abteilung für Mathematik und Physik	
Mathematik	2 Ord. Mathematik 1 EO Mathematik 1 Ord. Darstellende Geometrie 1 EO Instrumentelle Mathematik	1 Ord. Mathematik 1 EO für Spezialrichtung der Mathematik
Physik	1 Ord. Theoretische Physik 1 Ord. Experimentalphysik 1 Ord. Physik 1 Ord. Angewandte Physik	1 Ord. Theoretische oder Experimentalphysik
	1 Ord. — kw 131 GG — Kernphysik	kw-Vermerk streichen
	1 EO Röntgentechnik 1 EO Festkörperphysik	

insert

Fig. 2 Recommendations concerning the number of chairs in mathematics and physics at the *Technische Hochschule Stuttgart*, according to the science advisory board proposal

[160] "Recommendations of the science advisory board on the expansion of the scientific facilities".

[161] (Wissenschaftsrat, 1960).

[162] Cited after (Bartz, 2007), p. 68.

The resulting proposal for the department of physics at the *Technischen Hochschule Stuttgart* is shown in the illustration above.[163]

It was recommended that the physics department should get two more chairs. The sentence "kw-Vermerk streichen"[164] (cancel the kw-note) proved advantageous for Wolfgang Weidlich, because Professor Steinke, the holder of this chair, died unexpectedly in November 1963. Consequently, the chair was not cancelled, and the faculty took the chance to reorient and look for a theoretical nuclear physicist.

> "The reason is primarily that the "Radiation Physics Institute", performing experimental nuclear physics, is in urgent need of a theoretically oriented colleague with whom they can undertake the planned research, especially at the 4MeV-elementary particle accelerator. [...] In addition to this task, representing theoretical nuclear physics in research and teaching, the new "Ordinarius" should also contribute to the cycle of theoretical basic lectures. On top of this it was desirable that the candidate would not only be interested in his special subject but would also be interested in solid state physics, a strong research focus of the Stuttgart physics department".[165]

In describing the competence of Wolfgang Weidlich who was at the top of the recommendation list it was emphasised:

> "Herr Weidlich deals [in his works] primarily with applications of quantum field theory on problems of nuclear and solid state physics. His work on the theory of multi-channel scattering is especially remarkable".

And the referees highlighted:

> "that Herr Weidlich has a highly developed talent for the exact mathematical demonstration of physical principles by which the problems gain a special clarity.
> [...] his favourable understanding of physical problems and his substantial prowess in handling difficult mathematical methods".

Weidlich was awarded the chair. Within the scope of an internal reorganisation in the department this chair was renamed *II. Theoretical Physics Institute* in 1966. Restructuring of the department was completed with a third "Ordinariate", as

---

163 (Wissenschaftsrat, 1960), p.368.

164 kw = „kann wegfallen"; translates to „to be discontinued".

165 Berufungsvorschlag für die Besetzung des Lehrstuhls für Kernphysik vom 16.11.1965 (University Archive Stuttgart; Bestand 54, Berufungen (1949-1969)).

required by the *Wissenschaftsrat*. Finally in 1969 Max Wagner, a friend of Hermann Haken, was appointed head of the *III. Theoretical Physics Institute.*[166] In company with Risken this provided a fruitful constellation of colleagues being on friendly terms which each other, leading to an intense and, by and large, frictionless research atmosphere.

The expansion of the German universities led to an additional demand for professors to fill the posts and the different departments struggled to find the best experts. Thus Hermann Haken received calls offering him the chair of theoretical physics at the universities of Münster and Bonn. But Stuttgart did not want to lose Haken and immediately started retention negotiations that were settled successfully in December 1963. Haken's institute was staffed with another assistant post to be upgraded in the following year. It was also negotiated that the relocation costs would be paid for, if Max Wagner, being in the USA, accepted the position. Once again the funding of the guest professors had been confirmed and ring-fenced.[167]

Another two years later Haken again received a call from the *Technische Hochschule München*. He declined their request in December 1966. His institute was awarded an assistant lecturer, and furthermore a post for an extraordinary professor for theoretical solid state spectroscopy was created. The latter made it possible that he could keep Robert Graham, one of his best doctoral students, at the *University Stuttgart*, before he finally accepted the offer of the theoretical physics chair at the *University Essen* in 1975.[168]

Hermann Haken's research activities had been very wide. They comprised solid state physics and his specialty in excitons, laser theory, the theory of statistical dynamics and phase transition theory as well as synergetics in all its facets.

Haken's contributions to solid state physics during his years at Erlangen has been already described in the previous chapter. In the context of this book his additional works in the field will not be considered because we concentrate on his scientific development concerning the new field of synergetics. It should be mentioned, however, that Haken not only maintained his contact with Professor Serge Nikitine in Strasbourg, but extended it. From 1969 until 1975 he was appointed associated professor lecturing regularly on theoretical solid state physics at the University of Strasbourg.[169]

---

[166] Max Wagner (born 1931) studied physics in Stuttgart and at the University of Munich. After a 2 year term as post doc research fellow at the *Cornell University* and the *IBM Research Center in Yorktown Heights* (USA), he returned to Stuttgart in 1965 and then habilitated in theoretical physics. (Press release Nr.46/2001 of the University Stuttgart and private communication of Hermann Haken and Max Wagner).

[167] Erhaltungsvereinbarung vom 30.Juli 1963; University Archive Stuttgart, Personal file Haken.

[168] Erhaltungsvereinbarung vom 6. Dezember 1966; University Archive Stuttgart, Personal file Haken.

[169] See lecture notes of Hermann Haken. Archive Haken, Nr. 21.

His research on laser theory will be treated in detail in the following chapter. The theory of phase transitions and the resulting field of synergetics are discussed in Chapter 6.

On the basis of his exceptional position at the University Stuttgart – size of the institute, its financial equipment and its international reputation – and not least that he worked there for 35 years until his retirement in 1995, a large number of pupils and doctoral students originated from his institute. During our investigations we could identify 88 pupils finishing their diploma or doctoral dissertation with him. Their names and the topics of their work can be found in the Appendix. At least eighteen of his 63 doctoral students followed an academic career and received the status of professor.

Chapter 5
Hermann Haken and the "Stuttgart School" 1960 – 1970: Their Contribution to the Development of Laser Theory

5.1 General Introduction to the History of the Laser

The history of the laser has been written by scientists involved and historians several times. Great efforts to trace the lines of development were particularly made in 1985, when the science community celebrated the 25th anniversary of the "birth" of the laser by Theodore Maiman. The most important activity had been undertaken in the United States of America with the *Laser History Project* of the years 1982 to 1988. No less than four important scientific associations joined forces: the *American Physical Society*, the *Institute of Electrical and Electronics Engineer's Quantum Electronics and Applications Society,* the *Laser Institute of America* and the *Optical Society of America*. With regard to the jubilee they interviewed living contemporary witnesses in order to document the historical development. More than 80 interviews were recorded and resulted in the seminal book by Joan Lisa Bromberg:[170] "*The Laser in America*".

As the title declares, Bromberg concentrated on the development taking place in the United States. It was here that the pioneering work had been done, especially in the prehistory of the subject. Some Russian protagonists were also interviewed, as a result of the Nobel prizes that had been awarded meanwhile for their maser and laser research Parallel developments in Europe and in other countries worldwide were given a secondary role or no role at all in the project.[171]

Even before the publication of Bromberg's book the Italian experimental physicist Mario Bertolotti came out with his "*Masers and Lasers – an Historical Approach*"[172]. In 2005 it was succeeded by an enlarged edition.[173] Bertolotti's books are guided by the experimental approach to the laser, a subject in which he

---

The original version of this chapter was revised: Figures 6 and 8 have been updated. The erratum to the chapter is available at DOI: 10.1007/978-3-319-11689-1_11

[170] (Joan Lisa Bromberg, 1991)

[171] The records of the project are kept as *Sources for the history of Laser (SHL)* at the *Nils Bohr Library of the American Institute of Physics (AIP)* in New York.

[172] (Mario Bertolotti, 1983)

[173] (Mario Bertolotti, 2005)

B. Kröger, *Hermann Haken: From the Laser to Synergetics*,
DOI: 10.1007/978-3-319-11689-1_5

had been involved personally. There are many other historically relevant sources, mainly journal and magazine articles and the memories of physicists involved.[174] All these sources concentrate on the technical and experimental development of the laser. Theoretical research is only marginally mentioned.

Sole exception is the habilitation treatise by Helmuth Albrecht "*Laserforschung in Deutschland 1960 – 1970*", where he describes explicitly the development of the laser theory at the "Stuttgart School" led by Hermann Haken.[175] This description is enlarged and refined in the following, made possible by new interviews with the physicists involved and taking into account the personal archive of Hermann Haken.

It should be clear when evaluating the above sources it is noticeable that the development of the maser and the laser, having been accomplished mainly in the United States during the 1960s, is outlined in great detail. But (with the exception of Albrecht's work done in Germany) something is wanting: a description of the parallel developments in other parts of the world, especially in Europe. Another striking point is that the application of the laser in scientific research is given a wide berth. Very little notice is given to the application of the laser in medicine, industry and technology, which is at least equally important. Considering the importance of the laser for these fields this is amazing, and offers a vast open field for future historical research.

In the following chapter the theoretical development of laser research will be examined, starting with the seminal work of Townes and Schawlow in 1958.[176] The "Stuttgart School" and the two US-American theory schools of Willis Lamb Jr. and Melvin Lax will be investigated in detail, particularly in view of the competition between them.

The "Stuttgart School" was disregarded by American researchers and had to fight for a long time for recognition. This disappointed Hermann Haken and caused an element of bitterness.

We follow up the original published papers chronologically and try to shine light on the reasons for this temporarily oblivion.

5.2 From Maser to Laser[177]

One of the most fundamental discoveries of the 19th century was the detection and means of creation of electromagnetic waves by Heinrich Hertz. In the years to come the fundamental social and economic influence of this discovery became

---

[174] (Carroll, 1964), (Fischer, 2010), (M. Bertolotti, 1985), (C. M. Townes, 1999), (Hecht, 2005), (Lemmerich, 1987), (Joan L. Bromberg, 1988 (10))

[175] (Albrecht, 1997). Especially chapter 3.2.2.

[176] (A. Schawlow & Townes, 1958).

[177] For the most part the presentation in this sub-chapter follows (Mario Bertolotti, 1983), (Mario Bertolotti, 2005), (Joan Lisa Bromberg, 1991), (A. L. Schawlow, 1973), (C. M. Townes, 1999).

clear in the technological breakthroughs of radio, the telephone and television. The field of electromagnetic waves thus attracted many leading scientists, and in applied sciences the profession of electro-engineering was established.

Electromagnetic waves, created by oscillating charges, are mostly monochromatic and coherent. That means that they oscillate within a very small bandwidth and, at some distance from the source, are "plane waves". At the beginning of the 20th century it was possible to artificially create electromagnetic radiation with wave lengths of some hundred meters. Every emission and reception of electromagnetic waves is influenced (sometimes extremely negatively) by two types of noise: thermal noise (depending on the temperature) and "quantum mechanical noise" (depending on the number and distribution of photons). Engineers and scientists try to suppress this noise as much as possible because it disturbs their measurements. The continuous struggle against noise and fluctuations in these phenomena was well known in the scientific community.

In the years up to the Second World War technological advances led to shorter and shorter wave lengths, finally reaching the so called "meter-regime". Due to the phenomenon of diffraction of electromagnetic waves (they are "scattered" by obstacles) the radio engineers were not particularly interested in shorter waves. It was common understanding that mountains, houses, forests etc. would absorb or disturb the waves in such a way that the transmission of a signal over a long distance would not be possible. When it became apparent that this was a false conclusion the race for shorter and shorter waves was on.

It is characteristic of electromagnetic waves that they are deflected and reflected by metallic surfaces. Thus the signal emitted by a source and reflected by an object could be received and analysed. However the accuracy of the distance determination crucially depended on the power of the source and the wave length of the electromagnetic wave. Unsurprisingly this led to great efforts in the development of radar-technology during the Second World War. The goal was to create ever shorter wave lengths and higher power output of the source so as to detect aeroplanes and battle ships more precisely at ever greater distances.

The electromagnetic waves were amplified by cavity resonators, their size being adapted to the wave length. Typical dimensions had been just a few centimetres. Arthur Schawlow summed it up[178]:

> „One of the requirements for building an electronic oscillator to generate such short electromagnetic waves is the resonator to tune it. For microwaves, which have length ranging from millimeters to centimeters, tuning is usually achieved with some kind of cavity resonator whose dimensions are comparable to the wavelength. When the desired wavelengths are a small fraction of a millimeter, construction of cavity resonances becomes a difficult task."

---

[178] (A. L. Schawlow, 1973), p. 115.

Over time transmitter and receiver technology became ever more elaborate. Perhaps even more important is the fact that after the Second World War many scientists and experts were available who had genuine knowledge in the field of high frequency technology.

One of these scientists was Charles H. Townes. Born in South Carolina in 1915, he studied physics at the *California Institute of Technology,* where he took his doctoral degree with work on nuclear spin and isotope separation. The United States of America had not entered the war. Townes then joined the research laboratory of the *American Telephone and Telegraph Company (AT&T)* - the famous *Bell Laboratories* - located in Lower Manhattan (New York) at that time. He soon was assigned to radar work[179] and stayed associated with this subject even after the entry of the USA into the war. He was thus not involved in the development of the atomic bomb at Los Alamos, as were many of his colleagues.

After the second world war, Townes turned his scientific interest to molecular microwave spectroscopy, where he could profit from his experience in radar technology. Molecular beam spectroscopy had been developed by Isidor Isaac Rabi[180] during the late 1930s. It allowed high precision measurements by resonance phenomena triggered by radiation transitions of excited molecular beams into the ground state. In 1948 Rabi offered Townes a professorship at *Columbia University* (New York). Townes remembered fifty years later[181]:

> "During the 12 years I was a full-time member of the department, in addition to Rabi, Kusch, and Lamb, other professors there included T. D. Lee, Steve Weinberg, Leon Lederman, Jack Steinberger, Jim Rainwater and Hideki Yukawa; all were to receive Nobel Prizes. Rabi was the only one so recognized when I arrived. Students during that period included Leon Cooper, Mel Schwartz, Val Fitch, Martin Perl and Arno Penzias, my doctoral student who, in 1965, was co-discoverer (with Robert Wilson) of the cosmic background radiation (CBR), the relic photons from the big bang. All these were also to receive Nobel Prizes. Hans Bethe and Murray Gell-Mann were visiting professors there before receiving their Nobel Prizes. Then there were the young postdocs: Aage Bohr, Carlo Rubbia and my postdoc and close associate, Arthur Schawlow, now Nobel laureates."

---

179 (C. M. Townes, 1999).

180 Isidor Isaac Rabi (1898 - 1988) was an Us-American physicist of Austrian origin. In 1944 he received the physics Nobel Prize for the development of the molecular beam resonance method leading to the measurement of magnetic properties of the atomic nucleus.

181 (C. M. Townes, 1999), p. 48.

The further development of the maser and the laser was then highly influenced by the connection of Townes with Willis Lamb and Arthur Schawlow.

Amplification by cavity resonators was the conceptual hurdle required to advance to the electromagnetic wavelength millimetre and sub-regime, as was the usual way in radar technology. It seemed impossible to construct cavities of such a small size with the necessary precision to achieve the amplification.

> "The main problem [...] was that generating millimeter waves by conventional means required a very small resonant cavity. Only a wavelength, or a small multiple of a wavelength, in size. Making precise, delicate parts about a millimeter across is not easy. And to generate significant power one would have to pump considerable power through it, which wasn't easy. It would have to be strong and able to cope with a lot of heat"[182].

The idea for the solution of the problem came to him during a conference some 18 months later. Townes described the moment in his article of 1999[183]:

> "In musing over the problem and his frustration with it, he [Townes] suddenly realized that molecules could produce much more than thermal radiation intensities if they were not thermally distributed but had more molecules or atoms in an upper than in a lower state. Within about ten minutes he had invented such a system using a beam of ammonia and a cavity, and calculated that it seemed practical to get enough molecules to cross the threshold of oscillation. This meant that molecular-stimulated emission at a given radiation intensity would be greater than energy loss in the walls of the cavity".

Until then the whole radiation field, all frequencies, was amplified then a single frequency (or a small frequency band) was selected and the others suppressed. This technology had been highly complex and sophisticated. Townes knew Ansatz planned to amplify a single frequency with the help of the induced emission phenomenon.

The principle of induced emission will be outlined briefly. Normally atoms stay in the most energetic efficient state, called the ground state. When energy is supplied to the atoms they are stimulated to take on different and higher energy states. However, these excited energy states are only stable for a very short time, then the atoms return to the ground state, emitting a photon of the frequency that corresponds to the energy difference between the excited state and the ground state. This is true not only for electronic excitations but also for rotational and

---

[182] (C. M. Townes, 1999), p. 55.

[183] (W. Lamb, Schleich, Scully, & Townes, 1999).

other oscillations because there charges are also accelerated and thus radiate. In his spectroscopic research Townes himself had been occupied with the ammonia molecule. It consists of three hydrogen atoms and one nitrogen atom showing a pyramidal structure, the nitrogen atom being located at the top of the pyramid. Due to the symmetry of the configuration the nitrogen atom easily takes the lower position. There is of course an energy difference between the two states corresponding to a frequency of 23,786 gigahertz or a wavelength of 1.25 cm.

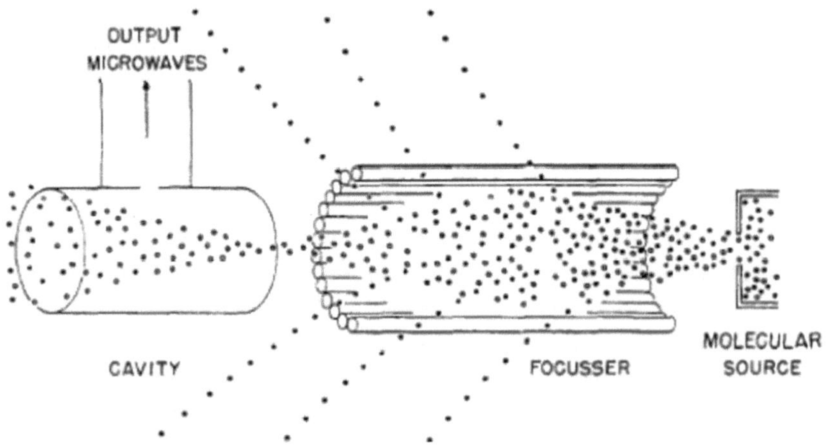

Fig. 3 Principle of the maser. (from (C. H. Townes, 1972), p. 62)

The principle of the maser is illustrated in the picture above. Ammonium rays are heated and thus excited, exiting the oven through a small hole. Atoms still in the ground are deflected away. Only excited molecules enter the cavity resonator. By means of stimulated emission a self-intensifying electromagnetic wave is created, leaving the cavity as a maser beam.

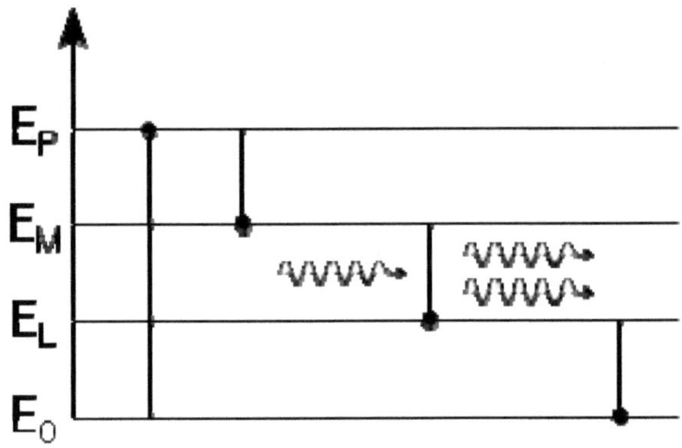

Fig. 4 Principle of the stimulated emission process (E_o ground state)

Stimulated emission is the decisive mechanism creating the maser and the laser effect. The principle of stimulated emission was introduced by Albert Einstein as early as 1916. He published his findings under the headline "Zur Theorie der Strahlung" in the mostly unread *Mitteilungen der Physikalischen Gesellschaft Zürich*[184], and shortly afterwards in the respected *Physikalischen Zeitschrift*.

How does stimulated emission work? Imagine an electron is in an excited ("upper") state. If a photon of the exact frequency corresponding to the energy difference between the upper state and the ground state is induced, the electron drops down to the ground state emitting another photon of the same frequency as the induced photon. Thus the intensity of this frequency is enhanced. Compared to spontaneous emission it is important that the stimulated photon oscillates not only in the same direction but also in phase with the primary photon. Coherent radiation is created.

It took another two and a half years until, in 1954, Townes was able to present, together with his doctoral students James Gordon and Herbert Zeiger, the first operational maser[185]. This team also created the name MASER, which is an acronym for **M**icrowave **A**mplification by **S**timulated **E**mission of **R**adiation.

Similar ideas were advanced by the Russian physicists Nikolai Basov and Alexander Prokhorov working at the *Lebedev Institut*e in Moscow, around the same time, but due to the language barrier their work has gone unnoticed in the United States.[186]

During the years to follow maser research flourished. It received an extra boost when Prokhorov, and Basov, as well as Nicolas Bloembergen, succeeded in constructing a solid state maser.

Only late in 1956 did Townes concentrate on maser research again, after taking a sabbatical working in Europe at the Institute of Alfred Kastler in Paris and in Japan at the *University of Tokyo*. During his stay in Japan he received a crucial stimulus about which he wrote in his scientific memoires[187]:

> "I settled down at the University of Tokyo,[…] As it happened, the faculty there included Koichi Shimoda, who had been a postdoc with me at Columbia and had participated in maser work.[…]
> Also on sabbatical there was another Columbia man, a biologist named Francis Ryan. We had known each other pretty well at Columbia. Naturally, we got to talking. He was studying an unusual paper by a British theoretical chemist, Charles Alfred Coulson, devoted to a treatment of microbial population growth. Coulson wanted to describe, quantitatively, the

---

184 Mitteilungen der Physikalischen Gesellschaft Zürich Nr. 18 (1916) and (Einstein, 1917).
185 (Gordon, Zeiger, & Townes, 1954).
186 The development of this work is described by (Mario Bertolotti, 1983), (Joan L. Bromberg, 1988 (10)), (Prokhorov, 1972).
187 (C. M. Townes, 1999), p. 84.

population fluctuations that occur when microbes are both dying and multiplying at the same time. In his paper, Coulson presented and discussed the solutions to an equation that allowed for both the probability of microbe multiplication by division and also a probability of death.

I recognized immediately that this was exactly the kind of mathematical formulation we needed to understand some aspects of the maser, in which photons are both dying (being absorbed) and being born (stimulated into existence) simultaneously, as the result of the presence of other photons. To Coulson's expressions, I knew I had to add another term to account for the spontaneous appearance of photons in a maser—which contrasts with the fissioning of microbial parents—since for microbes there is no chance of spontaneously creating life! But the basic approach, devised for a problem in a field far removed from physics, seemed just what was needed for a precise theory of noise fluctuations and amplification in a maser."

The drive for shorter wavelengths reaching down to visible light frequencies was widespread among physicists, but three strong reasons seemed to prevent its realisation. Firstly, in molecules the rate of spontaneous emission of energy (photons) increases with the fourth power of the frequency. To achieve higher frequencies and thus shorter wavelengths a disproportional amount of energy must be supplied to get the excited ("upper") states populated.

Secondly, according to the well-known formula,

$$E = h \cdot \nu = k \cdot T$$

(E = energy; h = Planck's constant; ν = frequency of the photon; k = Boltzmann's constant; T = temperature),

a high frequency corresponds to a high temperature. In gas – by means of the proper motion of the molecules – Doppler shift broadens the line width considerably. Last but not least it didn't seem possible to create a cavity resonator the size[188] of a few microns.[189]

Coming back from Japan Townes discussed the insights gained in Paris and Tokyo with his brother-in-law Schawlow[190]. At that time Schawlow was working on the subject of superconductivity at the *Bell Laboratories*. He convinced Townes that a cavity resonator of such a small dimension would not be necessary.

---

[188] One micron corresponds to one part of a million of a meter.

[189] (W. Lamb et al., 1999).

[190] In 1951 Art Schawlow had married the younger sister of Townes. He had got to know her during his time as post-doc with Townes.

A so called Fabry-Perot-Interferometer (two plane-parallel mirrors, mounted at some distance from each other), well know from optics, would do. In the period that followed both scientists occasionally worked on the problem, but discussions with colleagues and reading the literature convinced them that other groups were also working on the problem of an optical maser. On the other hand they knew it would take some time before they could realise a prototype optical maser, which they planned to realise with potassium vapour. They decided to publish their thoughts in an article (but only after the *Bell Laboratories* had submitted a patent application).[191] They submitted their paper titled "Infrared and Optical Masers" on the 26th August 1958 to the *Physical Review* journal, where it appeared in the 15th December issue. This article became one of the most quoted works in laser physics, attracting more than 1000 citations.[192]

In their paper Townes und Schawlow started their calculations by defining the number of atoms that had to be excited (energetically "pumped" to a higher state). The value they found was about 100 million atoms[193] leading to the comment "this number n is not impractically large". They then derived the energy necessary that needed to be fed into the system and declared:

> „the input power required would be [...] 10 milliwatts. This amount of energy in an individual spectroscopic line is, fortunately, obtainable in electrical discharges".

In the next step they tackled the problem of the size of the cavity resonator with its large dimensions compared to the wavelength of light and wrote[194]:

> „We shall consider now methods which deviate from those which are obvious extensions of the microwave or radio-frequency techniques for obtaining maser action. The large number of modes at infrared or optical frequencies which are present in any cavity of reasonable size poses problems because of the large amount of spontaneous emissions which they imply. [...] However, radiation from these various modes can be almost completely isolated by using the directional properties of wave propagation when the wavelength is short compared with important dimensions of the region in which the wave is propagated".

---

[191] Concerning the long-lasting „patent war" with Gordon Gould see (Hecht, 2005).
[192] Science Citation Index recalled 28. February 2011.
[193] (A. Schawlow & Townes, 1958), p. 1941.
[194] (A. Schawlow & Townes, 1958), p. 1943.

When calculating the mirror spacing of a Fabry-Perot interferometer they got a value of 10 cm that could be handled quite easily. Their publication ended up with a definite example of an experimental set-up for potassium vapour giving its atomic and spectroscopic values. As well as s the gaseous potassium vapour approach they dealt with the question of whether one could realise a solid state optical maser, and remarked:

> „There are good many crystals, notably rare earth salts, which have spectra with sharp absorption lines [...]"
> [but]
> „the problem of populating the upper states does not have as obvious a solution in the solid case as in the gas".

All the elements required to create an optical maser had been reported in this trend-setting publication. The door had been opened wide. Of course this was realised immediately by many colleagues and scientists working in the field and the "race" to build the first fully operational laser began.[195]

Scientific discussions do not only take place via publications. Sometimes what is more important is encountering people through conferences, workshops and guest lectures. To discuss and evaluate the new developments a conference "*Quantum Electronics – Resonance Phenomena*" was held at Shawanga Lodge, High View in the state of New York in September 1959. Financial support came from the American *Office of Naval Research*. Charles Townes acted as organiser and chairman. During preparation for the conference the subject matter had mainly been the maser. But the possibility of realising an optical maser (=laser) occupied much time, especially in private conversations. One of the conference participants had been Theodore Maiman[196] who gave a talk on "Temperature and concentration effects in a ruby maser". Maiman had been no outsider. He finished his doctoral studies with Willis Lamb Jr and then joined the *Hughes Laboratories* in Malibu (California). Certainly Schawlow was present at the conference. He reviewed the Schawlow-Townes paper from December 1958 now referring to it under the headline "Infrared and optical masers". The register of attendees shows 164 entries. Only 18 participants came from overseas. The former Soviet Union had sent A. Barchukov, N. Basov, L. Kornienko and A. Prokhorov. Germany was

---

[195] The details are given in (Joan L. Bromberg, 1988 (10)), (Joan Lisa Bromberg, 1991), (Mario Bertolotti, 1983), (Hecht, 2005)

[196] Theodore H. Maiman (1927 - 2007) was an US-American physicist. He was doctoral student with Willis Lamb Jr. and performed industrial research afterwards. His life is recalled in his autobiography (T. H. Maiman, 2000). It is hard to understand that Maiman did not receive the Nobel Prize for the first realization of the laser.

represented only by the scientists Helmut Friedburg[197] from the *University of Karlsruhe* and Christoph Schlier[198] coming from the *University of Bonn.*[199]

It took Theodore Maiman only nine months after the conference to realise and construct the first laser with the help of a contaminated ruby crystal.[200] Initially there had been some doubts about whether the effect could really be seen, but then the experiment was repeated at the *Bell Laboratories* and confirmed. Shortly afterwards the laser-effect was demonstrated with a helium-gas laser[201] and, one year later, in semiconductors.[202]

5.3 Semi-classical Laser Theory Until 1964

Meanwhile the theoreticians in the field had not been inactive. At the front line we have to mention the US-American physicist Willis Lamb Jr., born in 1913.[203]. Lamb was a disciple of Robert Oppenheimer, finishing his doctoral studies with a work on the theory of x-ray emission at the *University of Berkeley* (California) in 1938. He made important contributions to the theoretical and experimental development of quantum mechanics. During the Second World War he had been engaged in radar technology and thereafter did research in atomic hydrogen spectroscopy. In his experiments he provided evidence of the energetic displacement of the so called 2s and 2p level of the electron trajectories in the hydrogen atom. This effect can only be understood in terms of quantum mechanics. For his achievements Lamb was awarded the Nobel Prize in Physics for the year 1955. Theodore Maiman, who realised the first laser in 1960, had been a doctoral student with Lamb in the early 1950s, when Lamb had been Professor at the University of Stanford.

In the beginning Lamb had a hard time finding his place in the US-American science establishment. In part that might have been because he was married to the

---

197 Helmut Friedburg (1914 - 2007) was a German experimental physicist. He was a student of Wolfgang Paul and became professor at the *TH Karlsruhe* in 1958.

198 Christoph Schlier is a German physicist. He wrote his thesis on molecular beams at the *University of Bonn* in 1956. He got a chair in physics at the *University of Freiburg* early in the sixties.

199 (C. Townes, 1960).

200 (T. Maiman, 1960).

201 (Javan, Bennett, & Herriott, 1961).

202 (Basov, Vul, & Popov, 1959).

203 Willis Lamb Jr. (1913 - 2008) was an US-American theoretical physicist. In 1955 he was awarded the Nobel Prize in physics for the interpretation of the fine-structure spectrum of the hydrogen atom. Lamb, along with Melvin Lax, was the most important competitor of Hermann Haken in the development of the laser theory. Nevertheless they always have been on cordial terms. Life and work of Lamb Jr. are described in (Cohen, Scully, & Scully, 2009).

German emigrant Ursula Schäfer.[204] Despite being a renowned and respected physicist awarded with a Nobel Prize, Lamb was not offered a physics chair at a prestigious university, so in 1956 he accepted an appointment at the British *Oxford University*, where he lectured until 1962. During this time[205] "he pioneered the use of density matrix calculations", a mathematical tool that later on would become important in the formulation of his laser theory.

In retrospect Lamb's Oxford period cannot be judged as really satisfying. He had not a single doctoral student and could not gather a group of co-workers to perform experiments. Nonetheless, during this time he wrote the influential article "Theory of Optical Masers"[206] which only appeared late in 1964. Before its publication, in the course of 1962, he transferred to *Yale University,* staying there until 1974. During the Yale period he wrote the important articles on laser theory we are going to discuss.

The above mentioned article "Theory of Optical Masers", already written at Oxford, was a further development of his work on maser theory at the *University of Stanford* during the years 1954 to 1956. These results had only been published as a supplement to the doctoral thesis of his Stanford student J. C. Helmer.[207] It was only in 1960 that they were finally published in a peer-reviewed journal.[208] His previous student and co-worker Sargent III noted[221]:

> „The treatment [of the maser] utilized probability amplitudes for a two-level system and introduced the corresponding density matrix. It justified the popular rate equation method in appropriate limits and dealt with both weak- and strong-signal operation. Much of that maser theory applies directly to single-mode, homogeneously broadened laser operation."

After the laser effect had been shown by Maiman, Lamb remodelled his theory of the maser to a semi-classical laser theory of several modes (frequencies). This adapted theory showed an arbitrary broadening of the emitted frequency due to the "Doppler-motion" of the atoms. Lamb discovered that the intensity of the radiation decreases when the laser is tuned through the centre of the frequency line. This effect was later named, after him, "Lamb-dip".

---

[204] This might have been consequences of the notorious „McCarthy-era", when many scientists had been suspected of „non-American machinations". Famous is the "Case Oppenheimer".

[205] (Cohen et al., 2009), p. 9.

[206] (W. E. Lamb, 1964).

[207] Cited after (Sargent III, 1974).

[208] (Willis Lamb, 1960).

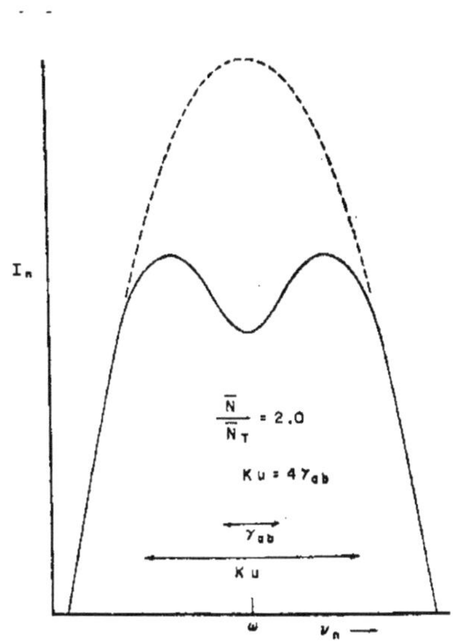

Fig. 5 "Lamb-Dip". Depending on the frequency, typical reduction of laser intensity. (from: (W. E. Lamb, 1964)).

Once again, as in the case of his maser theory, Lamb did not publish the results in a well-known journal. Instead in 1961, he informed his friends Walter Bennett[209] and Eli Javan working at the *Bell Laboratories* by letter, and asked them to search for the phenomenon. Subsequently the Lamb-Dip was detected by these researchers and their co-workers in August 1962.[210]

5.4 The Crucial Role of Laser Conferences

An exponential growth in experimental results was seen after the verification of the laser effect. This called for a *Second Quantum Electronics Conference*. From the 23rd until the 25th of March 1961 the meeting was held in Berkeley (California) under the heading "*Advances in Quantum Electronics*", and attracted not less than 448 participants. This was a threefold increase on the number of only eighteen months before.[211] But again the conference was dominated by American scientists who outnumbered all other nations with 436 attendees. Only two German researchers, H. Friedburg of the *Technischen Hochschule Karlsruhe* and G. Wiederhold of the *University of Jena* belonged to the small group of twelve participants from overseas. Another German physicist at the conference was

[209] (Bennett, 1962).
[210] (McFarlane, Bennett, & Lamb, 1963), (Szöke & Javan, 1963).
[211] The proceedings have been published in the same year: (Singer, 1961).

Wolfgang Kaiser[212], but he represented the *Bell Laboratories* at that time. The reports at the meeting were characterised by the dynamic evolution of experimental results. Ali Javan presented the first Helium-Gas Laser and Garrett, Kaiser and Wood gave a report on "Fluorescence and optical maser effects in $CaF_2:Sm^{++}$". Coming from the *Hughes Research Laboratory* in Malibu (California), where Theodore Maiman was working, W. Wagner and G. Birnbaum gave a short theoretical talk on "A Steady State Theory of the Optical Maser". In their paper of this semi-classical theory they cited the famous Schawlow and Townes publication of 1958 and referred to an article of their own that had appeared in the 1961 July issue of the *Journal of Applied Physics*.[213]

The next important conference took place in spring 1962, in Heidelberg.[214] The meeting was held in u of the retirement of the German physicist Hans Kopfermann and was titled "*Konferenz über optisches Pumpen*". Willis Lamb, in 1962 still working at the University of Oxford before moving on to Yale later that year, was one of the participants. It was here that Lamb and Hermann Haken met for the first time. Haken came from nearby Stuttgart and presented a talk on the "Theory of the laser and optical pumping".[215]

As described in the previous chapter Haken had been working as a visiting associate professor at *Cornell University* in New York and – simultaneously – at the research laboratory of the *General Electric Company* in Schenectady. During this time he received an invitation to stay at the renowned Bell Laboratories. He remembered his stay at Bell in one of the interviews:

> "I had been in the group eleven. Group head had been Phil Anderson who later on won the Nobel Prize. Other famous members of the group were Wannier, a solid state physicist, and coincidentally Melvin Lax, a solid state physicist as well. Anderson himself was an educated solid state physicist [...], this shows that I had been invited because I was viewed as a solid state theoretician. [...] First I continued my work on excitons, but then I recognised that laser was the up-coming subject. At

---

[212] Wolfgang Kaiser (born 1925) is a German experimental physicist who made important contributions to ultra-short-time laser spectroscopy. In 1952 he finished his doctoral studies at Erlangen where he came to know Hermann Haken. He worked at the Bell Laboratories at the time when Haken was visiting associate. A guest professorship offered by Haken in 1962 made it possible for him to habilitate in Stuttgart. Shortly afterwards he received a call from the *Technische Universität München* to a physics chair that he accepted in 1964.

[213] (Wagner & Birnbaum, 1961). Submitted 1. February 1961.

[214] The conference was held 26. – 28. April 1962 to honor the 67th birthday of Hans Kopfermann, (1895 – 1963).

[215] Published proceedings of the conference could not be identified. A list of participants, the names of the referees and the titles of their lectures as well as a short abstract of Haken's talk have been found in Haken's archive.

> Bell there was [...] Wolfgang Kaiser, a good friend of mine, and we spent whole nights discussing this subject. Later I had talks with Harry Frisch – I spent countless nights with him [laughs]. At Bell I wrote an article on the first phase of the operation of the laser, but discontinued this work later on. It is a fact that I was introduced to the subject of lasers by Wolfgang Kaiser".[216]

Back in Stuttgart Haken launched an ambitious programme on laser theory. He delivered first results on a semi-classical version of his theory at the above mentioned Heidelberg conference in spring 1962. Looking back he remembered:

> "At the conference I met Willis Lamb. I don't remember the name of my talk ... of course about the laser, but Lamb did not attend the lecture, he was out shopping. [loughs]. But we discussed the subject soon after, during the meeting. I told him what I was doing. And he said, well I did the same. We discussed how to handle the spontaneous emission ... and then I showed him my equations. He responded that he had found corresponding equations. But there had been a difference: from the very beginning Lamb had called his theory semi-classical, because he used the density matrix. I had called it quantum theory, because I used the second quantization formalism".[217]

Haken published his results under the heading "Nonlinear Interaction of Laser Modes". Early in 1963 the article appeared in the journal *Zeitschrift für Physik* (co-authored with his diploma student Herwig Sauermann) where it had been submitted on the 11th February. In a footnote to that article Haken recounted the discussions with Lamb in Heidelberg:

> „Prof. W.E. Lamb, jr. has kindly informed one of us (H.H.) in a private discussion at Heidelberg, spring 1962, that he has derived similar equations for the gas laser."

Lamb as well remembered this meeting:

> „I ran into Haken for the first time at Heidelberg. [...] I was pretty far along on the laser theory at that time, but I didn't talk about it at the conference. I talked on something else. But there was a talk by Haken about laser theory, and it seemed to me that he had some very good ideas, and I was a little upset, [...]

[216] Interview with Herrmann Haken 21.9.2010. (Haken-Archive).
[217] Interview with Herrmann Haken 21.9.2010. (Haken-Archive).

> because it seemed to me, that Haken might very well be a serious competitor, which in fact he certainly has been."[218]

In their publication Haken and Sauermann cited the article of Schawlow and Townes as well as the work of Wagner and Birnbaum[219]. The most important result of their calculations was that by taking into account non-linear effects a small deviation of the laser frequency occurs. This effect had been demonstrated in gas-laser experiments.

> „the main result of our analysis will be, that an increased pumping rate supports also off-resonance modes and leads to a repulsion of frequencies".[220]

The face-to-face exchange of ideas between the leading theoreticians was very strong during this time. Theoretical results were desperately needed because experimental results were , avalanche-like. An important event in early 1963 constituted the third international conference on "*Quantum Electronics*"[221], attracting more than 1000 participants from 15 countries. The meeting took place in Paris from 11 – 15 February 1963. The venue, this time in Europe, saw 74 attendees from Germany. Coming from Stuttgart were Hermann Haken, Herwig Sauermann and R. K. Sun [a guest professor]. The other German universities and *Technischen Hochschulen* sent 31 scientists, the German industry had 25 delegates and other research institutions were represented by 17 participants.

The topics of the conference were grouped into the following subjects[222]

- Theory of Coherence and Noise
- Optical Pumping and Magnetometers
- Molecular Beam Masers
- Gas Lasers
- Spectroscopy of solid state maser materials
- Solid State Masers
- Solid State Lasers
- Laser Modes and special techniques
- Non-Linear Optics
- Semiconductor and Photon Masers

---

[218] Interview Willis Lamb Jr. with Joan Bromberg from 7. March 1985. (AIP Niels Bohr Library & Archives. Call Number: OH 27491).

[219] (A. Schawlow & Townes, 1958), (Wagner & Birnbaum, 1961).

[220] (H. Haken & Sauermann, 1963b), p. 262.

[221] (Grivet & Bloembergen, 1964)

[222] (Grivet & Bloembergen, 1964)

It is striking that a special section on laser theory was missing. But, during the meeting in Paris, Haken and Sauermann obtained new and additional information on the current state of research in the United States. Therefore, while proof reading[223], they attached a note to their above-mentioned article[224]:

> "Note added in proof. After the present paper has been submitted for publication several talks (by W.E. Lamb, N. Bloembergen, D. McCumber, H. Statz) were given at the 3rd International Conference on Quantum Electronics, which consider the interaction of modes brought about by the nonlinear response of the atomic systems. From these papers the one of Lamb is most closely related with our present work, although the formalism and also the physical System are somewhat different from our case, Lamb treats moving atoms in gases with a Doppler broadened line, whereas we treat fixed atoms with a homogeneously broadened Lorentzian line. The two investigations give, however, similar results, for instance for the mode repulsion effect. On the other hand, Lamb's dipping effect has no analogue in our case. For a detailed comparison of results, however, the publication of Lamb's paper must be waited for.
> At the same Conference E. Snitzer reported results of Nd-doped glasses, which show additional modes appearing with higher pumping and also a repulsion of modes in good qualitative agreement with our analysis."

We have to mention that the talk at the Paris meeting given by Willis Lamb Jr. is not printed in the conference proceedings. The corresponding article only appeared in the Physical Review [225] (received 13th January 1964) in 1964. To claim some priority Lamb remarked:

> „The main results of the paper were reported at the Third International Conference on Quantum Electronics, Paris, February 1963. Lectures on some of the material were given at the 1963 Varenna Summer School."

Even before the publication of Lamb's article Haken and Sauermann extended their first paper with another contribution that they submitted on the 6th of June

---

223 The contributions mentioned in the note were titled: N. Bloembergen: "Optique Non-Linéaire"; D.E. McCumber: "Unified Theory of Steady State Cavity Masers"; H. Statz, C. Tang: "Zeeman Effect and nonlinear Interactions between Oscillating Modes in Masers".

224 (H. Haken & Sauermann, 1963b)

225 (W. E. Lamb, 1964).

1963.[226] In this article they analysed the frequency shifts in laser modes that occur due to the non-linear behaviour of the active laser material. It seems curious that they cited the paper of Lamb that had not been published at that point but presumably had been circulated as a pre-print.

Less than two month after the Paris conference had finished, the next "*Symposium on Optical Masers*" was organized for 16 – 19 April 1963, in New York. The publication of the proceedings was delayed and they appeared only in 1964.[227] The book presented articles by 92 different authors, among which only two German scientists could be found: K. Gürs and R. Müller of the research laboratory from the *Siemens und Halske* corporation. Not unexpectedly the majority of the attendees had been American scientists. Some Japanese researchers also gave talks but only referring to experimental results. The lecturers on laser theory had been N. Bloembergen, E. Wolf, E.C.G. Sundarshan, P.A. Grivet, B. Lax (on "Semiconductor masers"), H.A. Haus and J.A. Mullen, as well as G. Toraldo di Francia. Their contributions dealt with theoretical specialties, with no attempt at an encompassing theory such as those of Haken and Lamb.

In the period from 1963 until 1966 the meetings and contacts of the scientists involved in laser theory research followed each other in quick succession. From 1953 the Italian Physical Society organised regularly its famous physics summer schools at the Villa Monastero on the waterfront of Lake Como. They are named after Enrico Fermi, the renowned Italian physicist and Nobel Prize winner. Charles Townes headed the XXXI summer school that was held from the 19th until the 31st August 1963 dealing with "Quantum Electronics and Coherent Light".[228] These meetings had a twofold purpose. First they offered a genuine possibility for a selected bunch of young and eager scientists to learn about the newest theoretical developments. Secondly it was an invaluable chance to become acquainted with exceptional professionals in the field. The renowned lecturers at this summer school comprised, as well as C. Townes, the names W. Lamb, A. Schawlow, B. Lax, J. Gordon, N. Bloembergen, F. Arecchi and A. Javan. Hermann Haken was not on the list. In 1963 he was not yet known to the community as a laser theoretician. That would change quickly. About 60 doctoral students and post-docs had been invited. Among them were, coming from Stuttgart, were Haken's assistant Hannes Risken and his doctoral student Herwig Sauermann. Sauermann got the opportunity to give a talk on the results of the laser theory work he co-authored with Haken.[229] This is why, even though Haken had not been present, in the proceedings we find the contributions of Lamb "Theory of optical maser oscillators" and Haken/Sauermann "Theory of laser action in

---

[226] (H. Haken & Sauermann, 1963a)

[227] (Fox, 1964).

[228] The Proceedings were published in 1964: (Miles, 1964).

[229] Haken and Sauermann cited the work of Townes of the year 1961; Townes in his article referred to the paper of Haken and Sauermann in the *Zeitschrift für Physik* 173 (1963), 261.

solid-state, gaseous and semi-conductor systems“ in immediate succession. We can also assume that there had been a close dialogue between Lamb, Risken and Sauermann.

This contact was intensified when in the following year, 1964, a further summer school was held at the French mountain village Les Houches. Again Lamb and Sauermann were participants. These summer schools "*Ecole d'eté de physique théorique Les Houches*“, in close proximity of the *Mont Blanc Massive* had been held since 1951. Organised by the French physicist Cécile deWitt from the University of Grenoble these summer schools pursued the same intentions as the Italian ones. The proceedings of the meeting did not give a list of attendees, but the participation of Herwig Sauermann can be clearly derived from the following quotation in the report contributed by Willis Lamb:[230]

> "W. E. Lamb, Jr., Yale University: INTRODUCTION
> These lecture notes on the theory of optical masers were taken by Messrs. B. Decomps, M. Durand, B. Gyorffy and H. Sauermann, while assistance in their arrangement was given by Mme. A. Fouskova. For reasons mentioned below, the notes were not prepared in advance of the course, and I have not had sufficient opportunity to correct them. Because I declined to answer certain questions of an offensive nature, renewal of my passport was withheld by the U. S. Department of State. My lectures at the School were only made possible (on very short notice) by the Supreme Court decision in the case of Aptheker and Flynn vs. The Secretary of State."

Other speakers at Les Houches were Roy Glauber giving a lecture series on "Optical Coherence and Photon Statistics", Willis E. Lamb talking about the "Theory of Optical Maser", Ali Javan presented the latest results on "Gaseous Optical Maser“ and Nicolas Bloembergen lectured about "Non-Linear Optics“.

Hermann Haken, in his *Zeitschrift für Physik* article dated 4th July 1964, referred to the fact that its contents had been presented by his student Herwig Sauermann at the summer school in Les Houches. This seems plausible taking into account the duplicity of the events with the summer school in Varenna the year before.[231] In the proceedings there is no written evidence of this talk. The participation of Sauermann at the school is proven by the statement in Lamb's article.

---

[230] (deWitt, Blandin, & Cohen-Tannoudji, 1965), p. 331. Again another evidence of Lamb's problems with the US-American authorities concerning his „un-American“ activities.

[231] In an interview with Joan Bromberg as of 7th March 1985, Lamb confirmed the meeting with Sauermann. (AIP Niels Bohr Library & Archives. Call Number: OH 27491).

5.5 The Fully Quantum Mechanical Laser Theory

On 4th July, 1964 Haken submitted his seminal article to the *Zeitschrift für Physik* "A Nonlinear Theory of Laser Noise and Coherence I". This work constituted an important step forward towards a fully quantum mechanical laser theory. In this non-linear theory Haken explicitly mentioned the crucial difference in the nature of the laser radiation below and above the laser threshold.[232]

> „In contrast to linear theories there exists a marked threshold. Below it the amplitude decreases after each excitation exponentially and the linewidth turns out to be identical with those of previous authors (for instance WAGNER and BIRNBAUM), if specialized to large cavity width. Above the threshold the light amplitude converges towards a stable value, whereas the phase undergoes some kind of undamped diffusion process".

To demonstrate the situation at the threshold Haken chose the analogy of a potential. The illustration is given in Figure 6.

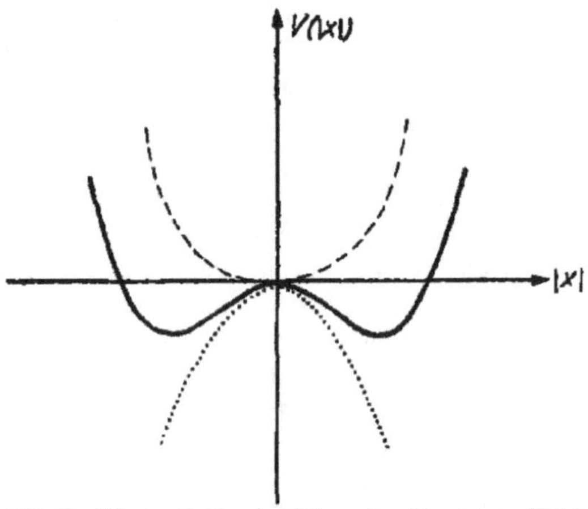

Fig. 2. Plot of "potential energy" versus light amplitude. – – – below threshold (*linear and nonlinear* theory); ········ above threshold, *linear* theory leads to instability; —— above threshold, *nonlinear* theory

Fig. 6 Behaviour of the potential below and above laser threshold for linear and non-linear theories

232 (H. Haken, 1964), p. 96.

The calculations showed that, pumping the laser above threshold in the non-linear quantum theory two stable minima of the potential exist, whereas in the linear theory no stable values occur. Because of the fact that this result depends strongly on the energizing pump-process nothing comparable could be found in maser theory. At the end of his article Haken summarised:

> „The main objective of our paper was to bridge the gap between linear and nonlinear theories of laser action. As we have shown linear theories represent a very good approximation at small inversion. On the other hand there is a marked threshold beyond which the system behaves qualitatively very differently from below threshold, its amplitude oscillating around a stable value".

Only three months later, on the 22nd September 1964, Haken published the second part of his work "A Nonlinear Theory of Laser Noise and Coherence II".[233] In the meantime Willis Lamb had published his nonlinear laser theory in the *Physical Review*[234] and Haken cited it. The second article expanded the first publication by the following aspects:

- Several laser modes are dealt with, not only a single mode,
- Standing waves were discussed and
- Non-complete cavity-resonance was calculated.

The noise source had been the optical transition of fluctuating dipols. It seems important to note that the methods for introducing noise (quantum mechanical, thermical, gaussian shape, etc.) into the calculations varied according to the particular authors. Different approaches often led to the need for a different mathematical treatment.

In autumn that year another possibility opened up for Hermann Haken to present his new theory at a meeting in Switzerland. "*Physik der Laser und deren Anwendungen*"[235] was the title of an international symposium held from 12th to 15th October 1964. This conference, organised on the initiative of the *Suisse Committee of Light- and Electro-optics* and by the *Institute of Applied Physics* of the *University of Bern,* attracted more than 250 participants from 22 countries.[236] The names of the attendees could not be found in the proceedings. Looking at the roster of talks given it seems reasonable to say that mainly non-American experiments and works was presented. Hermann Haken gave his lecture on "Nonlinear Theory of Noise and Coherence" at the session called "General Laser Physics".

---

233 (H. Haken, 1965).

234 (W. E. Lamb, 1964).

235 Translates into: Laser physics and its applications.

236 See "Introduction of the proceedings", printed in *Zeitschrift für Angewandte Mathematik und Physik Band 16* (1965). (Meyer, 1965).

During the years 1964 and 1965 laser theory developed dynamically. A race evolved between Hermann Haken and two US-American scientists, Willis Lamb Jr. and Melvin Lax. The previous exponents, Lamb and Haken, intensified their efforts by bringing in colleagues and students. Especially Hermann Haken was able to activate many co-workers (see Chapter 5f). A new competitor, the US-American theoretical physicist Melvin Lax, working at the *Bell Laboratories*, showed up. He had been working primarily on the theory of noise and coherence in classical and quantum mechanical applications. He was now going to apply this knowledge to the laser. The connection of the laser with noise and coherence phenomena is a consequence of the physical nature of the photons and the atomic structure of the laser active material (gas, solids, semi-conductors). In his article of 1964 Haken had already hinted at this connection.

The coherence theory of electro-magnetic waves, as well as non-linear optics, were a highly active research field at that time. Major contributions came from Roy Glauber and Nicolas Bloembergen who were later rewarded with the Nobel Prize for their work.[237] Because these segments had not been the focus of laser development we do not consider and evaluate them in depth in this article.

Until the end of 1964 laser theory had been advanced mainly by Willis Lamb and Hermann Haken, then a further group of theoretical physicists, Melvin Lax and his co-worker William Louisell at *Bell Laboratories* in Murray Hill (New York), joined the race.

For ten years Melvin "Mel" Lax had been a member of the scientific staff at *Bell Laboratories.* He headed the "Theoretical Physics Department" from 1962 until 1964.[238] Haken had met Lax for the first time during a solid state conference in 1958.[239] During his time as guest researcher at Bell in 1960, Haken and Lax intensified their acquaintance when simultaneously they were both members of the theory division 11 11 working under the leadership of Philip Anderson.

Lax was born in New York City in 1922 where he also spent his youth. Hampered by the Second World War he finished his doctorate at the *Massachusetts Institute of Technology* in 1947 and then moved to *Syracuse University* (New York State) where he stayed until 1955. During this time his research interest was focused on solid state physics, the field receiving a boost with the detection of the transistor effect. Whereas W. Lamb and C. Townes had been brought up scientifically with radar- and micro wave technology, Lax's education and focus on solid state physics showed some parallels to the scientific history of Hermann Haken. After the war, *Bell Laboratories* were regarded as the world's leading industrial research laboratory in solid state physics. When a separate theory division was installed at Bell in 1955, Lax unsurprisingly accepted a position as the first theoretician employed. Throughout his career he remained affiliated with Bell, even when he finally took a chair as physics professor at the

[237] Details on their work on coherence and quantum optics can be found in: Roy Glauber: (Grandin, 2006) and Nicolaas Bloembergen in: (Frängsmyr, 1993).

[238] (Birman & Cummins, 2005), see also (Millman, 1983).

[239] Hermann Haken (private communication).

City College at *University of New York* in 1971. Reading through his papers, and as also mentioned in his obituary, Mel Lax had "a deep interest in random processes".[240] This curiosity induced him to study the phenomena of fluctuations and quantum-mechanical noise in the laser process in detail. At the end of 1964 Lax gave up his direction of the theory division at Bell and concentrated again on scientific research. He cooperated with his colleagues J. P. Gordon and especially William H. Louisell[241], joining the race for a quantum-mechanical laser theory. In the course of eight years, from 1960 to 1964, Melvin Lax published not less than 18 articles on stochastic processes like noise and fluctuations.[242]

A first meeting of the exponents of the different theoretical laser schools occurred during the *Fourth International Quantum Electronics Conference*, held in San Juan (the state capital of Puerto Rico; an US-American territory in the Caribbean) from 28th to 30th June, 1965. The proceedings[243], repeatedly quoted afterwards, were published in 1966. The convention was dominated by American attendees because the venue was accessible only with great difficulty from abroad.[244] From the 257 scientists recorded on the roster not less than 222 came from the United States. France delegated 13 participants, the former USSR sent seven. Germany was represented by only two researchers: Hermann Haken from the *Technischen Hochschule Stuttgart* and Wolfgang Kaiser from the *Technischen Hochschule München*. This was sparse compared to the *Bell Laboratories* in Murray Hill with 25 employees and the *Massachusetts Institute of Technology* which was alone represented by 15 participants

We should thus not be astonished that laser theory and the theory of quantum electronics was presented mainly by American scientists. One talk each came from Great Britain, from the *Royal Radar Establishment* dealing with "Photon-counting statistics", and the Moscow *Lebedev Physics Institute,* speaking on the topic of "Dynamics of a Two-mode Operating Laser". According to Hermann Haken he also delivered a talk which was rejected for printing in the conference proceedings "due to its length".[245]

The other papers in the proceedings had been allocated to the other three known American theory groups:

- *Harvard University* in Cambridge (Mass.): Roy Glauber and his doctoral student Victor Korenman

---

[240] See (Birman & Cummins, 2005).

[241] William H. Louisell (born 1924) in Mobile (Alabama). He studied physics at the *University of Michigan* and received his doctoral degree in 1953. Thereafter he became member of the technical staff at Bell Telephone Laboratories (Murray Hill, N. J.), where he stayed until he assumed a physics professorship at the *University of Southern California* in 1967. (Source: Proceedings of the I.R.E. QE3 (1967), p. 97).

[242] For a summary see table 3, page .....

[243] (Kelly, Lax, & Tannenwald, 1966).

[244] Hermann Haken remembered that he crossed the Atlantic Ocean with a cargo vessel and went ashore in Venezuela. Then he took the airplane and flew into San Juan. (Hermann Haken, private communication).

[245] Hermann Haken, private communication.

- *Bell Laboratories* in Murray Hill (New Jersey): J.P. Gordon, Melvin Lax and William H. Louisell
- *Yale University* in New Haven (Connecticut): Willis E. Lamb and Marvin O. Scully

Attending the conference were Roy Glauber, J. P. Gordon, Willis Lamb, Melvin Lax, William Louisell, Marvin Scully and Charles Townes. During the three days of the meeting Haken had ample opportunity to speak with the most important American laser theoreticians.

Melvin Lax gave a talk titled "Quantum Noise V: Phase Noise in a Homogeneously Broadened Maser". As can be seen by reading the article the term "Maser" is equated with the word "Laser". At that time the Bell Lab scientists were under the impact of a patent dispute with Gordon Gould.[246] The latter had coined the word "Laser", whereas the Bell people used the name "Optical Maser". In the abstract of Lax's paper we find the following:[247]

„Our result for the full width at half power above threshold:

$$W = (\Delta\omega)^2\,(h\omega_o/2P_{tr})\,S(1 + \alpha^2)$$

where P_{tr} is the transferred power, is an improvement over Lamb, Haken, and Korenman in three ways: $\Delta\omega$ (....) depends on both cavity and atomic widths; S (...) depends on both photon and atomic noise sources (....); and α (....) includes the effects of detuning."

Furthermore Lax cited Haken's papers in the *Zeitschrift für Physik* volumes 181 and 182, as well as his article in volume 13 of the *Physical Review Letters*. He also mentioned a work of Hannes Risken that was in print. Willis Lamb was quoted with his contribution in the proceedings of the Les Houches lecture course of 1964.

Melvin Lax numbered his articles on noise and fluctuations, differentiating between classical and quantum-mechanical noise. Looking at the chronological order and numbering of his works one has to take into account a peculiarity. Articles published in the proceedings of a conference are mostly prepared in retrospect. Normally there will also be proofreading. It means that the last editorial intervention – the chance to quote the latest new articles – might be some month, up to a year, after the conference took place. It seems that this was the case with Lax's article at the Puerto Rico meeting. The title of his published talk runs under the number "Quantum Noise V". But his foregoing work on "Quantum Noise IV" was only published on the 6th of May 1966 and had been submitted on November 16th 1965 at the *Physical Review* , five months after the conference at San Juan.

---

[246] On this "patent war" see especially (Hecht, 2005).

[247] (Melvin Lax, 1966).

In "Quantum Noise IV. Quantum Theory of Noise Sources" Lax treated the fluctuations mathematically via the Langevin-equations, as Haken did with his Stuttgart school. Classifying and differentiating his approach from the publications of Haken and Sauermann of 1963 and 1964 he remarked:

> „Quantum noise sources have also been introduced into maser calculations in a heuristic way, or by methods similar to the perturbation techniques adopted here".

Then he gave reference to Haken and Sauermann in a footnote. In the time between submitting the paper on the 16th of November 1965 and its publication in May 1966, he noted that meanwhile the researchers in Stuttgart had also been active. Therefore he wrote a "Note added in proof":

> "After the completion of this manuscript (and after the results summarized in secs. 1 and 6 were presented at the 1965 Puerto Rico Conference) we have learned that several members of the Haken school have adopted a Markoffian approach closely related to our own. See H. Haken and W. Weidlich [Z. Physik 189, 1 (1966)]; C. Schmid and H.Risken [ibid. 189, 365 (1966)]. These papers treat the atomic fluctuations and lead to moments in agreement with ours. For the electromagnetic field, the noise sources are not derived by them but are taken from Senitzky— see, e.g., H. Sauermann, Z. Physik 189, 312(1966). Our procedure obtains the field noise sources by the same method as that used for the atomic noise sources, and moreover derives the independence of field and atomic sources."[248]

These works, cited by Lax, had been submitted by Haken and Weidlich as well as Weidlich and Schmid to the *Zeitschrift für Physik* on 17th of August 1965 and 7th of September 1965 respectively. The Puerto Rico meeting came to an end on the 30th of June 1965. It remains a question whether Haken and his co-workers had made use of the results given there, and "in a hurry" prepared fresh articles, or whether there had been preparatory work done before. The latter is indicated by the fact that Melvin Lax did not mention the work of Herwig Sauermann, which also appeared in volume 189 of the *Zeitschrift für Physik.* The article runs under the heading "Quantenmechanische Behandlung des optischen Maser (Dissertation)".[249] Submitted on the 17th of September it was overlooked by Lax, presumably because it was written in German. Sauermann recapitulated in detail the development of his considerations leading to his doctoral thesis:

---

[248] M. Lax: Quantum Noise IV. Pp. 120-121.
[249] (Herwig Sauermann, 1965)

> "In recent years one has been able to understand the essential classical properties of the laser on the basis of semi-classical theories. [...] These theories, where the atoms of the active material are described by a density matrix, the amplitudes of the laser oscillation by c-numbers, cannot explain, by their very nature, some essential quantum-mechanical properties of the laser. Among them range essential questions such as the line-width and the intensity fluctuations of the radiated laser light. [...]
> In the following period Lamb investigated the influence of the fluctuations in the cavity resonator using a semi-classical approach. Haken developed a non-linear quantum-mechanical theory of the fluctuations in the laser, where the pumping procedure is introduced explicitly as a stochastic process and the line-width, as well as intensity fluctuations, are calculated with the help of fluctuating dipoles. In contrast to Haken in our work we derive the fluctuations occurring in spontaneous emissions and pumping processes from "first principles" and introduce them totally symmetrically, in exactly correct quantum-mechanical fashion. Moreover the thermal noise of the resonator is taken into account. [...]"

At the end of his publication Sauermann added:

> *" After having finished the present publication the author gained knowledge of a talk by M. Lax given at the conference on quantum electronics in Puerto Rico, June 1965, where the phase diffusion of a laser mode was reported. To the extent a comparison with our results is deemed possible, the findings agree essentially".

Certainly this knowledge came from Hermann Haken who informed his co-workers and students about the papers delivered at the San Juan conference on his return. Nevertheless it also becomes clear that a doctoral thesis and other publications could not have been written in such a short time on the suggestion of Lax's results. They had a long prehistory in Stuttgart (see chapter 5f).

The other published theoretical contributions to the San Juan conference proceedings, those of Korenman as well as Scully and Lamb, did not quote the work of the Stuttgart School. They only cited the article published by W. Lamb in volume 134 of the *Physical Review* and the lectures of Glauber on coherence phenomena at the Les Houches summer school.

The race between Hermann Haken from Stuttgart with Melvin Lax continued in the following year. On May 2nd 1966 Lax submitted a paper titled "Quantum

Noise VII: The Rate Equations and Amplitude Noise in Lasers" to the *IEEE Journal of Quantum Electronics* that was accepted in a revised version on the 24th of October 1966. In turn, on the 21th of June 1966 Hermann Haken sent the fundamental article "Quantum Theory of Noise in Gas and Solid State Lasers with an inhomogeneously broadened Line I" to the *Zeitschrift für Physik*, written by him, his co-workers and students V. Arzt, H. Risken, H. Sauermann, Ch. Schmid and W. Weidlich.

Table 3 Articles published by Melvin Lax and co-workers on the theory of laser and noise

	Articles of Melvin Lax and co-workers	
	Classical Noise	
I	M. Lax: „Fluctuations from the Nonequilibrium Steady State", Rev. Mod. Phys. **32** (1960), S. 25	January 1960
II	M. Lax: „Influence of Trapping, Diffusion and Recombination on Carrier Concentration Fluctuations", J. Phys. Chem. Solids **14** (1960), S. 248	
III	M. Lax: „Nonlinear Markoff Processes", Rev. Mod. Phys. **38** (1966), S. 359	April 1966
IV	M. Lax: „Langevin Methods", Rev. Mod. Phys. **38** (1966), S. 541	July 1966
V	M. Lax: „Noise in Self-Sustained Oscillators", Phys. Rev. **160** (1967), S. 290	April 1966, published 23.2.1967
VI	M. Lax und R.D. Hempstead: „Self-Sustained Oscillators Near Threshold", Phys. Rev. **161** (1967), S. 350	published 28.3.1967
	Quantum Noise article	
Q0	M. Lax und D. R. Fredkin: „Oscillations of a Cavity Maser" (unpublished)	
QI	M. Lax: „Generalized Mobility Theory", Phys. Rev. **109** (1958), S. 1921	15.3.1958
QII	M. Lax: „Formal Theory of Quantum Fluctuations from a driven State", Phys. Rev. **129** (1963), S. 2342	11.10.1962; published 1.3.1963
QIII	M. Lax: „Quantum Relaxation, the Shape of Lattice Absorption and Inelastic Neutron Scattering Lines", J. Phys. Chem. Solids **25** (1964), S. 487	20.8.1963
QIV	M. Lax: „Quantum Noise: Quantum Theory of Noise Sources", Phys. Rev. **145** (1966), S. 110	16.11.1965
QV	M. Lax: „Phase Noise in a Homogeneously Broadened Maser", in (Kelly, et al., 1966), S. 735	June 1965
QVI	M. Lax und D. R. Fredkin: „Moment Treatment of Maser Noise (unpublished)	

Table 3 (*continued*)

QVII	M. Lax: „The Rate Equations and Amplitude Noise in Lasers", IEEE Journal of Quantum electronics **QE 3** (1967), S. 37	2.5.66; published 24.10.1966
QVIII	H. Cheng und M. Lax: „Harmonic Oscillator Relaxation from Definite Quantum States", in Löwdin, P.O. (ed.) „Quantum Theory of the solid State". Academic Press. New York 1966.	
QIX	M. Lax und W.H. Louisell: „Quantum Fokker-Planck Solution for Laser Noise", IEEE Journal of Quantum electronics **QE3** (1967), S. 47	2.5.1967; published 12.11.1967
QX	M. Lax: „Density Matrix Treatment of Field and Population Difference Fluctuations", Phys. Rev. **157** (1967), S. 213	
QXI	M. Lax: „Multitime Correspondence Between Quantum and Classical Stochastic Processes", Phys. Rev. **172** (1968), S. 350	15.1.1968
QXII	M. Lax und W.H. Louisell: „Density Operator Treatment of Field and Population Fluctuations", Phys. Rev. **185** (1969), 568	
QXIII	M. Lax und H. Yuen: „Six-Classical-Variable Description of Quantum Laser Fluctuations", Phys. Rev. **172** (1968), S. 362	19.2.1968

The above-mentioned articles of Lax and the Stuttgart School had only been published long after the subsequent *Quantum Electronic Conference* that had been held in Phoenix, Arizona (USA) from 12 until 15 April, 1966. Even though the publication date had been later, the contents of it and the results of the calculations had been presented and discussed in Phoenix. No book containing the proceedings of the conference had been edited, as with the former meetings, but many papers were reprinted in the second volume of the new journal "*IEEE Journal of Quantum Electronics QE 2(1966)*", in the April, August and September issues respectively. The programme of the conference, for instance, is given in the April issue. Under the heading "General Laser Theory" we read[250]:

„April 12, 1966: 8:00 p.m.
Session 3A: General Laser Theory
Pizzarro /A room

3A-1: Theory of Noise in Solid State, Gas and Semiconductor Lasers
H. Haken

[250] See IEEE Journ. Quantum Electronics QE 2 (1966, April), XIX (special conference program edition).

3A-2: Quantum Noise and Amplitude Noise in Lasers
M. Lax and W.H. Louisell"

Haken and Lax gave their talks consecutively in the evening session of that day. Prior to the conference there had been some discussion between Lax and Haken which Lax commented upon in his article "Quantum Noise VII"[251]:

> „Haken has recently [Zs. f. Phys. 190 (1966), 327] questioned the validity of the shot noise treatment of intensity noise because he finds that the dominant noise source is the off-diagonal random force F_{12} rather than (say) F_{22}. As has been shown in QIV, and by Haken [Zs. f. Phys. 189 (1966),1] F_{22} appears in atomic rate equations and yields atomic shot noise."

[...]

> "[...] the major source of noise in an optical laser is the off-diagonal atomic force F_{12}. [...] Haken now agrees with this viewpoint."

In a footnote he quoted as the source: "private communication at Phoenix Conference on Quantum Electronics". At least the face-to-face communication still holds.

Only four months later, the next encounter of the exponents of the competing laser theory schools took place in Kyoto (Japan). It was brought about by the local *Institute of Theoretical Physics* on the occasion of the "*Second Tokyo Summer Institute of Theoretical Physics*".[252] Haken lectured on "Dynamics of Nonlinear Interaction between Radiation and Matter", whereas Melvin Lax had chosen for his course the title "Quantum Theory of Noise in Masers and Lasers". The different headings of the lectures did not seem to have been selected purely by chance, as they reflected the different approaches to laser theory of the two scientists. Hermann Haken took the dynamics of many body systems as his starting point whereas Lax proceeded from the theory of noise and fluctuations. In the end, naturally, both theoretical laser theory approaches converge.

The Tokyo Summer School had been held in August 1966. Again it was a summer school that brought about the next meeting of the laser theorists. It took place in the context of the 42nd Summer School "Enrico Fermi" on "*Quantum Optics*", held from the 31st of July 1967 until the 19th of August 1967 at the *Villa Monastero* in Varenna at the shores of Lake Como (Italy). The director of the summer school had been Roy Glauber, bringing with him some of his American colleagues, Marlan Scully and William H. Louisell. Thus the three American laser

[251] (Melvin Lax, 1967a), page 37, 43, as well as page 46.
[252] (Kubo & Kamimura, 1967).

schools were present. Due to his illness and hampered by his work on the article for the physics “Handbook“ Hermann Haken was prevented from attending.[253] His place was taken by his colleague and co-worker Wolfgang Weidlich. He was accompanied by his doctoral students Fritz Haake and Heide Pelikan. Hartmut Haug and Karl Grob had been sent by Haken’s institute. In 1965 Grob had written his thesis on the theory of the stimulated Raman-effect supervised by Hermann Haken.

The Summer School was attended by 91 participants, mostly from Italy. But other European countries were also represented.

M. Scully gave a course on[254] “The Quantum Theory of a Laser“. He thoroughly discussed the work of Glauber, Gordon, Haken, Louisell, Shen and Weidlich, quoting the following articles:

V. Korenman: Phys. Rev. Lett. 14 (1965), 293
- M. Lax; W. Louisell: Journ. Quant. Elec. Q.E. 3 (1967), 47
- M. Lax: Phys. Rev. 157 (1967), 213
- C. Willis: Phys. Rev. 147 (1966), 406
- H. Haken: Zs. F. Phys. 190 (1966), 327
- H. Sauermann: Zs. F. Phys. 189 (1966), 312
- H. Risken, C. Schmid, W. Weidlich: Phys. Letters 20 (1966), 489
- J. Fleck Jr.: Phys. Rev. 149 (1966), 322
- J. Gordon: Phys. Rev. to be published (also 1967/68)

In comparison, Y. R. Shen coming from the Physics Department of the *Lawrence Radiation Laboratory* (USA) delivered a talk on “Quantum Theory of Nonlinear Optics”. He only cited the works of Lamb and Scully, neglecting the other contributors.

Roy Glauber, being the course director, referred to the papers of his colleagues in his introductory remarks “Coherence and quantum Detection“[255] :

> „When the fundamental equations of motion become nonlinear however, as they necessarily do in the case of the laser, the problem of finding the density operator assumes greater proportions. A number of lectures of our school have been devoted to this problem; it is discussed from various standpoints by Scully, Haken, Weidlich, Louisell and Gordon.”

Of some interest is the contribution of H. A. Haus[256], a physicist from the Massachusetts Institute of Technology, who recounted the measurement of the signal to noise ratio in lasers. He reviewed the theoretical situation as follows:

---

[253] See in detail pages .... of this book.
[254] (Marvin Scully, 1969).
[255] (Glauber, 1969), page 32.

> "While the experimental work on the intensity Fluctuations of lasers was progressing, Glauber obtained a quantum-theoretical description of radiation from coherent sources, a description better suited to deal with laser-radiation. The quantum theory of the laser oscillator proceeded apace. Even though the fluctuations observed in experiments could be satisfactorily explained by a semi-classical theory the need for such a quantum theory existed and was successfullly met by several authors [27-34]. The understanding of the measurement itself advanced along with the theoretical developments on the description of optical radiation and the description of fluctuations in optical masers."

The references given cite the work of Haken, Scully, Lamb, as well as Lax and Korenman, under the numbers 27-34.

In their talks Gordon and Louisell quote only their own papers (Bell-Lab-Team) and the first article of Scully and Lamb, which was in print. The work of the Stuttgart School was not even mentioned.

Finally Haken and Weidlich, although only Weidlich was present, could reap the fruits of Haken's labour on the "Handbook" article. In their lengthy lecture, comprising 49 printed pages, they analysed in detail the preparatory work on laser theory and structured the different phenomena. They made extensive references to all works of the different groups working on laser theory[257]:

> „At the 1963 Varenna summer school on the laser, two types of theories were presented. F. G. Gordon calculated the noise properties of the laser treating it as a linear device and gave a qualitative discussion of the nonlinear region. Lamb, Haken and Sauermann, Grasjuk and Orajewskij neglected laser noise but treated the laser quantitatively as a nonlinear system. Nonlinearity plays a decisive role in the stability of laser modes. Their coexistence and so on. These nonlinear theories, however, predicted no laser action at all below a certain threshold, and an infinitely narrow line above threshold. Thus there was a need for a theory which interconnected both aspects, nonlinearity and noise. Because laser noise stems primarily from spontaneous emission. which is a typical quantum-mechanical effect, this theory also ought to be quantum-mechanical. Since 1964 three essentially equivalent methods were developed to achieve this goal, namely

---

[256] (Haus, 1969), p.112.

[257] (Glauber, 1969), p.630.

a) the Heisenberg Operator equations with quantum-mechanical Langevin forces.
b) the density matrix equation for atoms and light field.
c) a (generalized) Fokker-Planck equation."

According to Weidlich and Haake[258] there had been intense discussions among the participating theoreticians at Lake Como during the three weeks of the seminar. One of the results had been that, after finishing his thesis, Fritz Haake joined Roy Glauber at *Harvard University* as a post-doc. There they co-authored some articles.

Thus we can ascertain that the leading scientists of the theoretical laser schools were in close contact and correctly quoted each other in their publications at the end of the year 1967. But other US-American researchers working in the field experimentally or theoretically had a tendency to cite only American sources.

During the years 1966 to 1968 a wealth of articles appeared that finally led to the definite formulation of a quantum-mechanical laser theory. Along with Haken and his Stuttgart School, Lax, and W. H. Louisell at *Bell Laboratories* Willis Lamb again came into the picture, this time publishing with his doctoral student Marlan Scully.[259] Scully was born in Casper (Wyoming, USA) in 1939, and was only 26 years old when he was awarded his doctorate, supervised by Willis Lamb at *Yale University* in 1965.[260]

As mentioned above the different groups closely followed the work of the competing groups and always cited their results. Thus we have a good understanding of the development in the subject. A synopsis of the time structure of the publications is given in Table 4, comprising the works from 1966 until 1970.

In reviewing the contents and focus of the different publications we essentially follow the analysis given by Haken in his seminal handbook article[261], as well as the contribution by Haken and Weidlich in the proceedings of the Varenna-meeting, published by Roy Glauber.

In 1964 the first move was made by Haken[262] using the Langevin-approach[263] (Heisenberg-picture) explaining the crucial difference in the characteristics of the laser light below and above the threshold. Below the laser threshold the light resembles that emitted by an ordinary lamp. It consists of different randomly distributed light waves emitted by spontaneous emission and amplified by stimulated emission. Above the threshold the laser acts as a self-stabilising oscillator, having steady c-number amplitude complemented by small fluctuations.

---

258 Fritz Haake, private communication.
259 (Schleich, Walther, & Lamb, 2000).
260 Besides being professor at the *University of Arizona* Marlan Scully had been working for many years at the *Max Planck Institute for Quantum Optics* in Munich.
261 (Hermann Haken, 1970). See also (Glauber, 1969).
262 (H. Haken, 1964).
263 See figure 8 on page ….

The phase of the laser radiation still shows non-damped diffusion.[264] Above the threshold the fluctuations of the phase and the amplitude can be calculated and resonance effects are observed.[265] Haken's theory predicted a decrease of the fluctuations in the vicinity of the threshold at increasing laser intensity.[266] This was soon independently verified by three different experiments.[267]

Choosing the density matrix "Ansatz" (Schrödinger-picture), mostly utilised by Weidlich and his co-worker Haake, as well as by Scully and Lamb, one is able to calculate the change of the photon statistics at the laser threshold.[268] Below the threshold laser light obeys

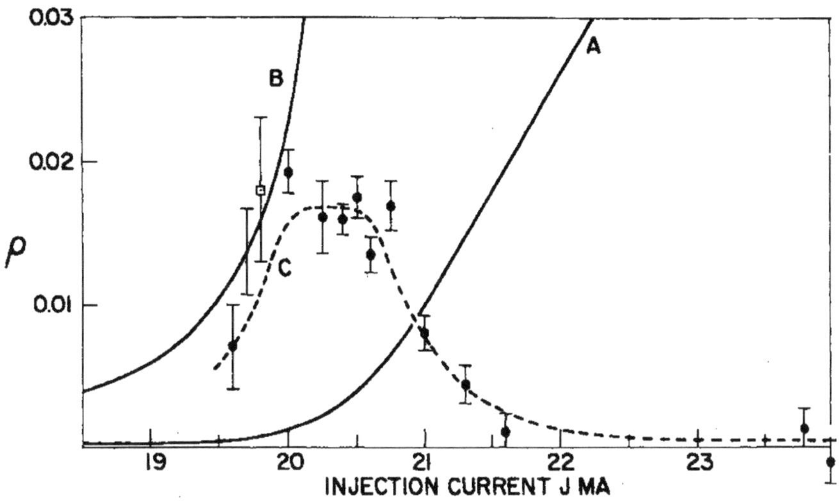

Fig. 7 Relative intensity fluctuation ρ measured for the strongest mode versus injected current J. (from: (Armstrong & Smith, 1965)

Bose-Einstein-statistics. Here the variance of the photon distribution reads

$$\Delta n^2 = n\,(n+1)$$

Above threshold we find a completely different variance subject to a Poisson distribution. Its value reads

$$\Delta n^2 = n$$

264 (H. Haken, 1964), p. 96.; Lamb in (deWitt et al., 1965); Lax in (Kelly et al., 1966); (H. Haken & Weidlich, 1966), p. 1.

265 (H. Haken & Weidlich, 1966).

266 (H. Haken, 1964).

267 (Armstrong & Smith, 1965), (F. T. Arecchi, Berné, & Bulamacchi, 1966), (Freed & Haus, 1965).

268 (Weidlich & Haake, 1965a), (Weidlich & Haake, 1965b), Scully and Lamb in (Kelly et al., 1966) p. 759 and (M. Scully & Lamb, 1966), (Melvin Lax, 1967b), p. 213, (Weidlich, Risken, & Haken, 1967).

Hannes Risken, Haken's assistant at Stuttgart, advanced a third approach, solving the problem of the quantum-mechanical laser theory, extending the method of Fokker-Planck equation. First he set up a semi-classical Fokker-Planck equation[269]. Then, above the threshold, by using quantum-mechanical defined dissipation and fluctuation coefficients he was able to solve the problem with a linearization "Ansatz".[270]

Lax and Louisell were then able to derive a c-number Fokker-Planck-equation for the electro-magnetic field and its fluctuations.[271] Other macroscopic variables such as the laser field, the collective atomic dipole moment and the total inversion rate were then calculated by the Stuttgart School in the Fokker-Planck-picture[272] as well as with help of the density matrix approach.[273]. J. P. Gordon from *Bell Laboratories* presented his solution in the course of the Varenna School.[274]

In Table 4 we document in chronological order the dates of the submission and publication dates of the most important articles concerning quantum-mechanical laser theory. As can be seen, the Stuttgart School often has a slight publication edge that diminishes in time. We would also like to mention that it is often not possible to assess the "real" chronological order because we do not know exactly when pre-prints had been distributed in advance of publication dates.

Table 4 Chronological order of publication dates of the articles on quantum-mechanical laser theory by the three laser schools of Hermann Haken, Willis Lamb and Melvin Lax.

	Date	H. Haken and the Stuttgart School	M. Lax and co-worker	W. Lamb and M. Scully
1	Submitted on 4.7. 1964	**H. Haken**: "A Nonlinear Theory of Laser Noise and Coherence I". Zs. für Physik **181** (1964), 96 - 124		
2	10. Oktober 1965	**H. Haken**: "Theory of Intensity and Phase Fluctuations of a Homogeneously Broadened Laser", Zs. f. Phys. **190** (1966), 327		
3	Submitted 21. June 1966	**H. Haken et al.**:"Quantum Theory of Noise in Gas and Solid State Lasers with an Inhomogeneously Broadened Line", Zs. für Physik **197** (1966), 207		
4	Submitted on 2.5.66; revised 24.10.1966; published January 1967		**M. Lax**: „The Rate Equations and Amplitude Noise in Lasers", IEEE Journal of Quantum electronics **QE 3** (1967), S. 37	

[269] (Risken, 1965), (Hempstedt & Lax, 1967).

[270] (H. Risken, C. Schmid, & W. Weidlich, 1966a), (H. Risken, Ch. Schmid, & W. Weidlich, 1966), (H. Risken, C. Schmid, & W. Weidlich, 1966b).

[271] (M. Lax & Louisell, 1967).

[272] W. Weidlich, H. Risken, H. Haken, 1967.

[273] W. Weidlich, F. Haake, 1965.

[274] J.P. Gordon, in: Glauber, 1969.

Table 4 (*continued*)

	Date	**H. Haken and the Stuttgart School**	**M. Lax and co-worker**	**W. Lamb and M. Scully**
5	Haken: Submitted 14.2.1967 Scully: 9.2.1967	**H. Haken, W. Weidlich, H. Risken**:"Quantummechanical Solutions of the Laser Masterequation I", Zs. für Physik **201** (1967), 396 - 410		**M. Scully, W. Lamb**: „Quantum Theory of an Optical Maser. I. General Theory" (thesis Scully), Physical Review **159** (1967), 208
6	Lax: 2.5.1967; published 12.11.1967 Haken: 3.5.1967;	**H. Haken, W. Weidlich, H. Risken**:"Quantummechanical Solutions of the Laser Masterequation II", Zs. für Physik **204** (1967), 223	**M. Lax und W.H. Louisell**: „QIX: Quantum Fokker-Planck Solution for Laser Noise", IEEE Journal of Quantum electronics **QE3** (1967), S. 47	
7	12.7.1966, revised 28.11.1966; published 10.5.1967		**M. Lax:** „QX: Density Matrix Treatment of Field and Population Difference Fluctuations", Phys. Rev. **157** (1967), S. 213	
8	Submitted 22.5.1967	**H. Haken, H. Haug**: "Theory of Noise in Semiconductor Laser Emission", Zs. für Physik **204** (1967), 262 – 275		
9	Submitted 30.5.1967	**H. Haken et al.**: "Theory of Laser Noise in the Phase Locking Region", Zs. für Physik **206** (1967), 369 - 393		
10	Submitted 10.7.1967	**H. Haken, W. Weidlich, H. Risken:** "Quantummechanical Solutions of the Laser Masterequation III", Zs. für Physik **206** (1967), 355		
11	Submitted 26.7.1967			**M. Scully, W. Lamb**: „Quantum Theory of an Optical Maser. II. Spectral Profile", Physical Review **166**(1968), 246
12	Submitted 31.1.1968			**M. Scully, W. Lamb**: „Quantum Theory of an Optical Maser. III. Theory of Photon Counting Statistics", Physical Review **179**(1969), 368

Table 4 (*continued*)

13	19.2.1968		**M. Lax und H. Yuen**: „QXIII: Six-Classical-Variable Description of Quantum Laser Fluctuations", Phys. Rev. **172** (1968), S. 362	
14	Submitted 6.4.1968	**H. Haken, R. Graham:** "Quantum Theory of Light Propagation in a Fluctuating Laser-Active Medium", Zs. für Physik **213** (1968), 420 – 450		
15	Submitted 4.5.1970			**M. Scully, D. Kim, W. Lamb:** „Quantum Theory of an Optical Maser. IV. Generalization to include Finite Temperature and Cavity detuning", Physical Review **A2** (1970),2529
16	Submitted 4.5.1970			**M. Scully, D. Kim, W. Lamb**: „Quantum Theory of an Optical Maser. V. Atomic Motion and recoil", Physical Review **A2** (1970),2529

5.6 Hermann Haken and the Stuttgart School

Willis Lamb essentially worked on laser theory only with his doctoral student Marlan Scully. At Bell Laboratories Melvin Lax joined forces with his two colleagues, J.P. Gordon and W.H. Louisell. But it was Hermann Haken, being the newly tenured Professor of Theoretical Physics at the *Technischen Hochschule* (later on *University of Stuttgart*) who had been able to involve a number of colleagues and assistants, doctoral as well as graduate students, into the laser project. This gave rise to the possibility that the three different approaches to quantum-mechanical laser theory could be executed simultaneously by the Stuttgart School. These are the Heisenberg-Ansatz using the Langevin-formalism, the Schrödinger-Ansatz by means of the density matrix equations and the route via the Fokker-Planck-equation. Haken outlined the different routes and presented them synoptically in graphical fashion in his seminal work[275] "Laser Theory". We show the illustration in Figure 6.

[275] (Hermann Haken, 1970).

In 1963 Haken appointed his first assistant Hannes Risken. He had studied physics at the *RWTH Aachen* and finished his studies with a thesis in solid state physics on “Zur Theorie heißer Elektronen in Many-Valley-Halbleitern“[276] in 1962. Risken won a grant from the *Deutsche Forschungsgemeinschaft* (DFG) and moved on to Stuttgart where Haken inspired him with laser theory. In company with Wolfgang Weidlich, Risken had been the most important collaborator of Haken in his first Stuttgart period. Risken stayed for eight years in Stuttgart and, after having written his professional thesis on “Zur Statistik des Laserlichtes“[277], received a call to the physics chair of the newly founded *University of Ulm*, which he accepted in 1971.[278]

Because Risken, with his DFG grant, was not on the institute payroll, Haken could offer another assistant position to Wolfgang Weidlich in 1963. This strengthened the research capacity in Stuttgart markedly. Haken and Weidlich knew each other from their time in Erlangen, where Weidlich had been assistant, after Haken had received the status of a “Privatdozent“. In 1959 Weidlich went back to Berlin, from which he had come, to finish his doctoral studies. Three years later he received his degree, writing a thesis on a topic in quantum field theory.[279] Then, early in 1963, Weidlich's scientific mentor, the theoretical physicist Günther Ludwig, accepted a call to the physics chair at the *University of Marburg*. Haken, who appreciated the personal qualities of Weidlich and was also looking for colleagues, approached him. He remembered the situation in an interview:

> "I was acquainted with Weidlich through our joint assistant time in Erlangen. He then went back to Professor Ludwig in Berlin. Afterwards, early in the 1960s, it was not clear how his professional career would develop. Thank god I had to fill in an assistant position in Stuttgart and, valuing Weidlich highly as a scientist and in his personality, I offered the position to him in [19]63, which he finally accepted. [...]
> Being in Stuttgart he worked his way into laser theory enthusiastically („he was on fire"). Coming from the United States in 1960 I started to develop laser theory. [...] I organised a regular seminar on this topic in my office at the Azenbergstrasse. Regularly Weidlich and others had been participants. That is how we started our collaboration. I persuaded the laser subject with the quantum-mechanical Langevin-equations, Risken, whom we mentioned earlier,

---

[276] “Theory of hot electrons in many-valley semi-conductors”.

[277] ”On the statistics of laser light“.

[278] Cited following (Albrecht, 1997), p. 236, Footnote 337 and Archiv-info des Deutschen Museums 1. Jg. Heft 1 (2000). Findbuch Nachlass Hannes Risken (1934-1994); Signature NL 131.

[279] Wolfgang Weidlich: „Die inäquivalenten Darstellungen der kanonischen Vertauschungsrelationen in der Quantenfeldtheorie“, Dissertation, FU Berlin 1962.

> interpreted it semi-classically with the Fokker-Planck-equation and Weidlich introduced an aspect of his own, the density matrix-equation or, as it is called sometimes, the master-equation".[280]

Haken had been called to the TH Stuttgart, being a well-known solid state theorist and specialist in exciton theory. We were therefore not surprised to learn that the first diploma students of Haken worked in this field.[281] Already in 1964 Herwig Sauermann (born 1938) had been the first to finish his studies with a diploma thesis in laser theory.[282] The cooperation with Herwig Sauermann, who later also did his Ph.D. with Haken[283], played an important part in Haken's work on semi-classical laser theory[284]. It had been Sauermann who represented the ideas of the Stuttgart School at the summer schools in Varenna in 1963, and at Les Houches in face of the US-American physicists.

Only three years after his appointment in Stuttgart Haken received two offers of physics chairs at the Universities of Münster and Bonn. In the course of retention negotiations that were finished at the end of the year 1963, Haken was able to allocate another permanent assistant position to his institute and to redouble the funds for scientific supporting staff.

The former assistant position was converted into a senior assistant position that was awarded to Hannes Risken in 1965. On top of that, the funds for guest professors were increased. The guest professorship position had been a matter very close to his heart.

Haken was deeply impressed by his experience of collaboration with Walter Schottky and his colleagues at the Siemens laboratory during his Erlangen time. Having seen the positive effects of the interaction between industry scientists and academia in the USA the ability to constantly exchange ideas with renowned scientists in his field was very important. As early as 1960 when he negotiated the conditions of his professorship at the *Technische Hochschule Stuttgart* he had ensured that funds for a semi-annual guest professorship position had been granted. In 1963 the funds were increased to 27,000 DM per annum. Haken used this money to invite well known scientists to Stuttgart, such as Rudolf Haag of the *University of Illinois* and Y. Toyozawa of the *University of Tokyo*, as well as his friend Wolfgang Kaiser.

---

280 Interview with Hermann Haken 16.11.2010, page 6.

281 Hartmut Haug (1963); Dieter Forster (1964); Roland Hübner (1964); Manfred Lang (1965). See also Appendix …

282 Herwig Sauermann (1964).

283 Herwig Sauermann: „Theorie der Dissipation und Fluktuationen in einem Zwei-Niveau-System und ihre Anwendung auf den optischen Maser", Dissertation TH Stuttgart 1965; (H. Sauermann, 1965).

284 See chapter 5d and 5e for details.

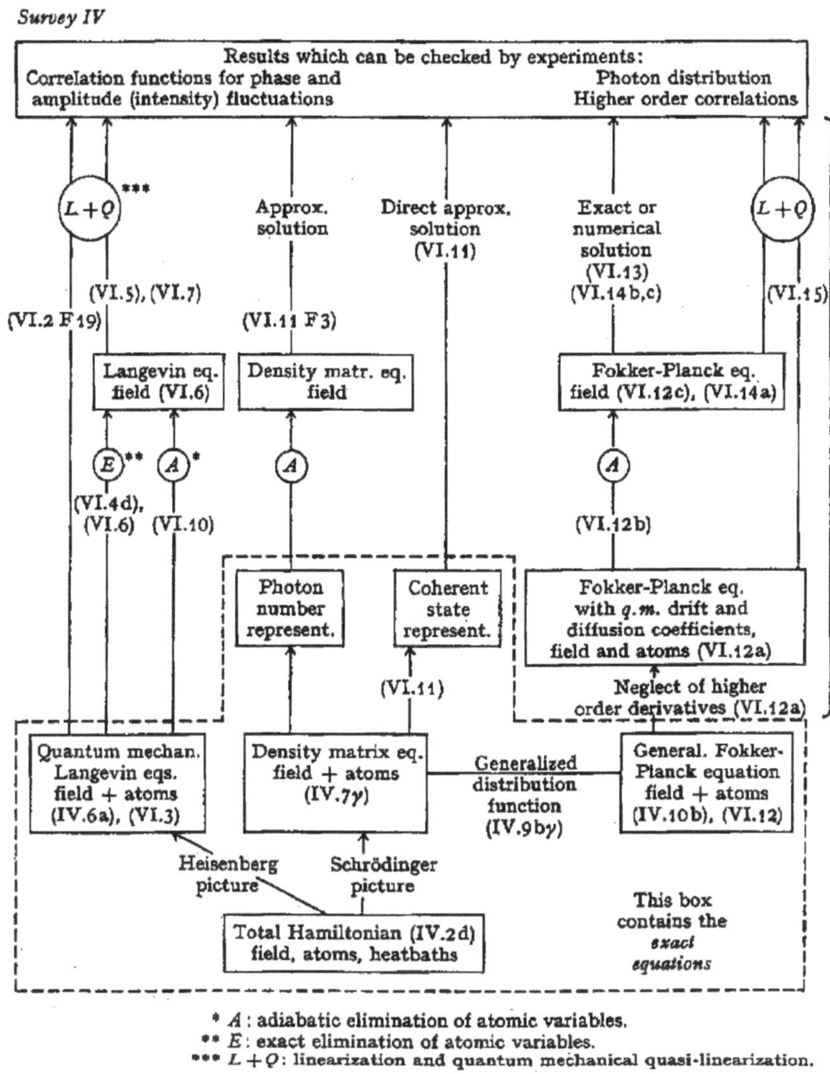

Fig. 8 Representation of the mathematical treatment of the quantum-mechanical laser theory. All three possible methods had been dealt with by the Stuttgart School. (from (Hermann Haken, 1970).

Kaiser has been one of the pioneering researchers in experimental laser physics. In consultation with Haken he used this time to finish his habilitation treatise. Immediately afterwards he received the offer of an experimental physics chair at the *Technische Hochschule München*. Haken remembered the situation:

> "During the six months he [Wolfgang Kaiser] had been in Stuttgart [being guest professor], he undertook two important

> experiments. And he had two graduate students[285], that was great. He gave a speech at the annual German Physical Society Association meeting. This talk had been attended by Maier-Leibnitz[286] from Munich. Afterwards he immediately called him to Munich".[287]

The better financial resources allowed for an expansion of his research activities. New students and co-workers were acquired: Christhard Schmid, Robert Graham, Fritz Haake and Robert Hübner reinforcing the team of Haken, Risken and Weidlich. Together with his collaborators, Haken now pushed his investigations concerning a fully quantized laser theory. His co-worker Wolfgang Weidlich recalled:

> "I had worked on relativistic quantum theory before: the so-called second quantization. It was important not to stop at the Dirac equation but to work on with the second quantization. Then you are dealing immediately with a multi-particle system. That is necessary if you want to make diversifications. I already knew how to deal with the problem in a relativistic way. It was very nice to see that Haken also mastered the second quantization. In terms of Heitler.
>
> [...] Haken knew all about it. And with the second quantization, with creation- and annihilation operators, he held the key to "quantize" the laser problem, which – I think – the Americans did not. [Using the second quantization formalism] one quantizes the atoms as well as the photons - where the relativistic "Ansatz" is inherent – and their interactions. In the Heisenberg picture you then see just how to introduce the dynamics, and the equations became clear. From a present day perspective this has been a straight forward procedure".[288]

In the years from 1963 until 1970 the Stuttgart School published no less than 58 articles on laser theory and related areas. Table 5 displays the wide range of the publications.

---

285 Dieter Pohl: „Einige Untersuchungen über die Ausstrahlungseigenschaften von Festkörperlasern", unv. Diplomarbeit TH Stuttgart 1964. („Some experiments on the emission properties of solid state lasers". Diploma thesis).

286 Heinz Maier-Leibnitz (1911 -2000) was a highly influential German nuclear physicist and long-time president of the German Research Association. Professor at the *Technische Universität München*. Retired in 1974.

287 Interview H. Haken 21.9.2010, page 23.

288 Interview W. Weidlich 18.1.2011, page 12.

Table 5 List of the publications of the Stuttgart School concerning Laser Theory during the years 1963 – 1970

Nr.	Year	Authors	Title and place of publication
1	1963	H.Haken, H. Sauermann	„Nonlinear Interactions of Laser Modes ". Zs. für Physik **173** (1963), 261 - 275
2	1963	H.Haken, H. Sauermann	"Frequency shifts of Laser Modes". Zs. für Physik **176** (1963), 47 - 62
3	1963	H. Haken, E. Haken	"Zur Theorie des Halbleiterlasers". Zs. für Physik **176** (1963), 421-428
4	1964	H. Risken	"Calculation of laser modes in an active Perot-Fabry-Interferometer." Zeitschrift für Physik **180** (1964), 150-169
5	1964	H. Haken	"A Nonlinear Theory of Laser Noise and Coherence I". Zs. für Physik **181** (1964), 96 - 124
6	1964	H. Haken, H. Sauermann	"Theory of laser action in solid-state, gaseous and semiconductor systems". in: (Miles, P.A. Hrsg.) Quantum electronics and coherent light, Proc. Int. Summerschool "Enrico Fermi" XXXI (1964), S. 111 - 155
7	1964	H. Haken	"Theory of Coherence of Laser Light". Phys. Rev. Lett. **13** (1964), 329
8	1964	H. Haken, E. Abate	„Exakte Behandlung eines Laser-Modells". Zs. für Naturforschung **19a** (1964), 857
9	1965	H. Haken	"A Nonlinear Theory of Laser Noise and Coherence II." Zs. für Physik **182** (1965), 346 - 359
10	1965	W. Weidlich, F. Haake	"Coherence-properties of the Statistical Operator in a laser model", Zeitschrift für Physik **185** (1965), 30-47
11	1965	H. Haken, Der Agobian, M. Pauthier	„Theory of Laser Cascades".Phys. Rev. **140** (1965), A 437
12	1965	W. Weidlich, F. Haake	"Master-equation for the Statistical Operator of solid State laser", Zeitschrift für Physik **186** (1965), 203-221
13	1965	H. Haken	„Der heutige Stand der Theorie des Halbleiterlasers". Advances in Solid State Physics **4** (1965), 1 - 26
14	1965	H. Sauermann	„Dissipation und Fluktuationen in einem Zwei-Niveau-System". Zeitschrift für Physik **188** (1965), 480-505

Table 5 (*continued*)

Nr.	Year	Authors	Title and place of publication
15	1966	H. Haken, W. Weidlich	"Quantum Noise Operators for the N-Level-System". Zs. für Physik **189** (1966), 1 - 9
16	1966	H. Haken	"Theory of Noise in solid state, gas and semiconductor lasers". IEEE Journ. of Quantum electronics **QE 2** (1966), 19
17	1966	H. Sauermann	„Quantenmechanische Behandlung des optischen Masers." Zeitschrift für Physik **189** (1966), 312-334
18	1966	C. Schmid, H. Risken	"The Fokker-Planck equation for quantum noise of the AZ-level System ." Zeitschrift für Physik **189** (1966), 365-384
19	1966	H. Haken	"Theory of Intensity and Phase Fluctuations of a Homogeneously Broadened Laser". Zs. für Physik **190** (1966), 327
20	1966	H. Risken	"Correlation function of the amplitude and of the intensity fluctuation for a laser model near threshold." Zeitschrift für Physik **191** (1966), 302-312
21	1966	H. Risken, C. Schmid, W. Weidlich	„Quantum fluctuations, master equation and Fokker-Planck equation." Zeitschrift für Physik **193** (1966), 37-51
22	1966	H. Risken, C. Schmid , W. Weidlich	„Fokker-Planck equation. Distribution and correlation functions for laser noise". Zeitschrift für Physik **194** (1966), 337
23		H. Haken, V. Arzt, H. Risken, H. Sauermann, Ch. Schmid and W. Weidlich	"Quantum Theory of Noise in Gas and Solid State Lasers with an Inhomogeneously Broadened Line". Zs. für Physik **197** (1966), 207
24	1967	R. Bonifacio, F. Haake	"Quantum mechanical masterequation and Fokker-Planck equation for the damped harmonic oscillator". Zeitschrift für Physik **200** (1967), 526-540
25	1967	H. Risken, H. D. Vollmer	"The influence of higher order contributions to the correlation function of the intensity fluctuation in a Laser near threshold." Zeitschrift für Physik **201** (1967), 323-330
26	1967	H. Haken, W. Weidlich, H. Risken	"Quantummechanical Solutions of the Laser Masterequation I". Zs. für Physik **201** (1967), 396 - 410

Table 5 (*continued*)

Nr.	**Year**	**Authors**	**Title and place of publication**
27	1967	H. Haken, W. Weidlich, H. Risken	"Quantummechanical Solutions of the Laser Masterequation. II". Zs. für Physik **204** (1967), 223
28	1967	H. Risken, H. D. Vollmer	"The transient Solution of the laser Fokker-Planck equation." Zeitschrift für Physik **204** (1967), 240-253
29	1967	H. Haken, H. Haug	"Theory of Noise in Semiconductor Laser Emission". Zs. für Physik **204** (1967), 262 - 275
30	1967	H. Haken, W. Weidlich	"A theorem on the calculation of multi-time-correlation functions by the single-time density matrix". Zs. f. Physik **205** (1967), 96 - 102
31	1967	H. Haken, W. Weidlich, H. Risken	"Quantummechanical Solutions of the Laser Masterequation. III". Zs. für Physik **206** (1967), 355 - 368
32	1967	H. Haken	"Dynamics of Nonlinear Interaction between Radiation and Matter". In: "Dynamical Processes in Solid State Optics" (ed. R. Kubo and H. Kamimura), Syodabo, Tokyo and W.A. Benjamin, Inc. New York , S. 168 - 194
33	1967	H. Haken, H. Sauermann, Ch. Schmid, H.D. Vollmer	"Theory of Laser Noise in the Phase Locking Region". Zs. für Physik **206** (1967), 369 - 393
34	1968	H. Haken, R. Graham	"Theory of Quantum Fluctuations of Parametric Oscillator". IEEE Journ. of Quantum Electronics **QE4** (1968), 345
35	1968	H. Haken, M. Pauthier	"Nonlinear Theory of Multimode Action in Loss Modulated Lasers". IEEE J. of Quantum Electronics **QE4** (1968), 454
36	1968	H. Haken, R. Graham	"The Quantum Fluctuations of the Optical Parametric Oscillator. I". Zs. für Physik **210** (1968), 276 - 302
37	1968	R. Graham	"The quantum-fluctuations of the optical parametric oscillator II ". Zeitschrift für Physik **210** (1968), 319-336
38	1968	H. Haken	"Exact Stationary Solution of a Fokker-Planck Equation for Multimode Laser Action". Physics Letters, **27A** (1968), 190

Table 5 (*continued*)

Nr.	Year	Authors	Title and place of publication
39	1968	R. Graham	"Photon statistics of the optical parametric oscillator including the threshold region. Transient and steady State Solution". Zeitschrift für Physik **211** (1968), 469-482
40	1968	H. Haken	"Exact generalized Fokker-Planck equation for Arbitrary Dissipative Quantum Systems". Physics Letters **A28** (1968), 286
41	1968	H. Haken, R. Graham, F. Haake, W. Weidlich	"Quantum Mechanical Correlation Functions for the Eletromagnetic Field and Quasi-Probability Distribution Functions". Zs. für Physik **213** (1968), 21 - 32
42	1968	H. Haken, R. Graham	"Quantum Theory of Light Propagation in a Fluctuating Laser-Active Medium". Zs. für Physik **213** (1968), 420 - 450
43	1968	K. Kaufmann, W. Weidlich	"Mode interaction in a spatially inhomogeneous laser". Zeitschrift für Physik **217** (1968), 113-127
44	1969	H. Haken	"Exact Stationary Solution of a Fokker-Planck Equation for Multimode Laser Action Including Phase Locking". Zs. für Physik **219** (1969), 246 - 268
45	1969	H. Haken, R. Graham	"Analysis of Quantum field statistics in Laser media by means of functional stochastic equations". Physics Letters **A29** (1969), 530
46	1969	H. Haken	"Exact Generalized Fokker-Planck equation for Arbitrary Dissipative and Nondissipative Quantum Systems". Zs. für Physik **219** (1969), 411 - 433
47	1969	F. Haake	"On a non-Markoffian master equation I. Derivation and general discussion". Zeitschrift für Physik **223** (1969), 353-363
48	1969	F. Haake	„On a non-Markoffian master equation. Application to the damped oscillator." Zeitschrift für Physik **223** (1969), 364-377
49	1969	F. Haake	„Non-Markoffian effects in the laser". Zeitschrift für Physik **227** (1969), 179-194
50	1970	H. Haken	"Theory of Multimode Effects including Noise in Semiconductor lasers". IEEE J. of Quantum Electronics **QE6** (1970) , 325

Table 5 (*continued*)

Nr.	Year	Authors	Title and place of publication
51	1970	H. Haken, R. Graham	"Functional Fokker-Planck Treatment of Electromagnetic Field Propagation in a Thermal Medium". Zs. für Physik **234** (1970), 193 - 206
52	1970	H. Haken, R. Graham	"Functional Quantum Statistics of Light Propagation in a Two-Level System". Zs. für Physik **235** (1970), 166 - 180
53	1970	H. Haken, R. Graham, W. Weidlich	"Flux Equilibria in Quantum Systems far away from Thermal Equilibrium". Physics Letters **32A** (1970), 129
54	1970	H. Haken	„Laserlicht - Ein neues Beispiel für eine Phasenumwandlung?". Schottky's Festkörperprobleme **X**, Pergamon, Vieweg 1970, S. 351 - 365
55	1970	H. Haken, R. Graham	"Laserlight - First Example of a Second-Order Phase Transition Far Away from Thermal Equilibrium". Zs. für Physik **237** (1970), 31 - 46
56	1970	H. Haken	"Laser Theory". Handbuch der Physik (Flügge, S. Hg.) Band XXV/2C. Springer, Berlin-New York 1970.
57	1970	H. Haken, R. Graham	"Microscopic Reversibility, Stability and Onsager Relations in Systems far from Thermal Equilibrium". Physics Letters **33A** (1970), 335
58	1970	H. Haken	"Quantum Fluctuations in Nonlinear Optics". Opto-Electronics **2** (1970), 161 - 167

In 1965 Haken took over the task of writing a detailed review article for the new edition of the renowned *Handbuch für Physik*. His personal aim was to present a encompassing survey of laser theory in all its facets: rate equations, semi-classical theory and quantum-mechanical laser theory, referring to the original works.

This turned out to be a very time consuming and exhausting effort. Haken and his collaborators were very much absorbed in this task during the years 1966 and 1967. Weidlich remembered his feelings:

> "To write the book responsible-minded Hermann Haken ordered the file of the original laser articles that had been published since 1960. This file comprised about 6,000 publications to which about 100 articles were added monthly. Even the strongest would be knocked down by this.

Whilst concentrating on the most important contributions, to which belonged his own works and that of the Stuttgart Laser School – about 25 doctoral thesis – he would not listen to his wife. Writing the book, night after night he drank too much coffee, until a "vegetative dystonia" set in, for which an impolite euphemism is burnout. But his strong human nature and the recreational surroundings of the Black Forest made it possible: Hermann Haken recovered rather quickly"![289]

Survey II

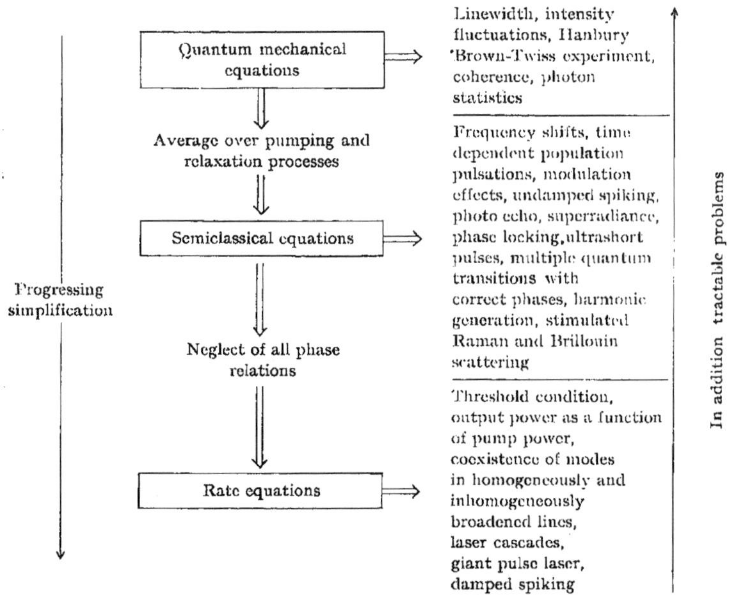

Fig. 9 Haken's survey on the increasing correctness of the different approaches to laser theory and the parameters that could be determined and checked experimentally.[290]

The "file of the original works" mainly included four collective volumes: the anthology "*Laser Literature" – A permuted Bibliography 1958 – 1966*" by Edward Ashburn[291], the contents of which had been previously published in the issues of the "*Journal of the Optical Society of America*" from May 1963 until

[289] W. Weidlich in his „Laudatio inofficialis. Unpublished manuscript written on the occasion of Hermann Hakens 80th birthday. (Archive Haken).

[290] (Hermann Haken, 1970), p. 8.

[291] (Ashburn, 1967).

January 1967; the book "*Laser Abstracts*" edited by A. Kamal[292], "*The Laser Literature – An Annotated Guide*" by K. Tomiyasu[293] and the German book "*Laser – Lichtverstärker und –Oszillatoren*" written and compiled by D. Röss.[294]

While Haken was working on the handbook article he received another offer of the theoretical physics chair of the *Universität München*.[295] Again, in the course of the following retention negotiations, he was able to extend his institute personally and financially. Another assistant lecturer position, more funds for guest professors as well as a new tenured professorship on theoretical solid state spectroscopy (including two assistant posts) were allocated. At the same time, in summer 1966, due to the "Hochschulreform" (reform of the universities) Wolfgang Weidlich was appointed to the newly constituted Second Chair of Theoretical Physics at the TH Stuttgart. It had been these positive developments that allowed Haken to provide appropriate positions to his most able co-workers Risken, Graham, Haake, Sauermann and Weidlich, thus linking them to his institute.

At last Haken had to pay attention to his diverse activities. Apart from his stressful retention negotiations and his normal duties as professor he had been traveling to and participating at the 4th Quantum Electronics Conference in Arizona (USA) in April. Only three months later he lectured in Japan and then plunged into work for the handbook article on laser theory. These activities finally surpassed his physical and mental strength. During the winter 1966 he faced burnout and had to take a forced break, so that the manuscript for the handbook article could not be finished completely.

To bring the manuscript to a close Haken applied for a research sabbatical during the summer term 1967.

> "Hereby I ask for a research sabbatical during the summer term to come, keeping my full earnings. [...] Since September 1960 I have been appointed to my chair [of theoretical physics] at the *Technischen Hochschule Stuttgart* and have not availed myself of several offers for guest professorship, taking a research sabbatical. For some time past I have been occupied in writing an article on the laser for a handbook, having by far overrun the manuscript deadline due to strain. Moreover I greatly have fallen behind with my research on laser theory. On these two grounds I am badly dependent on having a focus on writing the handbook article and on the other work during the summer term".[296]

---

[292] (Kamal, 1964).
[293] (Tomiyasu, 1968).
[294] (Röss, 1966).
[295] Personal file Hermann Haken; University Archive Stuttgart.
[296] Letter to the TH Stuttgart dated 8.2.1967; Personal file Haken.

This application was supported by the director of the mathematics and physics section and was forwarded to the Dean of the Faculty of Natural Sciences and the Humanities. The latter added:

> "Herr Haken had expressly avoided naming his illness as the reason for his application. One should be aware of the fact that Herr Haken has had a severe breakdown due to a circulation disorder and exhaustion during this term, from which he has not yet recovered completely".[297]

After the request was initially turned down, with the renewed intervention of Wolfgang Weidlich it was finally granted.[298]

Immediately after his recovery in spring 1967 Haken continued his research activities. He worked jointly with Weidlich and Risken on the seminal article "*Quantummechanical Solutions of the Laser Masterequation*" that appeared in a series of three contributions in the *Zeitschrift für Physik* in the same year. In 1967, the fundamental papers of Melvin Lax, as well as that of Marlan Scully and Willis Lamb, were published, indicating the consolidation of the first phase of the quantum-mechanical laser theory that was finally achieved during the year 1968.

5.7 Synopsis of Laser Theory in Book Form; ca. 1970

As we have shown in the preceding chapter, the essential elements of quantum-mechanical laser theory had been arrived at in the summer of 1968. Hermann Haken finally finished the manuscript on "*Laser Theory*" for the handbook in February 1969. Its delayed release due to his illness had at least one advantage: he could include the latest developments of the years 1967 and 1968 in the manuscript:

> „The main part of the present article had been completed in 1966, when I became ill. I have used the delay to include a number of topics which have developed in the meantime, e.g. the Fokker-Planck equation referring to quantum systems and the theory of ultrashort pulses."[299]

The raw manuscript for the laser article had been ready in its essential parts in spring 1967. Robert Graham, describing the situation, confirmed this in his interview:

---

[297] Letter dated 16.2.1967; Personal file Haken.

[298] Letter from Wolfgang Weidlich to Reg. Dir. Kammerer dated 10.4.1967; Personal file Haken.

[299] (Hermann Haken, 1970) Preface.

> "As I wrote my diploma thesis [during the winter term 1966/1967] Haken became seriously ill. ... The article [Handbook – article Laser Theory] existed in its first version – in a raw copy – with which I could start. He [Haken] had already given it to me when I started with my diploma work. Of course, in the beginning, I had to plough through [the manuscript]. It was the first text from which I learned the subject".[300]

Even some time before 1969 Melvin Lax had left the race. The "*Brandeis Lectures on Statistical Physics, Phase Transition and Superfluidity*" written in 1968, constituted the first detailed review of his works.[301] The Brandeis Summer School in Theoretical Physics had been held at the *Brandeis University* in Waltham/Boston (Massachusetts) in June and July 1966. In his "article", comprising not less than 200 pages, Lax summarised the results of his six "noise"-and, until then – twelve "quantum-noise" papers.[302] The most recent reference was to a research paper that had appeared in volume 172 of the *Physical Review.* The issue in question had been published in January 1968, therefore the manuscript of the Brandeis Lectures could have not been finished before that date. Comparing his results to the work of the other laser schools Lax came to a well-rounded conclusion:[303]

> „Lamb and Scully made a calculation of the density matrix field + atoms and adiabatically eliminated the atoms at the start. [...] When our equation [...] is translated into an equation for the density matrix [...] we find (see for example QX) complete agreement with corresponding equations of Scully and Lamb. [...] Haken and coworkers have introduced a quantum Langevin procedure similar to ours. However, they proceeded to introduce a classical Fokker-Planck equation which ignores the commutators. [...] Risken repeated this work for the lowest four modes obtaining results in excellent agreement with ours.
> In summary, no disagreements seem to remain, in the discussion of noise and coherence in masers [optical masers = laser]. Our work, moreover, provides a bridge between the Lamb-Scully and Haken treatments."

The comprehensive review of his Brandeis lectures constituted a kind of legacy from Lax with respect to the subject of noise and coherence. Afterwards he again

---

300 Interview with Robert Graham 29.3.2011. (Archive Haken (University Archive Stuttgart)).

301 (Melvin Lax, 1968).

302 See the table on the Lax „noise"-articles on page 73–74

303 (Melvin Lax, 1968), pp. 295-296.

turned towards the problems of phonon transport in solid state physics. This subject was of great interest for semi-conductors and was investigated in detail experimentally at the *Bell Laboratories*. After having worked for sixteen years with *Bell Labs*, Lax accepted a physics chair at the *City College* of the *New York City University* in 1971.

Finally it was his long time co-worker William H. Louisell who wrote a monograph of their mutual work in 1973.[304] In the preface of his book, titled "*Quantum Statistical Properties of Radiation*", Louisell wrote:

> „The invention of the laser was directly responsible for a tremendous development in the field of nonequilibrium quantum statistical mechanics. Although many people have made important contributions, W. E. Lamb, Jr., of Yale University, H. Haken of the Technische Hochschule in Stuttgart, Germany, and M. Lax of Bell Telephone Laboratories and City College of New York and their collaborators have been the trail blazers."

In 1974, as the last book in this "series", the textbook "*Laser Physics*" written by Lamb, Scully and Sargent III was published.[305] Written from the very beginning as a text book for university use it had a significant impact in English speaking countries. Within only three years it achieved three print runs and is regarded as the standard reference on laser theory, especially in the United States. Having the character of a textbook and, as different from Haken in his handbook, there are only few references to the original publications. This had been done deliberately and the authors justify themselves in the preface of the book:

> „[...] The laser theory discussed in this book tends to follow the approaches of the Lamb school. Parallel work of the Bell Telephone Laboratories group has been presented in the book by W. H.Louisell, "Quantum Statistical Properties of Radiation", (John Wiley & Sons, New York, 1973), while H. Haken's contribution, "Laser Theory," in Encyclopedia of Physics, Vol. XXV/2c, edited by S. Flügge, Springer-Verlag, Berlin, 1970; also Chap. 23 in Laser Handbook, gives a very complete account of the Stuttgart work.
> In keeping with the text format, no uniform attempt is made to assign credit to the original papers."

In addition to these three comprehensive monographs by the leading laser theoreticians, the publication of a multi-volume encyclopaedia, the "*Laser*

---

[304] (Louisell, 1973).
[305] (Sargent III, Scully, & Lamb, 1974)

Handbook" commenced in 1972.[306] It was the first effort to cover the broad experimental and theoretical results of laser research. The proposal for such an undertaking had come from Hermann Haken.[307] Haken contributed the article on "The theory of coherence, noise and photon statistics of laser light" and Sargent III and Scully wrote on the "Theory of Laser Operation".

The achievements in laser theory by Hermann Haken and his Stuttgart School were finally acknowledged by the American scientists. They manifested themselves not least by the fact that Haken was elected into the programme committee of the 3rd Rochester-Conference "*Coherence and Quantum Optics*", held in 1973. Other members included, among others, Nicolas Bloembergen and Melvin Lax.

5.8 Summary of Laser Theory

The history of the development of the quantum-mechanical laser theory has, as yet, not been described in detail. Bromberg and Bertolotti[308], in their comprehensive overviews of the history of the laser, nearly make no mention of it. Not only did they ignore Hermann Haken and the Stuttgart School, even the American theoreticians Lax, Louisell and Scully were not mentioned. According to their narrative, laser development occurred on an experimental basis alone, unrelated to theoretical considerations. This could not have happened, because many physics effects of the laser are only to be understood by means of quantum mechanics. Quantum-mechanical effects play a decisive role in spectroscopy precision measurements.

This can be shown by a succinct piece of information in Bertolotti's book from 1983:[309]

„ Later, in 1966, Lamb et al. finally established the quantum theory of lasers"

that completely led astray too.

One can distinguish four different stages in the chronology of laser theory development, as we have briefly shown.

The first phase was that of maser research. This preliminary work had been dominated by the American physicists Townes and Schawlow, as well as the Russian scientists Basov and Prokhorov. The decisive stimulus, theoretically and experimentally, was laid by the seminal work of Townes and Schawlow in 1958.[310] When in 1960 Maiman demonstrated that the laser effect could be achieved by many other laboratories, the Bell Labs in particular demonstrated its existence in different materials. In this second step the theoretical efforts concentrated on classical rate equations. Here the occupation numbers of the different atomic levels play the crucial role.

---

306 (F. Arecchi & Schulz-Dubois, 1972).

307 (F. Arecchi & Schulz-Dubois, 1972), Vol. 1 Page XI.

308 (Joan Lisa Bromberg, 1991), (Mario Bertolotti, 2005).

309 (Mario Bertolotti, 1983), p. 238.

310 (A. Schawlow & Townes, 1958).

From 1962 until 1964 Willis Lamb Jr. and Hermann Haken developed, independently of each other and nearly simultaneously, a semi-classical laser theory, denoting the third step. These theoretical approaches were then discussed intensely at different conferences and "summer schools".

The final stage then was triggered by Hermann Haken in 1964, pointing out and explaining the crucial difference in the radiation behaviour of the laser below i.e. above laser threshold. Together with his co-workers and students, Haken rapidly investigated different aspects of laser radiation especially at and above threshold. Line-width, fluctuations and photon number of the radiation were determined theoretically. This had been done in competition with Melvin Lax and William Louisell, American theoreticians from Bell Lab. In the past they had made important contributions to the theory of noise and fluctuations, now applying their results to the laser where they played an important and crucial role. At the same time, and in parallel, the scientific subjects of non-linear optics were developed by Nicolas Bloembergen and Emil Wolf and his colleagues at *Rochester University* as well as the theory of coherent radiation. The latter was especially advanced by Roy Glauber. During the 1970s and 1980s these efforts converged. The mathematical methods to describe the quantum-mechanical and stochastic effects of these fields are quite analogical and finally coalesced into the field of research called non-linear quantum optics.

Analysing the chronological order of the publication of the different papers, one can understand Haken's statement:

> "Back then it was a strong competition ... but one has to say explicitly, and we are still proud of it, that in Stuttgart we had always been the first to publish the theories, something the Americans will not admit voluntarily or not at all".[311]

Haken, Risken and Weidlich, representing the Stuttgart School, consistently published some months prior to their American colleagues. But it is also true that in between submitting the papers, revising and the final publication date there was often overlap.

There is one thing that is entirely singular and exceptional in the work of the Stuttgart Laser School, and that is dealing simultaneously with laser theory through three different approaches: the Heisenberg-model, the density matrix or master equation "Ansatz" and using the Fokker-Planck equation. This can only be ascribed to the rare liaison of three distinguished physicists: Hermann Haken, coming from solid state physics and exciton research; Wolfgang Weidlich, who brought from Berlin his knowledge of the mathematical tools of quantum field theory, and Hannes Risken immersing himself in the use of the Fokker-Planck formalism. A comparable situation of this character did not happen in the United States.

---

[311] Cited after (Albrecht, 1997), p. 241.

If we ask ourselves why the results and efforts of the Stuttgart Laser School did not subsequently receive the attention it deserved, we are left with speculations. One factor that should not be underrated might be its final publication in book form. Haken published his detailed review in the new edition of the renowned "*Handbuch der Physik*", thoroughly citing all relevant papers. These handbooks normally are very expensive, one will find them in libraries and often you cannot borrow them. Therefore students only rarely consult them in depth and they do not use it in their usual studies. Scully, Sargent III and Lamb on the other hand published their work in the English language, explicitly written as a textbook for students. Of course, the book was widely disseminated. Even though they mentioned the different laser theory schools in their foreword, as time went by this awareness was lost in the users of the book.

In the years before data based publishing was common in the physical sciences, bibliographies played an important role, when a scientist wanted to follow the development of a special area of research. Therefore it seems striking that in the early years of laser research the German journal *Zeitschrift für Physik* had not been referenced in one of the leading American bibliographies, and thus the articles that appeared in this journal had not been cited. In comparison, the publication of Kiyo Tomiyasu[312] had been the basis of five comprehensive bibliographies on lasers that had been published by the renowned "*IEEE Journal of Quantum Electronics*" from 1965 until 1967. Subsequently the developments of the Stuttgart Laser School, regularly publishing in the *Zeitschrift für Physik,* went unnoticed by some American laser scientists.

Nevertheless the work of Hermann Haken and his co-workers was clearly recognised in the circle of experts, especially after the publication of his review article in the *Handbuch der Physik.* This can also be seen in the many tenured professorships in theoretical physics that his students later received. Notably Fritz Haake and Robert Graham made significant contributions to the fields of quantum optics and stochastics. Graham finally was honoured with the *Max Planck* medal of the *German Physical Society*, the highest award a theoretical physicist can receive.[313]

Laser theory set the course of Hermann Haken for his further research programme. Late in the 1960s, recognition that the character of the laser radiation was completely different below and above threshold convinced him that the laser is a paradigm for phase transition.[314] This led him to intense activity with the theory of self-organising processes, far away from thermodynamic equilibrium. These so called (thermodynamically) open processes are dominant in nature and Haken finally led to the foundation of the "new science" of synergetics. This Greek word means a "theory of working together". Nevertheless it still holds true that the laser is the "trailblazer of synergetics" as he declared in a talk presented at the 4th Rochester conference in 1977.[315]

---

312 (Tomiyasu, 1968).
313 Haken received the Max Planck medal already in 1990.
314 (Graham & Haken, 1970).
315 (Mandel & Wolf, 1978), pp. 49-62.

Chapter 6
Early Years of Synergetics: 1970 – 1978

Writing the manuscript of his article "*Laser Theory*" in the *Handbuch der Physik,* Hermann Haken became overworked. In the winter of 1966-1967 he fell ill, showing the burnout he experienced, but recovered soon after staying some weeks at the medical wellness clinic, *Bühlerhöhe* near Baden-Baden in the Black Forest.[316] Chastened by his illness, he applied for a research sabbatical to finalise the book. It was during this period that for the first time he came to know scientific contacts outside of solid state and laser physics, a period that strongly influenced him in his subsequent research activities. The conferences in which he participated dealt with the question of the causal connection between physics and biology, a question of eternal historical and philosophical tradition.

6.1 The Versailles Conferences from 1967 to 1979

Later on these meetings were nicknamed "Versailles Conferences" after the Hotel Trianon at Versailles, the place where they occurred. French physician, Maurice Marois[317] and the French aristocrat, Francois de Clermont-Tonnerre developed the meetings; Marois being the driving force. In 1960, the starting point had been the creation of the *Institut de la Vie*[318] setting for itself the goal[319]

1. To stimulate the interest of scientifical, philosophical, spiritual and political circles, the interest of statesmen and other persons of responsibility in labor and economic fields, and the interest of the public in general in the problems raised by the maintenance and the development of Life, and primarily, of human Life.
2. To conduct research and study, and to facilitate exchanges of ideas concerning the problems of life between persons of

[316] Private communication Hermann Haken.

[317] Maurice Marois (1922 – 2004) was a French physician. Founder of the *Institut de la Vie*. For biographical data and a review of his works see `http://www.mauricemarois.net/Textes/biographie/index.html`, recalled 20.03.2013.

[318] (Marois, 1998).

[319] (Marois, 1997a), p. 35

B. Kröger, *Hermann Haken: From the Laser to Synergetics*,
DOI: 10.1007/978-3-319-11689-1_6

different disciplines and cultures, to promote a scientific approach of all the questions concerning life and to analyse their implications in economic, ethical and educational fields.

3. To prepare and present all kinds of studies or suggestions in relation with the aims of the institution, and to help Governments, international institutions, private associations or all organizations concerned, in order to contribute to the maintenance and development of life and mankind.

Looking back these somehow pretentious goals should be reached at organising international conferences with renowned scientists. Unfortunately, Marois was still lacking the stirring subject that he needed to attract the "scientific celebrities" he courted. He finally found this topic somehow accidently during a conference at the Alp resort Alpbach in Austria. There he met Hermann Haken's longtime friend Herbert Froehlich and his wife, Fanchon Froehlich. The biographer of Herbert Froehlich has described the encounter:[320]

> "Around 1965, Fröhlich and his wife were in Alpbach (Austrian Tyrol), where she was attending a Conference on science and life, while he indulged his love of mountain-climbing. Quite by chance, she there met Maurice Marois, a professor of Medicine at the Sorbonne and founder in1960 (together with Francois de Clermont-Tonnerre) of l'Institut de la Vie in Paris, and duly embarked on a discussion of the relation of physics to life, casually telling him that her husband was a famous theoretical physicist to whom she later introduced him.
> Keen to pursue the contact, Marois suggested that they meet in Paris, where, during a lunch, Fröhlich's wife happened to mention that, according to Wigner, 'life' was impossible from the point of view of quantum mechanics. At this, Marois became excited, and asked Fröhlich what could be done to bridge the gap between physics and biology. At the time, he was rather reluctant to get involved, since not only had he never really been interested in this question, but also, because he was then immersed in pure theoretical physics from which he did not wish to be deflected. Marois, however, persisted and eventually Fröhlich agreed to help organize what was to be the first of many successfull international Conferences on theoretical physics and biology that were to be held, under the auspices of l'Institut de la Vie, at the Trianon Palace Hotel in Versailles. These conferences which continued, biennially, until

320 (Hyland, 2006).

> 1988, were attended by highly eminent physicists and biologists, including such people as Onsager, Prigogine, Crick, Edelman, Cooper and Wigner himself."

The general topic of the *Versailles Conferences,* initiated by the *Institut de la Vie* has been "From Theoretical Physics to Biology." The meetings were aimed at facilitating the discussion between biologists, chemists and theoretical physicists on the subject of whether life can be explained by the laws of physics or if other aspects are needed.

Of course that discussion has not been a new one. For hundreds or even thousands of years philosophers and other thinkers had pondered this topic. The dispute has serious and profound aspects concerning philosophy, theology and the natural sciences. There are questions such as, if a creative god exists, if man has a special role in nature, the dispute on the theses of Charles Darwin and not least, the question of if we need a "Deus ex machina" in nature to explain certain features of life. For many natural scientists, the discussion had been revived in 1944 by a small book written by the Austrian physicist Erwin Schrödinger. Being one of the creators of quantum theory, Schrödinger in his work titled "*What is life?*"[321] took the view that the laws of the natural sciences are sufficient to explain the phenomenon of life on earth:

> „How can the events in space and time which take place within the spatial boundary of a living organism be accounted for by physics and chemistry? The preliminary answer which this little book will endeavor to expound and establish can be summarized as follows: The obvious inability of present-day physics and chemistry to account for such events is no reason at all for doubting that they can be accounted for by those sciences.."

After the Second World War Schrödinger's book was the reason for many natural scientists to devote themselves seriously to this question, enormously advancing the areas of biophysics and biochemistry. A big step forward was also the deciphering of the genetic code and the triple helix formation of the DNS by Francis Crick and Charles Watson in 1953. Nevertheless, grave doubts remained because DNS and the biological cell in particular are extremely complex systems with highly interconnected dynamical relations. There was no idea of how this had come into being. The Hungarian born American theoretical physicist and Nobel Prize winner Eugene P. Wigner had dealt with that question in an article written in 1961 having the title "*The probability of the existence of a self-reproducing unit.*"[322] He arrived at the conclusion that the probability was infinitely small that so many parts (being in place in a cell) could have assembled spontaneously to an ordered system. It is before this background that Franchon Froehlich made her

---

[321] (Schrödinger, 1944).
[322] (Wigner, 1961).

statement to Maurice Marois at Alpbach explaining that (according to Wigner) from the point of quantum mechanics, the coming into being of life is not possible.

The organising committee of the first conference that was held from the 26th until the 30th of June 1967 in Versailles, comprised only five people: the physicists Herbert Froehlich, Léon Rosenfeld[323], André Lichnerowicz[324] as well as the biologist Pierre-Paul Grassé[325] and the physician Maurice Marois. Seventy-two scientists had attended, among them eight Nobel Prize winners. The topics encompassed discussions on subjects like "dissipative systems," "non-equilibrium thermodynamics," "information in biology" and "physiological mechanism and their theoretical description."

Hermann Haken benefited from his longstanding acquaintance with Herbert Froehlich and was already invited to the very first meeting. Likewise, the Russian born Belgian chemist Ilya Prigogine[326] who, at the time, was working in the field of nonlinear thermodynamics and so called "dissipative structures" played an important role; he gave the introductory presentation on "Structure, Dissipation and Life." Prigogine had studied chemistry at the *Free University of Brussels.* During his life, he had occupied himself with the question of how systems can be described theoretically that are not in thermodynamic equilibrium. He was especially interested in the question of which role was played by the irreversibility of processes in nature.[327] He also coined the word "dissipative systems," describing processes that lead to stable configurations in systems where energy and matter are fed in and out. Ludwig von Bertalanffy had said these systems to be in a state of "Fließgleichgewicht" (flux equilibrium). What is more, the laser is such a dissipative system.[328] In his investigations, Prigogine came up with a theorem he called "minimum entropy production." Using this theorem, he tried to apply classical thermodynamics to systems not in thermal equilibrium.[329]

In the discussion that followed at the end of the presentations of the first day of the conference, Hermann Haken introduced the laser as a system being far from thermodynamic equilibrium that shows a transition from total disorder to total

---

323 Léon Rosenfeld (1904 - 1974) was an eminent Belgian theoretical physicist working in close connection with Nils Bohr. He was the longtime responsible editor of the leading journal *Nuclear Physics,* therefore being very well interconnected within the physics community.

324 André Lichnerowicz (1915 - 1998) was a French mathematician and physicist, Professor at the *College de France* and member of the *Academie des Sciences.*

325 Pierre-Paul Grassé (1895 - 1985) was a French biologist, Professor of Zoology at the Université de Paris. In 1967 he had been *Président de L'Académie des Sciences.*

326 Ilya Prigogine (1917 - 2003) was a Russian-Belgian chemist holding a chair at the Université Libre de Bruxelles and simultaneously, beginning in 1967, at the *University of Texas* in Austin. In 1977 he won the Nobel Prizes in chemistry for his work on dissipative systems.

327 In the basic equation of physics in principle all processes are time reversible.

328 A comparison of Prigogine's theory of dissipation with the work on synergetics of Hermann Haken is given in more detail in chapter 9d of this book.

329 (Prigogine, 1955), see as well (Prigogine, 1962).

order. He demonstrated that the light field could be described by a potential of the form

$$V = - \text{const}'' \cdot \frac{1}{2} |b|^2 + \text{const}''' \cdot \left\{\frac{1}{4} |b|^4\right\}; \; \text{const}'' = \sigma_0 \cdot \text{const} - \chi$$

σ_0 being a measure for the pumping intensity (energy that is fed into the system). The potential is represented by the following graph

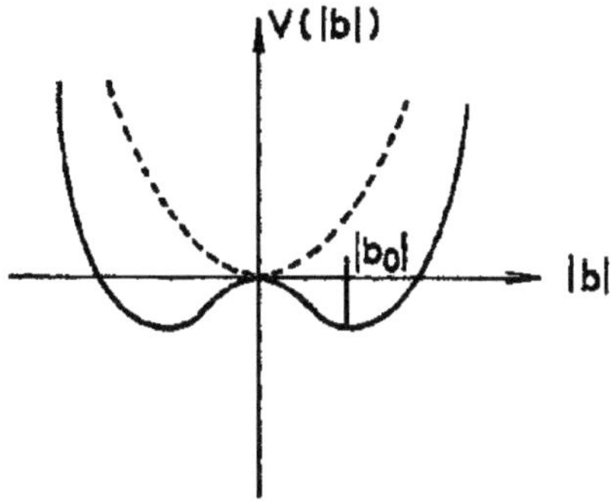

Fig. 10 Graphical display of the laser potential below and above threshold

Haken had been well acquainted with this picture by his works of the year 1964. For most attendees of the meeting, it must have been quite new. The important point was that by feeding in energy randomly above a certain value (the threshold) a stable ordered state occurred.

Haken concluded[330]:

> „If σ_0 is small, the potential looks like the dotted line [... the lightfield] performs a damped oscillation. On the other hand, if the parameter σ_0 becomes bigger and bigger we find the potential given by the solid line. That would mean that the lightfield is now stabilized at a certain value b_0. This is a case of more complete order."

At the first Versailles Conference Hermann, Haken came to know Manfred Eigen, a German physicist from Göttingen where he did biophysical chemistry research at the Max Planck Institute. In autumn of the same year, Eigen received the chemistry Nobel Prize for his work *"for their studies of extremely fast chemical reactions, effected by disturbing the equilibrium by means of very short pulses of energy."* During the following years Manfred Eigen became an important dialogue partner for Hermann Haken and the scientists formed a lifelong friendship. Having the new status as a Nobel laureate, Marois immediately elected Eigen into the organising committee of the second *Versailles Conference* that

---

330 (Marois, 1969), pp. 94-97.

again took place in Versailles in 1969. The organisation panel of that meeting had grown rapidly. In 1967, there had been five scientists, now the list comprised not less than 24 members.[331] They met in advance to discuss the list of participants and the speakers. Studying the names of the organising committee, one can feel the efforts Marois has undertaken to attract "reputable" scientists. The aim to enhance the significance of the conference and, by that, the significance of the *Institut de la Vie,* can clearly be seen.

After the successful performance of the first meeting, every two years Marois was able to organise the Versailles Conferences on "*Theoretical Physics and Biology.*" In total there were eight conferences until 1981; then the frequency slowed down. The last two meetings were held in 1984 and 1988.[332] We like to stress the fact that Hermann Haken participated at each of the first nine reunions, leaving out only the last one, because the spectrum of subjects had grown too much apart from his research interests.

The dominating characters in the first conferences had been Ilya Prigogine and Manfred Eigen. They delivered fundamental lectures and/or presided over the discussion sessions. Eigen spoke about the early phases of evolution: "Alkali ion carriers: specificity, architecture and mechanisms" (jointly with Ruthild Winkler; 2nd Versailles Conference),[333] "Introduction to the first steps of Evolution and the nature of life" (3rd Versailles Conference), "The origin of biological information" (4th Versailles Conference). Prigogine and his colleagues from Brussels spoke about "Dissipative structures in biological systems" (2nd Conference), "Fluctuations and the Mechanism of Instabilities" (with G. Nicolis, 3rd Conference), and "Models for cellular communication" (with G. Nicolis and R. Lefever, 4th Conference).

As can be seen in the conference proceedings, Haken lively engaged in the discussions. Finally, he got the chance to lecture at the third meeting in 1971. The title of his presentation "Cooperative phenomena far from thermal equilibrium" fit seamlessly into a whole range of other publications from this time.[334] But we should note that Haken did not mention the word synergetics during his talk. The equations that Haken used to describe the self-organising effect of the laser radiation baffled Manfred Eigen:[335]

---

331 (Marois, 1997b), p. 28. The names of the organizing committee of the 2nd conference: P. Auger (F), S. Bennett (USA), S.E. Bresler (USSR), G. Careri (I), E. Cohen (USA), A. Cournand (USA), M. Eigen (D), A. Fessard (F), H. Fröhlich (GB), A. Katchalsky (ISR), M. Kotani (JPN), R. Kubo (JPN), A. Lichnerovicz (F), H. Longuet-Higgins (GB), P. Löwdin (SWE), F. Lynen (D), O. Maaloe (DK), M. Marois (F), K. Mendelssohn (GB), R. Mulliken (USA), I. Prigogine (B), L. Rosenfeld (DK), S. Sobolev (USSR), A. Szent-Gyorgyi (USA).

332 (Marois, 1997b). Conference proceedings only were published for the first four gatherings: (Marois, 1969), (Marois, 1971), (Marois, 1973), (Marois, 1976).

333 (M. Eigen & Winkler, 1971).

334 (Haken, 1974a). (Haken & Wagner, 1973).

335 Interview with Manfred Eigen and Ruthild Winkler-Oswatitsch from 24.5.2011 in Göttingen, page 6 (Archive Haken; University Archive Stuttgart).

> "I got to know him [Haken] in Paris. [...]: I had to give a presentation, after the one he delivered, and had been late. ... Then I spoke about my findings and was going to write the equations down at the blackboard. I stopped short and said "they are already there, how does this come about?" [Interviewer: ... the rate equations...] Yes, it have been the ones from Haken, but very much alike, that is to say the autocatalytic term."

Events like these must have been a great stimulation for Haken. His acquaintance with Manfred Eigen deepened over the years. For instance for more than 25 years he had been participating in the annual "Winterseminare" (winter seminars) organised by Eigen in the Swiss ski-resort Klosters. These fortnightly seminars had been (and are) the perfect place to exchange, in an informal way, ideas on questions dealing with early stages of molecular and evolutionary biology. That is to say, how ordered and complex structures evolve.

It is hard to say how great the influence of the *Versailles Conferences* had been on Haken's thinking about synergetics. In the early years, until about 1975, at the *Versailles Conferences* many subjects had been addressed that, later on – but in a quite different context – had been dealt with at the *ELMAU Conferences* organised by Hermann Haken. A comparison of the topics at the first three Versailles Conferences with the topics of the first *ELMAU Conferences* should illuminate this:[336]

> **Versailles 1967:**
> Theoretical concepts: statistical mechanics, dissipative structures, neurocybernetics; physical chemistry of life, non-equilibrium thermodynamics; information in biology; Physiological mechanism.
> **Versailles 1969:**
> Order in physical systems; structure and the functions of life; self-organzation; dissipative structures in biology; generation of biological molecules; mutation and evolution processes; Information and biological systems; iformation storage in the central nerveous system.
> **Versailles 1971**:
> Physical aspects of the order in biological systems; the first steps of evolution and the nature of life; systems of order recognition and of recognition; systems of sensorial analysis; neurophysiological aspects of vision; perespectives of theoretical physics and biology and their social implications.
> **Elmau 1972:**
> Mathematical and physical concepts for cooperative systems; instabilities and phase-transition-like phenomena in physical systems far from thermal equilibrium; biochemical kinetics and population dynamics; biological structures; general structures.

---

[336] Translation of the French original by the author.

During the early conferences in Versailles the discussions centred on topics like self-organisation and on phenomena related to the thermodynamics far from thermal equilibrium. Subsequently, the subject matters developed towards biological specialities[337] that had been of no special interest to Hermann Haken. In the seventies and eighties, he focussed on questions of the physical and mathematical nature of self-organisation.[338] On the other hand, besides the subjects under discussion, the personal contacts at the *Versailles Conferences* had been invaluable. The meetings created a network of connections that had a positive bearing on Haken's future work in synergetics. Over the course of time, not less than 37 participants of the *Versailles Conferences* also attended the *ELMAU Conferences* organised by Haken. That shows the intense interaction and fruitful exchange of ideas between the members of the two series of events.[339]

6.2 The Foundation of Synergetics: Analogies and Phase Transition

Let us now have a look at the developments that motivated Hermann Haken at the turn of the year 1969 - 1970 to create the new research area of synergetics. Of course, the starting point had been his work on the laser. The phenomena at the laser threshold had especially fascinated him from the very beginning. A detailed examination of the mathematical structure of the equations at this point caused him to make the prediction that the laser light above threshold would differ completely from the light emitted below threshold. Soon after, this prediction was experimentally verified.

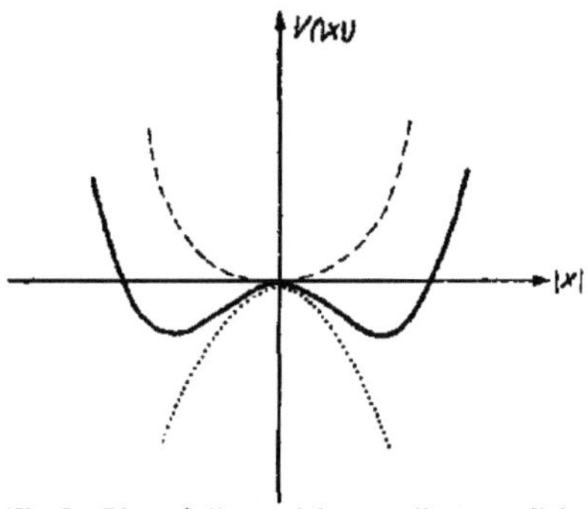

Fig. 2. Plot of "potential energy" versus light amplitude. – – – below threshold (*linear and nonlinear* theory); ······· above threshold, *linear* theory leads to instability; ——— above threshold, *nonlinear* theory

Fig. 11 Plot of the potential behavior below and above laser threshold (from (Haken, 1964))

[337] See for instance the topics of the 1977 conference: „Proteins and small molecules: structure, catalysys and dynamics; proteins and lipids in membranes; antigenes and antibody detection in cells; the development of synapses; modification of synapses".

[338] See especially the chapters on the *Elmau Conferences*: 6d, 6h, 7d, 7e and 8a.

[339] Appendix 4: List of participants to both series of events in annex ….

In his calculations, Haken introduced a potential that changes with the amount of energy applied. At threshold spontaneous symmetry breaking occurred and above threshold two new energy minima appeared, enabling stable states of the system.

Symmetry breaking of this kind arises also in phase transitions and is well known to physicists by the Landau theory of ferromagnetism.

In 1969, Robert Graham, then a doctoral student with Haken, was working on the quantum mechanical formulation of the laser phenomenon with infinite degrees of freedom. In his dissertation he came across an equation, the structure of which reminded both of them of the theory of superconductivity formulated by Vitaly Ginzburg and Lev Landau.

Robert Graham remarked in retrospect:[340]

> "The very specific had been that there is no resonator from the very beginning. Instead imagine having a laser medium that is pumped [with energy] and solely by the light amplification factor one gets the laser activity. And then one can deal with the whole subject mathematically not only with discrete modes but one treats the system also as a spatially extended continuous space-time field.
>
> That had been very instructive for the analogy of phase transitions. At that time the phase transition theory best understood had been the one from Ginzburg-Landau on super conductivity. When I chose this ansatz for the laser with spacially extended propagating fields we found a probability functional that showed a form resembling the one well known from the Ginzburg-Landau theory. In this respect that had been enlightening."

Even before the publication of their paper in the *Zeitschrift für Physik* Haken presented the results at the Spring Meeting of the *German Physics Association,* early in 1970 in Freudenstadt.[341]

Haken and Graham detected the following analogue structures in the corresponding equations:

[340] Interview with Robert Graham from 29.3.2011, page 10.

[341] The meeting was held from 6th till 11th of April. See (Haken, 1970).

Laser equation:

$$f = N \exp\left(-\frac{B}{Q}\right)$$

$$B = \int \left\{\alpha\, |E(x)|^2 + \beta |E(x)|^4 + \gamma \left|\left(\frac{d}{dx} - i\frac{\omega_o}{c}\right) E(x)\right|^2\right\}\, dx$$

Ginzburg – Landau theory of superconductivity:

$$f = N \exp\left(-\frac{F}{kT}\right)$$

$$F = \int \left\{ \alpha |\psi(x)|^2 + \beta |\psi(x)|^4 + \frac{1}{2m} \left|\left(V - \frac{2ei}{c} A\right) \psi(x)\right|^2\right\} dx$$

Fig. 12 Comparison of the mathematical structure of the laser with the basic notions of the Ginzburg-Landau theory of super-conductivity

When asked about this by the author, Hermann Haken commented:[342]

> "When Graham and I look at this expression in more detail we strongly recalled the Ginzburg-Landau theory on super-conductivity. In this theory, as in the whole theory of thermodynamics, one can determine - by means of the free energy - the probability to find certain configurations of the pair wave function. The expression from Ginzburg-Landau has the following form: a quadratic term, a term of the 4th order and a term having a spatial derivation. [...] A comparison of our expression for the laser with the expression of Ginzburg-Landau shows perfect analogy. Even the vector potential A is represented by the constant ω_o in one dimension. Therefore we stated that the laser must show the same statistical properties at the phase limit as a super-conductor does."

But what is so very special about this finding? Until then, all phase transitions looked at in physics, for instance the phase transition from the solid phase to the fluid phase (ice to water) or the transition within ferromagnetism or super-conductivity occur in systems that are in thermal equilibrium.[343] In the laser

[342] (Haken, 1970), p. 362.
[343] (Haken, 1970), p. 352.

system that is not the case. This system, pumped heavily by the energy applied, is in a state far from thermal equilibrium. Of course, symmetry breaking is not the only feature of phase transitions; other factors also apply, as Haken and Graham were fully aware:[344]

> "Notwithstanding that laser light is a system far from thermal equilibrium, at laser threshold we observe all the characteristic features of a phase transition: symmetry breaking, instability of a "hard mode," critical fluctuations, symmetry restoration through excitation, "off-diagonal long range order," existence of an order parameter. The electric field probability distribution of the laser light shows a perfect analogy to the pair wave function given by the Ginzburg-Landau theory of super-conductivity. Therefore, we have proven that laser light undergoes a second order phase transition at threshold."

At the same time, and independently from the Stuttgart scientists, this analogy of the laser light undergoing a second order phase transition at threshold had been noticed by Marvin Scully and Vittorio deGiorgio.[345] It should be mentioned, however, that in one of our interviews, Hermann Haken remarked as an aside that during the 1967 summer school at Varenna, where Marvin Scully and Vittorio deGiorgio had been attending, Wolfgang Weidlich presented their concept of symmetry breaking. It might be that Scully and deGiorgio could have had this concept "in the back of their minds."[346]

Being an unsolved theoretical problem, phase transitions was a subject of research interest by a greater physicist's community at that time. To give an example: Siegfried Grossmann, a theorist coming from Marburg, gave a talk on "Analytical properties of thermodynamical functions and phase-transitions" at the DPG spring meeting in 1969 in Munich. In his presentation he dealt with systems in the immediate vicinity of the threshold where phase transition occurs, but not directly at this point.[347] It was in June 1969 that Haken then partook at the 2nd *Versailles Conference*. In autumn of the same year "phase transitions" was the topic of an invited talk[348] by H. Thomas, at that time with the IBM-research laboratory in Zurich.[349] At this annual DPG meeting the Stuttgart theory school played a prominent role. Haken gave an invited talk on "Non-linear optics," while

---

344 (Haken, 1970), p. 351.

345 (Scully & deGiorgio, 1970). The article had been submitted on the 29th December 1969, but was published only in the October issue 1970.

346 Interview with Hermann Haken 20. 4. 2011, Pages 4/5.

347 Using classical mathematical tools there will be divergent terms.

348 Die 34. Annual physics meeting was held from 29. September till 4. October 1969 in Salzburg.

349 Later on H. Thomas has been professor of theoretical physics at the University of Frankfurt.

Hannes Risken lectured on "Ultra short impulses" and Robert Graham presented the results of his thesis on "The theory of the parametric oscillator."

This up-to-date topic and the results from the dissertation of his student Robert Graham, induced Hermann Haken to give a university lecture course during the winter term 1969-1970 and in the summer term 1970 titled "Fließgleichgewichte, Phasenübergänge und Fluktuationen in Quantensystemen weit weg vom thermischen Gleichgewicht".[350] In this lecture, Haken used the name "Synergetics" for the first time. Finding this name, Haken tells a nice anecdote:[351]

> "I had searched the dictionary of my father, who had studied Greek, and found another word "Symkamnetik." The meaning of it was "to struggle about something." Well, I thought, that's quite nice. But later on I learned that one could also translate it into "to sleep together" [...] and there it goes ... I went to my close friend Hans Christoph Wolf who had been educated at the "Latina" [in Halle] there being taught Greek and discussed the subject with him. He reckoned that Symkamnetik would be not so nice and proposed the word Synergetics. I owe the word to Hans Christoph Wolf, but we searched for it together. I had been on the wrong track with Symkamnetik."

The scheme of the lecture course and part of the notes are preserved in the archive of Haken. The course was partitioned into 12 chapters where the different segments were attributed to the referents Haken (H) and Graham (G).[352] A transcription of the handwritten notes read:

§ 1. Voraussetzungen und Plan (H) [*assumptions and scheme*]
§ 2. Klassisch: Langevin – Fokker-Planck (H) [*classic: Langevin-Fokker-Planck*]

[350] "Flux equilibria, phase transitions and fluctuations in quantum systems far from thermal equilibrium". See also the footnote in their article (Haken & Graham, 1971a) „Der vorliegende Artikel basiert auf einer von den Autoren an der Universität Stuttgart im Sommersemester 1970 gehaltenen Vorlesung." (This article is based on a lecture given at the University of Stuttgart during the summer term 1970).

[351] Interview with Hermann Haken from 21.09.2010, page 28; Archive Haken. Since 1972 Haken also used the word „synergetics" in his English written publications. Because of the fact that the American architect Buckminster Fuller published in 1975 a comprehensive work about "geometric figures in architecture", things often get mixed up. Especially performing keyword search, one has to keep apart the two concepts because they describe two completely different subjects. See (Fuller, 1975).

[352] Whether the lecture course had been given by the contributors in the respective sequence cannot be reconstructed anymore.

§ 3. Quantenmechanisch: Dichtematrixgleichung (G) [*quantum-mechanical: density-matrix-equation*]
   a) Elementar [*elementary*]
   b) Agyres u. Kelley [*Agyres and Kelly*]
§ 4. Quasiverteilungsfunktion (H) [*quasi-distribution function*]
§ 5. Q.[uanten]M.[echanisch] Fokker-Planck: Bose (G)
§ 6. Q.[uanten]M.[echanisch] „ „ : beliebige (G) [*any*]
§ 7. Fliessgleichgewichte, Definition (H) [*flux equilibria, definition*]
§ 8. Bedingungen für Fliessgleichgewichte (H) [*conditions for flux equlibria*]
§ 9. Phasenübergänge und deren Klassifizierung (G) [*phase transitions and their classifications*]
§ 10. Laser, Parametrische Prozesse (H,G) [*laser, parametric processes*]
§ 11. Biologische Prozesse (G,H) [*biological processes*]
§ 12. Ordnungshierarchien (H,G) [*hierarchies of order*]

Looking at the scheme of the lecture we can deduce that in Chapters 1 – 9 we are concerned with a theoretical lecture on statistical physics that proceeded from "first principles" dealing with the elementary parts of a quantum mechanical system like photons, electrons or molecules. Only after the equations and the problem-solving approach had been constructed in Chapter 10 to 12, examples of applications in the macroscopic world are discussed. We guess that at that point of the course Hermann Haken introduced the word "Synergetics."[353]

Whilst the lecture course was given, and shortly after the spring meeting in Freudenstadt, Graham and Haken, together with their colleague Weidlich, mathematically extended the concept of phase transition over and above the transition point. It has been Weidlich who called attention to Haken, that a stationary state in an open system far from thermal equilibrium (energy introduced and dissipated) had been named "flux equilibrium" by the biologist Ludwig von Bertalanffy (1901 - 1972). In May 1970, they submitted a paper to *Physics Letters*

[353] No lecture course with this title is to be found in the university calendars of the TH Stuttgart. In the winter term 1969/70 Hermann Haken is mentioned with lectures on „*Quantenfeldtheoretische Methoden in der Festkörperphysik II (mit Übungen)*" and with a seminar on „*Energietransport in biologischen Systemen*", performed jointly with W. Weidlich and M. Wagner. In the summer term 1970 the seminar had been named „Energieübertragung in physikalischen und biologischen Systemen", again with his colleagues. The name of Robert Graham did not appear. There seems to be a discrepancy between the titles printed in the university calendars (printed early in advance) and the lectures that had been actually held. Specifically asked to that point Hermann Haken as well as Robert Graham confirmed explicitly that the above mentioned lecture „Fließgleichgewichten..." had been given.

titled "Flux Equilibria in Quantum Systems far Away from Thermal Equilibrium" describing their motivation concerning this "Ansatz:"[354]

> "The concept of flux equilibria ("Fliessgleichgewichte") was originally coined by a biologist [1]. Recently it was applied by Weidlich to nuclear reactions [2]. This concept applies to the following situations: A system, in particular a quantum system, is coupled to reservoirs at different temperatures, so that there is a flux of energy through it. This energy flux may cause new stable configurations of the system, which are not present in complete thermal equilibrium. By studying specific examples, e.g. the laser, we have found [3] that there may be several different stable configurations which are quite analogous to the well-known phases of systems in thermal equilibrium [4]. It is even possible to study the analogue of phase-transitions [5]."

And, at the end of the article:

> "We expect further applications not only to active devices in solid state physics (e.g. the Gunn oscillator), but also in astrophysics and biology."

The situation must have been an "Aha-experience" for Hermann Haken. His considerations concerning the symmetry breaking at the laser threshold and his subsequent search for the creation of emergent stable states in systems far from thermal equilibrium, as they are common especially in biology, could be "naturally" combined and achieved a solid with a mathematical explanation. This indicated a solution to the problems that had been dealt with at the *Versailles Conferences* in 1967 and 1969: namely, how theoretical physics could help to explain biological processes.

At the same time, a new vein opened up how, out of a random, only statistically describable motion, a new ordered state could emerge.

This track had to be followed and qualified mathematically. In quick succession, further articles appeared[355] that worked out the connections between thermodynamics and the statistics of the underlying microscopic processes:

---

[354] (Graham, Haken, & Weidlich, 1970) The citations are [1]L. v. Bertalanffy, Theoretische Biologie, Vol. 1 and 2 (Berlin 1932 and 1942). [2] W. Weidlich, Z. Physik 222 (1969) 403. [3] R. Graham and H. Haken, Z. Physik 213 (1968) 429; [4] D. Ter Haar, Elements of thermostatistics, (Holt, Rinehart and Winston, New York, London 1969); [5] R. Graham, to be published.

[355] (Graham & Haken, 1971b), page 290. Submitted 6th of January 1971 and (Graham & Haken, 1971a), pages 151 and 152.

> "The inseparable connection between macroscopic thermodynamics and microscopic statistical theories is a very old and extremely successful branch of physical study. Outstanding marking points in the development of this connection are e.g. Boltzmann's proof of the H-theorem, Onsager's theory of microscopic reversibility and the many sophisticated mathematical devices, which allow to bridge the broad gap between the microscopic and the macroscopic theory. [...]
> This [Laser-]theory is based on a microscopic Hamiltonian and it is worked out by extensive use and even new development of quantum statistical methods. It came as a big surprise to many physicists, however, that many of the central results of this theory could be consistently explained and understood in terms of a purely macroscopic, thermodynamic theory: the Landau theory of phase transitions.
> One of the difficulties to fully accept this connection between laser theory and Landau theory comes from the fact that the Landau theory is based on the analytical properties of a thermodynamic potential, which can only be defined in a thermodynamic equilibrium, whereas such an equilibrium cannot be invoked for a laser, even if it is in a stationary state.
> [...]
> Nevertheless the analogue of a thermodynamic potential could be explicitly constructed for the laser case.
> ...Nevertheless the connection between "thermodynamic potentials" and probability densities turned out to be the same in both theories. This seemed to indicate that there should exist some basic physical properties, which systems far from thermal equilibrium like the laser and systems in thermal equilibrium should have in common."

The existence of a "potential condition" is central for the application of the well-known Landau-theory of thermodynamics. To their great surprise, Graham and Haken found an unexpected solution:[356]

> „Recently, we found the unexpectedly simple answer to this question: Within the framework of a Fokker-Planck equation the potential conditions in their most general form are equivalent to the condition of detailed balance."

---

[356] (Graham & Haken, 1971b), page 291.

If one wants to deal with irreversible systems far from thermal equilibrium, one could use the well-known methods and results from equilibrium-thermodynamics, if one can prove that the system is in "detailed balance." A relation of this kind was indicated by the IBM-physicist, Rolf Landauer, at the same time.[357]

Only two months later, Graham and Haken deepened their theory mathematically. In their paper "Fluctuations and Stability of Stationary Non-Equilibrium Systems in Detailed Balance," submitted on the 23rd of March 1971, they declared:

> "In our work we have abandoned the usual procedure of irreversible thermodynamics, to start from equilibrium and look for approximations, valid in its vicinity. Instead we used as a starting point the stationary state of a system which may be very far from equilibrium. Our basic assumption was that this stationary state has the property of detailed balance."

In this way it was possible to describe systems far from thermal equilibrium by the integration of methods from "classical" thermodynamics. These symmetry-breaking transitions occur quite often in different fields of physics. Graham and Haken mention the Gunn-effect, electronic information processing and hydrodynamics, as well as other areas of science: examples from chemistry and biology. For Hermann Haken, by education a mathematically oriented physicist, the linking condition had been detected:

Look out for and find systems that show phase transitions far from thermal equilibrium and that are in a state of detailed balance.

6.3 The UMSCHAU-Article of 1971

Hermann Haken intended to present the subject matter and its cross connections to a larger audience. He surprised his co-author Robert Graham with a manuscript he prepared for the popular science magazine *Umschau in Wissenschaft und Technik* edited by the Umschau Publishing house in Frankfurt am Main. In the recollection of Robert Graham, Haken had chosen this magazine with its broad readership[358]

„because it had become clear to him that the main impact of these ideas might not even be in physics but rather in other fields like chemistry, biology or oecology, to name just a few. Graham added:[359]

---

[357] Cited from (Graham & Haken, 1971b) Footnote 8 on page 291: "A relation between the validity of detailed balance and the existence of a thermodynamic potential was conjectured by Landauer, R.: IBM Research RC 2960, 1970."

[358] (Graham, 1987) p. 4.

[359] Interview with Robert Graham from 29.3.2011, p. 14 (University Archive Stuttgart (Archive Haken)).

> "For the most part the [Umschau-article] was written by Haken. Well, one day he showed the manuscript to me saying, "how about doing this"? Even than he had put me down as leading author on the typed manuscript. But I changed that because I thought "Nope! That's really his idea writing this article." The paper really showed common things that were not new to me at that time."

At that time, Graham was working on his habilitation thesis dealing with the "Theory of non-equilibrium systems in stationary states."

In the Umschau-article titled, "Synergetik – die Lehre vom Zusammenwirken"[360] March 1971, Haken used the word *Synergetics* for the first time in print. He expressed his motivations very clearly in the introduction:[361]

> "Recently a new field of research had developed that allows, under a common perspective, to address phenomena in quite different areas like physics, chemistry, biology and even sociology. As a matter of fact one succeeds to reveal common regularities for instance between problems like the following: to which rules obeys laser light, the structure of a forest, the formation of an enzyme, the creation of life, the development of language? The mutual problem is the following: Although the examination objects, i.e. a body in physics or a cell in biology, consist of very many subsystems, they act to the outside world as a characteristic entity, where, in most cases, its properties are not just the accidental superposition of its subsystems. [...] Because in physics a number of rankings or states of order had been explored in detail, it seems obvious to begin with these. We like to emphasize that by no means we try to reduce to physics biology or even sociology. It is merely that the relatively simple physics phenomena can inspire us to develop concepts and methods that allow for new insights into quite different fields."

In the beginning, Haken explained the phase transition phenomena of physical systems in thermodynamic equilibrium. He used the examples of super-conductivity and of ferro-magnetism. Subsequently he showed with the laser, his paramount example that phase transitions occur as well in systems far from thermal equilibrium. Following these fundamental argumentations he made a

360 „Synergetics – the science of working together."

361 (Haken & Graham, 1971a), p. 191.

plunge to the fields of chemistry and oecology, as can be seen in the following figure, adapted from the original paper:[362]

Abstraction	Realization		
	1. example	2. example	3. example
Scientific area	Physics	Oecology	Chemistry
System	Laser	Forest	System of different molecules
Subsystem	Atom	Tree	Molecule
Time of process	Between exitation and emission	Life time of the tree	Reaction time
Statistical processes	Excitation, spontaneous emission	Sprouting, dieback of the tree	Spontaneous creation
Order parameter	Lightfield	Density of trees	Density of molecules
External parameter	Electric current	e.g. climate	Pressure, temperature, energy applied (light, energetic molecules

Fig. 13 Examples on the realisation of abstract synergetics concepts (after (Haken & Graham, 1971b))

A completely new development in biology reassured Hermann Haken in his approach. The first time he heard about it was in a talk given by Manfred Eigen in Göttingen in December of 1970.[363] Eigen had presented a model that created an autocatalytic cycle in prebiotic molecular biology. Hereby, he could explain important steps in the self-organisation of matter and in the evolution of biological macromolecules. This should emerge as a decisive step forward, explaining the creation of life out of inanimate matter. In his paper, Haken was able to demonstrate that Eigen's kinetic equations of autocatalytic self-organisation of molecules and his equations for the laser phenomenon have the same analogue structure.[364]

$$\frac{dn_j}{dt} = n_j\ (\alpha_j - \beta_j) + F_j\ (t) \qquad j= 1,\ldots, N$$

Physics: Laser n_j: phonton number in direction j
Chemistry: auto-catalytic reaction n_j: number of molecules of class j
Molecular biology: auto-catalytic reaction n_j: number of molecules of class j

Fig. 14 Examples of master equations in physics, chemistry and biology. (from (Haken & Graham, 1971a)) .

362 (Haken & Graham, 1971a), p. 194 fig. 3.
363 Published in October 1971 in Zeitschrift für Naturwissenschaften. (Manfred Eigen, 1971) (submitted in May).
364 (Haken & Graham, 1971a), p. 194 fig. 4.

Then he proceeded:

> "In our interpretation order parameter would be the subsystems of the different groups of molecules. [...]. The difference between the creation rate and annihilation rate would act as a "filter" (*Eigen's*, Wertfunktion). In this case evolution shows up as a cooperation of systematic selection, described by $(\alpha_j - \beta_j)$, and stochastic processes, described by F_j.
> [...]
> Of course, the theory of evolution is not confined to molecules but is as well true for living things – therefore the possibility to describe mathematically the concepts of Darwin is given."

In his view, that example showed a connection between animate and inanimate nature that could be described mathematically. It now seemed possible to treat these, up to now, as completely disparate phenomena. That is why he finished off with the optimistic words:[365]

> "To summarise, we may say that a variety of widely different phenomena can be treated from a unified perspective with the help of a few notions. The seemingly mysterious order principles that govern the cooperation of the different parts of a large system prove to be feedback systems created by the subsystems. Unexpectedly sudden changes in these order principles are caused by phase transitions. A new vein of mathematical coverage of these phenomena seems to open up."

Haken was well aware of the fact that these ideas were "revolutionary." He had himself geared up for a lot of criticism and incitements. But what happened? – Nothing![366]

6.4 The First ELMAU-Synergetics Conference 1972

In winter of 1971-1972, Hermann Haken, motivated by the *Versailles Conferences* and the annual meetings of the *German Physical Association,* made plans for a meeting on Synergetics of his own devising. Right after receiving a subvention by the *Bavarian Ministry of Culture*[367] he selected *Schloss Elmau* in the Bavarian Alps as conference venue. At that time it was not a luxury resort as it is today. The general idea of the meeting, he formulated quite distinctly in the preface of the conference proceedings that appeared in 1973:[368]

365 (Haken & Graham, 1971a), p. 195.
366 Private communication with Hermann Haken.
367 Interview with Hermann Haken from 21.9.2010, p. 30.
368 (Haken, 1973c)

At a first glance the reader of this book might be puzzled by the variety of its topics which range from phase-transition-like phenomena of chemical reactions, lasers and electrical currents to biological svstems, like neuron networks and membranes, to population dvnamics and sociology. When looking more closely at the different subjects, the reader will recognize, however, that this book deals with one main problem: the behavior of systems which are composed of many elements of one or a few kinds. We are sure the reader will be surprised in the same way as the participants of a recent Symposium on synergetics, who recognized that such systems have amazingly common features. Though the subsystems (e. g. electrons, cells, human beings) are quite different in nature, their joint action is governed by only a few principles which lead to strikingly similar phenomena. ....Though the articles of this book are based on invited papers given at the first International Symposium on Synergetics at Schloß Elmau from April 30 to May 6, 1972, it differs from usual conference proceedings in a distinct way. The authors and subjects were chosen from the very beginning so that finally a well-organized total book arises."

In this statement two facts were emphasised to which Hermann Haken always attached great importance: from the beginning, his conferences had always been devised multidisciplinary. He was searching for analogies in all kind of sciences and did not confine himself to physics, chemistry or biology. On the other hand, the invited speakers at his *ELMAU Conferences* and the topics they dealt with had been "hand-picked" by him, giving the chance to gain a comprehensive survey of the latest state of research in a science field.

We would like to mention that, if in the following we speak about the participants of the *ELMAU Conferences,* always meant are the speakers and lecturers, normally about twenty individuals. But the meetings had about eighty attendees, their names not being printed in the proceedings.[369]

The first *ELMAU Conference* may be regarded as the starting point of Hermann Haken's activities on Synergetics. In what follows, we therefore analyse more precisely its group of participants and contents. The lecturers of the meeting originated from five different sources:

- Former participants of the *Versailles Conferences*[370]
- Invited speakers from the recent meetings of the *German Physics Association*[371]

369 Unfortunately the list of participants had not survived, neither in Haken's archive nor at the *Volkswagen Foundation* that supported these conferences by grants. Only one list of attendees of the 1981 conference was found in the papers of the *Volkswagen Foundation.* It shows that this conference had had 25 invited lectures and 75 attendees in total. (A copy can be found in the Archive Haken (University Archive Stuttgart)).

370 H.Fröhlich, B. Julesz, R. Kubo, R. Lefěver, T. Matsubara, W. Reichardt, H.R. Wilson (placed by J.D. Cowan?), H. Kuhn (placed by M. Eigen?), E.W. Montroll.

- Japanese solid state theoreticians, working on stochastic problems in multicomponent systems[372]
- Scientists performing research on phase transitions[373]
- Members and students of the Stuttgart theory school[374]

Speakers of the late *Versailles Conferences* amounted to the largest group of attendees and by far the most prestigious. Furthermore, noticeable is the number of mathematically oriented theoreticians. The topics of the symposium had been structured into segments of highest interest to Haken at that time:

- Mathematical and physical concepts of cooperative phenomena
- Instabilities and phase transition like phenomena in physical systems far from thermal equilibrium
- Biochemical kinetics and population dynamics
- Biological structures
- General structures

In his introductory talk (the published version, of course, was written after Haken had received the manuscripts of the other lecturers, i.e., in June and July, 1972) he concentrated on the analogies between the different systems showing phase transitions and their order parameters respectively

Table 6 Synergetic systems and their order parameters compared by Hermann Haken at the first *ELMAU Conference*

Science	System	Subsystem	Orderparameter
Physics	ferromagnet	Elementary magnets (spin)	Mean field
	superconductor	Electron spin	Pair wavefunction
	laser	Atoms	Lightfield or photon number
Chemistry	Chemical ensembles	Molecules	Number of molecules
Biology	Biological clocks	Molecules	Number of molecules
	Neural network	Neurons	Pulserate
Ecology	Group of animals	Individual animal	Number of animals
	Forest	Individual plants	Density of plants
Sociology	Society	Human beings	Number of people of given opinion

371 R. Landauer, H. Thomas, W. Reichardt, F. Schlögl.

372 H. Mori, K. Tomita, F. Yonezawa.

373 L. Kadanoff, G. Adam, (placed by W. Weidlich?).

374 M. Wagner, R. Graham, W. Weidlich.

Let us have a short look into the order parameter concept. Originally coming from the Landau theory, Haken adapted and extended it to systems far from thermal equilibrium. This concept has a distinctive specialty: in large systems of different but equal parts it allows one to abstract from the single subsystem, let it be electrons, molecules or neurons, and concentrate on one or fewer parameters that define the macroscopic action of the total system. A classic example is a gas with its order parameter temperature or pressure. In his introductory notes, Haken discussed the mathematical formulation of order parameters. He quoted two examples that he had, until then, not presented in his publications: for one thing, the famous example from population dynamics first formulated and named after Alfred Lotka[375] (1880 - 1949) and Vito Volterra[376] (1860 - 1940) that consists of a predator and a prey. Looking at the dynamic evolution of sych, a model system on oscillatory state occurs between predator population and prey population. Oscillation of that kind, is observed as well in autocatalytic reactions and in biological clocks. The second example related to the so called Zhabotinsky-Belousov reaction,[377,378] a self-excitatory oscillating chemical system showing periodic change of colours. Ilya Prigogine and his co-workers invested research efforts into this area. From the mathematical point of view, the occupation numbers in the system play the dominant part. Systems of this type are common in many fields, having the consequence that a variety of cases of application can be treated by the same mathematical formalism.

Haken then turned to the phenomenon of symmetry breaking that occurs in phase transitions at threshold bringing in his laser example. He used this phenomenon to classify other conference contributions:[379]

> "A very similar problem occurs for the parametric oscillators and the tunnel diode which has been treated by Landauer in the connection with the theory of the computing process (see his article).Because „symmetry breaking instabilities" play an important role in synergetics, we mention a few further examples: Molecules which differ only in one property, e. g. optical dichroism, but having identical a, ß, and F, or DNA or RNA, which differ by the arrangement of their constituents, but again having the same production factors a etc. Another example is provided by molecules, whose concentration is

[375] (Lotka, 1925 (reprinted 1956)).

[376] (Volterra, 1931).

[377] (Zhaboutinsky, 1964) On the history of the Belousov-Zhabotinsky reaction see (Zhabotinsky, 1991) and (Winfree, 1984).

[378] (Vavilin, Zhabotinsky, & Zaikin, 1968).

[379] (Haken, 1973c), p. 16/17.

> space dependent. In the broken symmetry case e.g. spatially periodic molecule concentrations may occur. Such systems and related ones have been studied in detail by Prigogine and coworkers and were called "dissipative structures" (see also the article of Prigogine and Lefever in this book). Analogies with respect to instabilities include a large variety of systems, e. g. plasmas [14], or current instabilities (see Thomas' article). A further example is provided by different states of interstellar matter [15]. Instabilities (or multi-stability) of exactly the same type as treated here are also found in visual perception (see the article by Reichardt) and in sociological models (see Weidlich's contribution). The ingenious experiments by Julesz are a challenge for a similar interpretation and considerable progress has been achieved in the model of neural networks by Wilson and Cowan".

If we compare the topics described by him in this conference with the items of the first publication the year before, the following extensions are noticeable: in addition to the quoted examples ferro-magnetism, super-conductivity, the laser and general systems of chemistry, – especially the theory of self-organising biological macro-molecules according to Manfred Eigen and the formation of public opinion within a society (after Wolfgang Weidlich), new topics showed up.

- Electric current instabilities (Gunn-effect)
- Cooperative processes in information processing
- Population dynamics according to Lotka-Volterra
- The Belousov-Zhabotinsky-reaction (chemical oscillations)
- Pattern recognition of the visual system
- Cooperative phenomena of the nervous system
- Other cooperative biological systems

Besides these more concrete phenomena there were some theoretical contributions on the statistical aspects of phase transitions that were of high significance to Hermann Haken. The conference had been a great success:

> "In the end my Japanese colleagues said, - there had been four of them -, they had been very sceptical in the beginning, not knowing, what it was all about. But because we had a good relationship they said: "Well, we will take part." At the end of the meeting they mentioned: "Now we have understood that it is a reasonable matter." That has been the quintessence of the conference."

Rolf Landauer of IBM, the computer and software manufacturer, had been one of the invited lecturers. Many years later he still remembered the meeting and honourably mentioned the proceedings:[380]

> „Hermann Haken in 1972 had the first interdisciplinary meeting [...]
> Interdisciplinary meetings existed before then, but they typically had a high component of flaky stuff. Probably not everything presented at Haken's session in 1972 has stood up, but all of it was serious and represented real intellectual depth and effort.
> Participation was a breath-taking experience for me; for the first time I found myself among people with comparable interests and a comparable sense of values. I was no longer an orphan! And the meeting had another earmark of a good conference: The conference had broad representation in its selection of speakers; it was not dominated by the organizer and his close associates."

Stimulated by the positive response, Haken looked for ways to organise additional conferences of this type. But he did not instantly succeed. Finally, starting in 1976, he got the chance by a generous grant of the *Volkswagenwerk Foundation* promoting *Synergetics*. (See Chapter 7b for details).

Looking back, Haken valued the first conference as the inception of a new interdisciplinary cooperation that also made an impact on other fields of research:

> "On the other hand, we recognise a downright tipping point. Whereas, until not long ago it seemed that science as a whole would disintegrate into ever smaller subfields that did not (or not much) take notice of each other, today we see the advent of many conferences and new journals that aim at searching for deep lying connections between the different disciplines, - in the sense of Synergetics. I think I may say that 25 years ago the Synergetic Conference had been one of the first, if not really the first one, that pursued the aim of finding common principles."[381]

---

[380] (Landauer, 1988), p. 392.
[381] (Haken, 1999).

Although many arguments are in favour of this appraisal, some other activities went unnoticed by Haken at that time. We would like to mention the activities of the research groups from Ludwig von Bertalanffy and Heinz von Foerster.[382]

6.5 The Startling Discovery of a Common Mathematical Basis: Laser – Bénard-Effect – Brusselator

Hermann Haken got another chance to promote his ideas on Synergetics on the occasion of the retirement of his friend and colleague Herbert Froehlich. In 1973, he and Max Wagner edited, with the Springer publishing company, a "Festschrift" entitled "*Cooperative Phenomena*" encompassing many renowned authors. As is usual on these special occasions the "Festschrift" reflected Herbert Froehlich's broad range of scientific interests. The articles had been arranged in the following order: „Quasi-particles and their interactions; Superconductivity and superfluidity; Dielectric Theory; Reduced density matrices; Phase Transitions; Manybody Effects and Synergetic Systems as well as Biographical and Scientific Reminiscences".

Haken grabbed the chance to create a chapter called "*Synergetics.*" Besides his own article he assigned articles to I. Prigogine, M. Wagner, G. Careri, B. Holland and C.B. Wilson to this segment.[383] In terms of content, his contribution "Synergetics – Towards a new Discipline" constituted only a slightly modified version of his introductory address to the *ELMAU Conference* in 1972, the printed version of it appeared too in the beginning of 1973. In his book chapter, Haken emphasised the laser equations with respect to their selective function. This could be the consequence of discussions with Manfred Eigen at the "Winterseminars."

Aside from these topics published before, a new association aroused. Jacques Monod, French chemist, Nobel Prize winner and participant of the Versailles Conferences, in 1970 had published a provocative book titled "*Le hasard et la necessité*"[384] that immediately raised controversial discussions.[385] In his book,

---

[382] For instance see the interdisciplinary conference on „Self-Organizing Systems" from 1959. In the preface the editors justified the necessity of such a meeting. They wrote „On the one hand the psychologist, the embryologist, the neurophysiologist and others involved in the life sciences were attempting to understand the self-organizing properties of biological systems, while mathematicians, engineers, and physical scientists were attempting to design artificial systems which could exhibit self-organizing properties. Accordingly, the Information Systems Branch of the Office of Naval Research together with Armour Research Foundation, decided to sponsor a conference enabling the workers in the many disciplines involved to meet together and discuss their research activities and to explore common problems, mutual interests, and similar directions of research." (Marshall Yovits & Cameron, 1960) A second conference was held in 1962. (M. Yovits, Jacobi, & Goldstein, 1962).

[383] (Haken & Wagner, 1973).

[384] The English version appeared in 1971: (Monod, 1971a).

[385] (Monod, 1970).

Monod holds the view that the creation of life finally had happened by chance, as extremely improbable it might have been. Following this singular event, life has developed according to the natural laws. Manfred Eigen, who wrote the preface to the German edition of the book, explained:[386]

> "Necessity" enjoys equal rights to "Chance," as soon as for an event there exists a probability distribution, and – as it is possible in the physics of macroscopic systems – it can be described by *large numbers*. The title of the book expresses this equal status unambiguously. But Monod has to stress "Chance" because of the fact that "Necessity" is voluntarily accepted by everyone. [...]
>
> Thereby, the deep caesura between the inanimate world and the biosphere disappears, that philosophy, ideology and religion had attached such a great importance to. The creation of life that is the development of the macro-molecule towards micro-organism is only one step forward among many others: such as from the elementary particle to the atom, from atom to molecule ..."

Haken took up that issue. It has been one of his special strengths to consider new developments in different fields of research, whether or not they were analogous to research in his own field, especially the laser. This had been the case in the analogy between the laser equations and the Ginzburg-Landau equations of super-conductivity, as well as in the "Wertefunktion" ("value" function) of Eigen's selection theory. Haken looked at the order parameter equation of the single mode laser, which in his notation can be written as:

$$dE/dt = \alpha' E - \beta' |E|^2 E + F(t)$$

where α' denotes the saturated gain, β' the saturation constant, E is the field amplitude and F(t) represents the time dependent fluctuation. This equation consists of two terms: the first expression describes the interaction of the light field with the electrons of the laser medium. The second term models the spontaneous emission of photons by excited electrons that can not be influenced or prohibited. It is a quantum-mechanical effect. Therefore, Haken concluded:[387]

> [This equation] "is one of the simplest examples, but very instructive of the interplay between fluctuating forces [...] and systematic forces [...] or, in Monod's words, of the interplay between "chance and necessity"".

[386] (Monod, 1971b) p. XIV-XV.

[387] (Haken & Wagner, 1973) p. 269.

Later on we find the idea of this interplay that represents an interaction of chance and (mathematically speaking) selection rules in Haken's comprehensive review on Synergetics of 1977.[388]

It may seem as if Hermann Haken had been occupied only with phase transition phenomena and Synergetics during the years 1971 to 1973. But this impression would be wrong. On the contrary, at that time Haken had been actively involved in his cooperation with Serge Nikitine (1904 - 1986) at the *University of Straßburg* concerning the theory of excitons in solids. He delivered several lectures in French, which are part of the manuscripts surviving in his archive. These activities should also be judged in the context of his book "*Quantenfeldtheorie des Festkörpers*" published in 1973.[389] It can be fairly stated that the research activities of his institute during this period concentrated on the theory of the laser, statistical mechanics and on solid-state theory. From 1971 until 1974, not less than seventeen articles devoted to solid state physics had been published (Haken being author or co-author). Furthermore, from 1971 until 1977, eleven diploma and doctoral thesis on this subject were completed.

At that time, *Synergetics* was a specialty of Hermann Haken; it had not yet arrived at the curriculum and research activities of the institute.

Due to close co-operations in laser research Hermann Haken cultivated good contacts with his Italian colleagues; especially, Rodolfo Bonifacio and Tito Arecchi (born 1933) from Milan. The summer schools "Enrico Fermi" held in Varenna at Lake Como played an important role. In the early seventies in laser theory, the questions and concepts of short and ultra-short laser pulses, the research on coherence of the light wave as well as the topic of "Superradiance" kept the theoreticians busy. Superradiance means the collective emission of radiation after the atoms had been exited coherently. Particularly, Bonifacio worked on this cooperative quantum effect.[390] One of the outstanding annual activities of the Italian physicists was the organisation of the summer schools, not only at Lake Como but also in Erice (Sicily). In 1963, physicist Antonio Zicchici founded, and was chairperson of, the annual meeting at the "Centro Ettore Majorana." The main topic was elementary particle physics, but a parallel "*Summer-School on Quantum Electronics*" was held. In 1974, Arecchi charged Rodolfo Bonifacio and Hermann Haken with a lecture course on "Cooperative effects in multi-component systems." The proceedings of the meeting were edited by Hermann Haken and published in the same year.[391]

The summer school consisted of two different parts. There was one section on "Quantum optics" prepared and presented by the Italian physicists Arecchi, Bonifacio and Narducchi along with Roy Glauber, Fritz Haake and E. Courtens from the IBM laboratory in Zürich. The other part had been devoted to "Cooperative Phenomena" organised by Hermann Haken. The lecturers were

388 (Haken, 1977b).
389 (Haken, 1973a).
390 (Bonifacio, 1971).
391 (Haken, 1974a).

nearly exclusively participants of the first *ELMAU Conference* in 1972. The list consisted of L. Kadanoff, F. Schlögl, H. Thomas, M. Wagner, B. Julesz, H.R. Wilson, H. Froehlich and W. Weidlich. Prigogine's Brussels school was represented by G. Nicolis, replacing R. Lefever. Only two new names - P. G. de Gennes[392] and G. Brettschneider[393]- completed the roster. Therefore, it is not surprising that only a few new arguments were presented. The meeting served mainly to develop further the discussions that started in 1972. In his lecture that was highly mathematical, Haken presented the well-known analogies between phase transitions and Ginzburg-Landau theory and derived the order parameter concept in detail. Nevertheless, there was one new subject in his talk, the analogy between the laser and the Bénard instability only recently found.

Haken took notice of the Bénard phenomenon occurring in hydrodynamics by reading the book by Glansdorff and Prigogine,[394] in which the structure formation had been explicitly. What is the Bénard phenomenon about? As early as in 1900, the French chemist Henri Bénard (1874 - 1939) had detected that by heating a fluid uniformly from below dynamic stable structures develop.[395] This effect can be observed by those who boil water, but perhaps is done without reflecting about the subject.

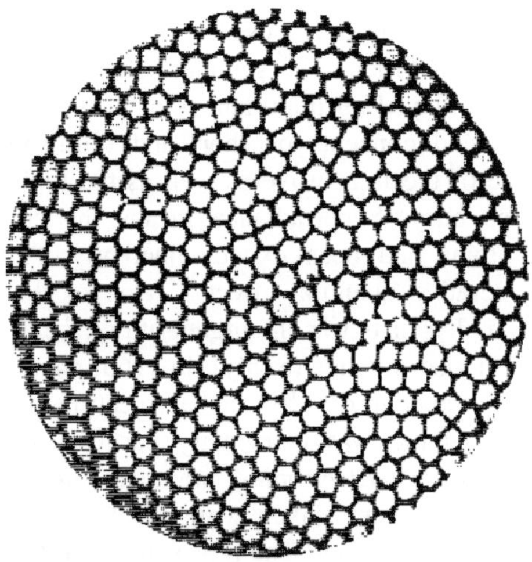

Fig. 15 Cell structure (Bénard rolls) of the Bénard instability[396]

[392] Professor at the *College de France*. He gave a lecture on "Gravitational instabilities of liquid crystals".

[393] Scientist coming from Siemens AG in Munich. His talk was titled „Cooperative Phenomena in telephony". This subject at the first *ELMAU Conference* had been dealt with by the IBM researcher Rolf Landauer.

[394] (Glansdorff & Prigogine, 1971).

[395] (Bénard, 1901) and (Bénard, 1900).

[396] Figure taken from (Chandrasekhar, 1961).

Concerning the Bénard instability Haken's co-worker Robert Graham had written in 1973:[397]

> „The Bénard convection instability [...] is probably the simplest nontrivial example of an instability in fluid dynamics. By the same token, the appearance of convection cells at the Bénard point is also a very clear-cut example of a "dissipative structure". Such structures are well known to appear in many systems if they are driven by some external force into a nonlineare domaine far from thermal equilibrium. The very abrupt appearance of dissipative structures, like convection cells, [...] has sometimes been compared to phase transitions in equilibrium systems."

The Bénard instability shows all the typical features of a dissipative structure: on the one hand, heat (i.e., energy) is supplied uniformly and dissipated on the other side by convection and thermal radiation. This is analogue to the laser phenomenon: in that case energy is supplied by "pumping" (electrically or optically) and dissipated by the laser beam and heating of the cavity. Of course, Haken immediately recognised this analogy and checked with Graham regarding whether the mathematical tools they had developed for the laser would also hold for the Bénard instability. During the summer 1973, they published two articles where they derived a thermodynamic potential function that describes the stability, the fluctuations and the dynamics of the system near the instability threshold. The methods and concepts for the laser also worked for the Bénard instability:[398]

> „In the Bénard and Taylor problem the conditions are fulfilled for our theorem [...] on the exact solution of the Fokker-Planck equation (which was originally developed for applications to laser theory)."

Therefore, Haken could present the following analogies between the laser and the Bénard phenomenon at the Erice Summer School:

Table 7 Analogies between laser and Bénard instability (from (Haken, 1974b), p. 19)

Analogies between the laser threshold and the Bénard instability		
	Laser	Bénard instability
external parameter	pump strength (or inversion d_o)	Temperature gradient (Rayleigh number R)
instability at	d_c	R_c

[397] (Graham, 1973), p. 1479.
[398] (Haken, 1973b), p. 193.

Table 7 (*continued*)

behaviour below d_c, R_c	No stimulated emission, averaged field zero	No motion, averaged velocity zero
Behavior at d_c, R_c	One (or several) field modes get unstable (depending on geometry) Soft mode (critical slowing down) Critical fluctuations Principle of detailed balance holds	One or several modes get unstable (depending in geometry) Soft mode (critical slowing down) Critical fluctuations Principle of detailed balance holds
Behaviour above d_c, R_c	Stabilization of "unstable modes" via damped modes	

With respect to the Bénard instability, Graham and Haken had found another example of a dissipative structure to which they could apply the Synergetics methodology.

The topics of phase transitions and non-equilibrium thermodynamics had been the dominating issue at the spring meeting of the German Physics Society in Freudenstadt (Black Forest), April 1974. Invited lectures were given by H. Thomas from Basel speaking on "Structural Phase Transitions" as well as Paul Glansdorff from Brussels on "Non-linear Thermodynamics."[399] Not less than three special sub-meetings had been addressed to the theory of phase transitions and two more meetings were arranged on the topics of "Non-equilibrium thermodynamics and the theory of stochastic processes." Haken played a dominant part, giving three different lectures that were highly connected. The first talk considered the subject "Stability and fluctuations of many particle configurations in the non-linear regime of the convection instability," an old problem of the theory of turbulence. Generalising the solution he read about "Generalized Langevin-equation for systems far from thermodynamic equilibrium" and demonstrated in his third contribution the "exact stationary solution of the master equation far from thermal equilibrium in detailed balance." There is a good caste to believe that at this conference Hermann Haken had discussions with Paul Glansdorff who, three years earlier (in 1971) jointly with Ilya Prigogine, had published the influential book on "*Thermodynamic Theory of Structure, Stability and Fluctuation*."[400] It became apparent that the later so called "Brussels School" of Prigogine and Glansdorff grew to be the major scientific competition to Haken in the theory of dissipative systems.

Only some weeks later, another reunion provided the chance to continue the discussions. The symposium "On cooperative phenomena in equilibrium and

399 See „Verhandlungen der Deutschen Physikalischen Gesellschaft", Reihe VI Band 9 (1974) (Physik Verlag Weinheim).

400 (Glansdorff & Prigogine, 1971).

nonequilibrium" was held at the *Monastery Gars* at the banks of the Inn river in Bavaria. This Summer School of the *German Physical Society*[401] (DPG) was organised by Friedrich Schlögl, professor of theoretical physics at the *RWTH Aachen.* Schlögl also led the topical committee regarding "Thermodynamics and statistical mechanics" of the *DPG.* Aachen is situated only about 150 kilometers from Brussels and Schlögl was in close contact to the group surrounding Prigogine at the *Free University of Brussels.* Therefore, he invited Grégoire Nicolis and René Lefever to give lectures at the gathering. Coming from Stuttgart, Haken and Weidlich joined the meeting. Scanning the roster of participants, we again find the names of Rolf Landauer, J.S. Nicolis, E. W. Montroll, S. Grossmann and H. Thomas,[402] all of them having attended the first ELMAU-Conference in 1972.

Up to that time Hermann Haken had advocated his ideas and results on *Synergetics* at conferences and meetings. He now imagined that the time had come to present the results in an extended review article. During the summer of 1974 he wrote the fundamental and detailed report "Cooperative phenomena in systems far from thermal equilibrium and in non-physical systems" that appeared in the January issue, 1975, of the renowned American journal *Review of Modern Physics.*[403] Amounting to 50 pages, this publication presented in great detail, mathematically as well as by examples, the results of the research on cooperative behaviour of systems far from thermal equilibrium. In comparison to the *Zeitschrift für Physik* that, at that time, contained articles written in English as well as in German and was not read throughout the scientific world, the *Review of Modern Physics* is indispensable for physicists and can be found worldwide in nearly every scientific library. The impact of this article is made clear by the fact that it earned 650 citations,[404] the second highest value any of Haken's publications ever attracted.

Quoting Haken, the goal of the presentation was to demonstrate how subsystems work together to create order on a macroscopic scale. This should be exemplified by examples from different scientific fields, showing the common mathematical basis for these different systems.[405] In every detail, he elucidated the examples now common to the reader: the laser and its different instabilities being the paradigmatic example, non-linear optics, the Gunn instability of the tunnel diode, chemical oscillations creating the famous Belousov-Zhabotinsky reaction, instabilities of hydrodynamics, neural networks of the brain and interacting social groups.

In the second part of his work, Haken developed the mathematical tools of statistical mechanics showing the different approaches (Langevin equation, density matrix ansatz, Fokker-Planck equation) and their solutions for the most

[401] No proceedings of this symposium appeared in print. An invitation leaflet including the invited speakers was found in the Archive Haken.
[402] The exception had been J.S. Nicolis.
[403] (Haken, 1975c).
[404] Science Citation Index, recalled 8 June, 2011.
[405] (Haken, 1975c), p. 68.

import special applications. While reading this section on the mathematical methods one is reminded strongly of Haken's presentation in his handbook article "*Laser Theory*" from the years 1967-1970[406] where this exposition of the laser problem is shown in every detail. Another important point is the analogy found by Graham and Haken between systems far from thermal equilibrium and phase transitions. Here the occurrence of order parameters and critical fluctuations (at the threshold) play a central part.

> „This concept of order parameter also sheds new light on the problem of self-organization: the subsystems themselves create fictitious or real quantities which via feedback loops organize the behavior of the subsystems."

It is important to note that Haken subsequently decidedly referred to the fact that the correct treatment of the examples shown was only by a statistical (stochastic) approach and it is not permitted to argue with mean values. If one makes use of mean values, fluctuations would get lost (being averaged out). But these fluctuations (often called noise) play an important role if the system approaches threshold. On top of that, at threshold non-linear effects occur leading to the effect that all "linear" mathematical approaches would fail. Using this ansatz, Haken set himself apart from the approach of the Brussels school of Prigogine and Glansdorff, something he phrased expressly:[407]

> "In conclusion, a word should be said of the relation of the approach presented in this article to approaches made within irreversible thermodynamics, or to still more advanced thermodynamical approaches like that by Glansdorff and Prigogine. In our approach we start from stochastic equations of motion either for microscopic systems or for systems described by order parameters. The thermodynamic approach begins with the assumption that there exists local thermodynamic equilibrium; this allows us to define quantities like entropy, so that the principle of excess entropy production (Glansdorff and Prigogine) can be applied. While this principle proves to be a useful tool in the "linear regime" its applicability to the "phase-transition" region which requires a truly nonlinear treatment seems to require further investigation."

Meanwhile the subject of chemical oscillations, i.e., the Belousov-Zhabotinsky reaction, had become a prevailing research subject in chemistry. Not least by a so-called "Faraday-Symposium" organised by the *Royal Institution* under the heading

---

[406] See figure 8 page. 58
[407] (Haken, 1975c) p. 69.

"*The Physical Chemistry of Oscillatory Processes*" in London, 1974.[408] Among the participants, we find from the Brussels school Ilya Prigogine, G. Nicolis and their co-workers Agnes Babloyantz and René Lefever. From Germany, Friedrich Busse (born 1936), Benno Hess (1922 - 2002) and Otto Rössler (born 1940) attended the meeting. Haken had sent his young doctoral student Arne Wunderlin.

While still working on the article for the *Review of Modern Physics,* Haken came to realise that the subject of phase transitions, far from thermal equilibrium and his Synergetics, presented a rich field of future research. He felt the strong desire to devise the necessary mathematical-physical theories. Thus in November 1974, for the second time, he applied for a research sabbatical in the summer term of 1975 at the *Baden-Wuerttemberg State Ministry of Culture.* His goal was to "develop general methods for the treatment of cooperative phenomena."[409]

Within the two-month period between application and authorisation of the sabbatical he scored a great success: using his synergetic approach, Haken presented a solution solving part of the questions of chemical oscillations. On 3 February 1975, only a few days after the extensive review contribution had appeared in "*Review of Modern Physics,*" he submitted a further article in the *Zeitschrift für Physik.* The work, titled "Statistical Physics of Bifurcation, Spatial Structures, and Fluctuations of Chemical-Reactions"[410] solved, via the order parameter concept, the problem of chemical oscillations that was investigated by Nicolis, Lefever and Prigogine. Relating his research to the work of the Brussels school was important to him, as we can see also in his paper "Statistical Physics of a chemical reaction Model," submitted to *Physics Letters* on 22 January 1975. There we can read in the abstract:[411]

> „We give a nonlinear theory of fluctuations at chemical instabilities using the Prigogine-Lefever-Nicolis model. Our results apply to arbitrary dimensions, a mode-continuum and temporal oscillations. Striking analogies to laser phase transitions and hydrodynamic instabilities are found."

The so-called "Brusselator"[412] covers a chemical reaction of the type

$$\begin{aligned} A &\rightarrow X \\ B + X &\rightarrow Y + D \\ 2X + Y &\rightarrow 3X \\ X &\rightarrow E \end{aligned}$$

---

[408] (Institution, 1974) See the list of participants. On the history of chemical oscillations see (Tyson & Kagan, 1988).

[409] Application for a research sabbatical by Hermann Haken from 6.11.1974 and authorization from 27.1.1975. (Personal file Haken; University Archive Stuttgart).

[410] (Haken, 1975f).

[411] (Haken, 1975e).

[412] Named after *Université libre de Bruxelles,* the place where Prigogine and his colleagues did their research.

A and B represent solutes being applied continuously from the outside, the amount kept constant (dissipative system), whereas X and Y are reactants that are variables in space and time. Thus, the "Brusselator" deals with two chemical sorts of molecules that show time- and space dependent structures comparable to the famous Belousov-Zhabotinsky reaction. These structures are controlled by the concentrations of the other reactants introduced into the "reactor" (that is the container where the mixing takes place). The degree of concentration of the reactant, in the words of Haken, then takes on the role of the order parameter. If the degree of concentration is smaller than the critical threshold value, the system will show no structure. Approaching threshold, the system gets unstable and shows big fluctuations. At even higher values "bifurcation" occurs, the system oscillates between two states. Previous theories could not provide for an explanation at threshold or did not take into account the fluctuations. Haken succeeded by using the "Ansatz" he found in laser theory:[413]

> "In our paper we give a novel approach to the bifurcation problem which includes fluctuations and thus seems promising to replace hitherto used bifurcation theory. As an explicit example we treat the Prigogine-Lefever-Nicolis model […], which we incidentally generalize to two and three dimensions and to a mode continuum. […].The stationary solution of it [the equations] is the well-known Ginzburg-Landau functional of the theory of superconductivity and of the continuous mode laser. This allows us to interpret the present chemical instability as a quasi-phase transition including symmetry-breaking (bifurcation)."

Quasi in an accessory clause, Haken gave a mathematical solution for the Bénard-instability:[414]

> „Here we just mention that in a thin layer, the solutions of […] are identical to those of the hexagonal Bénard cells or rolls in hydrodynamics."

Haken than detailed his theory and published it at length in the next volume of the *Zeitschrift für Physik* under the title "Generalized Ginzburg-Landau equations for phase transition-like phenomena in lasers, nonlinear optics, hydrodynamics

[413] (Haken, 1975g), p. 414.

[414] (Haken, 1975g), p. 414. Already two years before Haken had given the corresponding equations in an article submitted to the journal *Physics Letters*. See also (Haken, 1973b).

and chemical reactions." Phenomena in all these quite different looking fields could be dealt with by the same mathematical formalism: [415]

> „The recently found close analogies between the continuous mode laser, the Bénard instability, and chemical instabilities with respect to their phase transition-like behavior are shown to have a common root. We start from equations of motion containing fluctuations. We first assume external parameters permitting only stable solutions and linearize the equations, which define a set of modes. When the external parameters are changed the modes getting unstable are taken as order parameters. Since their relaxation time tends to infinity the damped modes can be eliminated adiabatically leaving us with set of nonlinear coupled order parameter equations resembling the time dependent Ginzburg-Landau equations with fluctuating forces. In two and three dimensions additional terms occur which allow for e.g. hexagonal spatial structures.
> [...]
> Our procedure has immediate applications to the Taylor instability, to various chemical reaction models, to the parametric oscillator in nonlinear optics and to some biological models. Furthermore, it allows us to treat analytically the onset of laser pulses, higher instabilities in the Bénard and Taylor problems and chemical oscillations including fluctuations."

The method of the order parameter and the generalised Ginzburg-Landau formalism made it possible for Haken to analytically solve the phase transition behaviour of quite different phenomena: the Taylor instability of hydrodynamics, chemical reaction models, the so-called parametric oscillator in non-linear quantum optics and some biological models. He had the certainty to be on the right track.

It was only in a small footnote at the end of his article that Haken referred to a close analogy between the description of the laser and a fluid, being a formal identity of the so called "Lorenz-equations" from atmospheric physics with the single mode equations of the laser.

6.6 The "Lorenz – Equations"

The Lorenz-equations[416] initiated a paradigm-shift in many areas of natural and human sciences in the late twentieth century. They had been the starting point for chaos theory leading to new vistas of natural phenomena in different fields,

[415] (Haken, 1975d), p. 105.
[416] (E. N. Lorenz, 1963).

commencing in the 1980s.[417] Edward Lorenz also created the word "Butterfly-effect" that meanwhile has become common language. To better understand the importance of the relationship between the Lorenz-equations and the laser equation we need to briefly explain the genesis of the Lorenz-equations.

Edward N. Lorenz was born in 1917 in West Hartford (Connecticut). He studied mathematics and meteorology at the *Harvard University* and at the *Massachusetts Institute of Technology* (MIT). In 1941, caused by the entry of the United States to the war, the young meteorologists at MIT had been kept busy with weather forecasting, being of great importance for warfare.[418] This was the origin of Lorenz's research, which he maintained his whole life. Modelling the atmosphere, and taking into account its interaction with the surfaces of the earth and oceans, the influence of wind and many other parameters are the subjects of a dynamic meteorology. Starting with a large set of measurements as initial values and by means of many differential equations using the physical laws, one tries to model the future development of the weather. Until the 60s, meteorologists assumed that, the differential equations and the physical laws were deterministic; the development of the initial values should then also be deterministic and thus predictable. It seemed to be only a matter of calculating power. Of course, the atmosphere is a gigantic physical system. It is a gas consisting of an inconceivable number of atoms. To be able to start a calculation at all, it was necessary to use mean values of parameters like air temperature, wind force and direction, heat capacity of air, etc. The number of measurement points is also decisive. To create a network-model of earth's surface that has knots (measurement points) in a horizontal dimension of 100 kilometres (km) and a vertical direction of 2 km would need more than a million measurement points. Without powerful computers, not available in the 1960s, a mathematical solution to this task is hopeless. Therefore, meteorologists work with systems comprising a reduced set of equations. Quite early, the question arose regarding how far the systems of equations could be simplified and reduced, thus, still producing intelligible results. Edward Lorenz devoted a large part of his scientific career to this question. In the beginning, he was working with a set of thirteen equations. In the early 1960s, reducing this set again and again, he finally arrived at three non-linear equations having only three degrees of freedom that modelled the large-scale energy transport of air circulation in the atmosphere.[419] These deterministic equations can be written as:

$$\frac{dX}{dt} = -\sigma X + \sigma Y,$$
$$\frac{dY}{dt} = -XZ + rX - Y,$$

[417] (Aubin & Delmedico, 2002).
[418] (E. N. Lorenz, 1984).
[419] (E. N. Lorenz, 1963).

$$\frac{dZ}{dt} = XY - bZ$$

X, Y and Z denote the convection and the temperature gradient of the air. The constants σ, r and b contain the so-called Prandtl number and the Rayleigh-constant.[420] Performing the calculations, Lorenz expected that these deterministic equations would evolve in a stable and predictable way, because every step in the calculations follows deterministically from the previous step. Surprisingly, he could not find stable solutions. On the contrary, the equations were highly sensitive to the initial input values. It had been the early times of computing. Lorenz used a computer of the Royal McBee LGP-30[421] type that needed about a second per calculation step of X, Y and Z. Therefore, to work through 6,000 cycles of this simple system would take about 100 minutes. It was inevitable that the evaluation sometimes was interrupted. Lorenz remembered: [422]

> „I stopped the computer, typed in the old state [outprinted numbers of the former run], and set the computer running again. Upon returning after an hour, I found that the solution was quite different from the one, which the computer had previously produced."

He soon figured out that the computer internally worked with six digits, while the numbers on the print out consisted only of three decimals. Thus, typing in these three digit numbers altered the former results by some tenth of a percent. This difference in initial values led to completely different final values. More importantly, he found that there were no periodic (recurring) solutions. That explains the title of his publication "Deterministic non-periodic flow." Sketching the solutions of X, Y and Z in a phase-space diagram, as is common for dynamic systems, a highly complex pattern occurred.

---

420 The Prandtl-number and the Rayleigh-constant are specific values depending on the material used in the theory of turbulence of gases and liquids.

421 A computer of the size of a cupboard that, from 1957 until the midst 60s, was manufactured by the Royal McBee Corporation from Port Chester (N.Y.). Its cost has been about. 40.000 Dollar (cited after `http://webdocs.cs.ualberta.ca/~smillie/ComputerAndMe/Part19.html`, recalled 19.01.2012.

422 (E. Lorenz, 1979) p. 105.

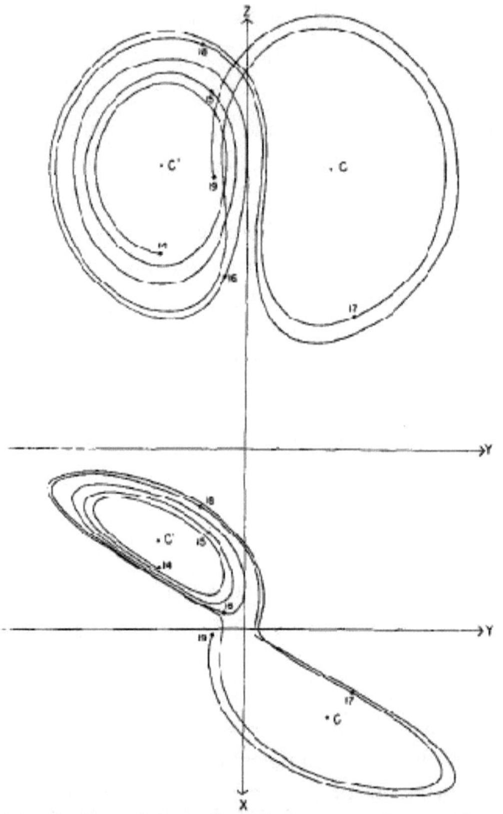

FIG. 2. Numerical solution of the convection equations. Projections on the X-Y-plane and the Y-Z-plane in phase space of the segment of the trajectory extending from iteration 1400 to iteration 1900. Numerals "14," "15," etc., denote positions at iterations 1400, 1500, etc. States of steady convection are denoted by C and C'.

Fig. 16 Graphic representation of the numerical values of the so-called Lorenz-attractor (from (E. N. Lorenz, 1963))

The mathematicians David Ruelle and Floris Takens who "rediscovered" the work of Lorenz in the beginning of the 1970s, dubbed the famous name *"strange attractor."*[423] Lorenz came to the conclusion that long-term weather forecasts would not be possible because of the fact that initial values entering into the calculations, which can only be measured with limited precision.

When Haken took notice of Lorenz's publications, he immediately recognised the following analogy: in an atmosphere, consisting of nearly indefinitely many

[423] (Ruelle & Takens, 1971) In the language of dynamic systems theory an attractor denotes the subset in a phase-space confining the solutions in its time development. Simplest example is a point in phase-space to which all solution converges irrespective of the starting value of the system.

molecules, the three variables in the reduced set of equations constitute order parameters. In addition, as Haken discovered for the laser, Lorenz showed that, varying the parameters, his system passed through different "phase transitions" before entering the irregular non-periodic state. Therefore, the variables in Lorenz' system were equivalent to order parameters, whereas the air molecules constituted the dynamic subsystem. It lent itself to a test regarding if the laser equations in a modified form could also be used for the hydrodynamic system of Lorenz.

It took Haken only a few days to devise the solution of this analogy and he submitted his article on 10 April 1975, to *Physics Letters*, where it was published on 19 May, 1975.[424] Haken again used the formalism of the single mode laser to describe the Lorenz instability of liquids. The basis of his considerations had been an article of J.B. McLaughlin and Paul C. Martin that appeared in the November 1974 issue of *Physical Review Letters*.[425] McLaughlin and Martin studied the development of turbulence of a gas having low a Prandtl-number, as it is typical for air.[426] They aimed at mathematically describing the transition towards non-periodic fluctuations in an ideal thermal convection system. In their work, they referred to the abstract mathematical approach that Ruelle and Takens had used in their publications where they dubbed the name "strange attractor."[427] The development of turbulent behaviour in gas and fluids is characterised by two dimensionless constants. The Prandtl-number describes the ration of kinematic viscosity to thermal conductivity of the material researched, while the Rayleigh-number denotes the heat transfer in a fluid.[428] For a boundless fluid the relations read:

Rayleight-number $R \equiv g \epsilon H^3 \frac{\Delta T}{\kappa\nu}$ and

Prandtl-number $\sigma \equiv \nu / \kappa$

(g is the gravitation-constant, ϵ denotes the heat expansion coefficient, H describes the thickness of the fluid layer, ΔT denotes the temperature difference within the layer, κ the temperature conductivity and ν the kinematic viscosity.)

McLaughlin and Martin demonstrated that for Reynold-numbers with R high enough, instabilities occur. These transitions depend on the Prandtl-numbers: low σ show a transition to wave-like convection rolls, as one can often see in clouds in the sky. Increasing R, other transitions occur, finally resulting in a non-periodic motion. McLaughlin and Martin then showed that at high σ there exists a dependence on the bifurcation theorem found by Hopf. This theorem leads to two solutions. Within the realm of an inverted bifurcation, the limit cycle is not stable and leads to an instability of the system. That had been the case for Edward Lorenz. McLaughlin and Martin then remarked:[429]

---

424 (Haken, 1975b).

425 (McLaughlin & Martin, 1974).

426 The Prandtl-number of air has the value of about 0,72.

427 See footnote 357.

428 Fluid = gas or fluid.

429 (McLaughlin & Martin, 1974) p. 1190.

„The three-mode model of time dependence in high-Prandl number convection, which was studied by Lorenz (E. N. Lorenz, 1963), is an example of inverted bifurcation. In this model there is an immediate transition to a complicated, nonperiodic motion."

Taking "normal" bifurcation into account, Ruelle and Takens already gave a qualitative answer: a range of instabilities should occur and, starting with the fourth instability, the motion should become chaotic, finishing up at a "strange attractor."

Haken seized this idea that connects phase transitions with bifurcation theory. Here, he saw another system the structure of which he could describe the mathematical methods of *Synergetics*. If there really were common roots of these non-linear systems far from thermal equilibrium, one should be able to map the mathematical structures onto each other; and he succeeded. The following figure shows the analogous equations:

Lorenz - Functions	Laser - Functions
$\xi = \sigma\eta - \sigma\xi$	$E = \kappa P - \kappa E$
$\eta = \xi\zeta - \eta$	$P = \gamma ED - \gamma P$
$\zeta = (b(r - \zeta) - \xi\eta$	$D = \gamma(\Lambda + 1) - \gamma D - \gamma\Lambda EP$

Identity relations

$$t \rightarrow t'\sigma/\kappa\,,\ E \rightarrow \alpha\xi\,,\ \text{whereby } \alpha = [b(r-1)]^{-1/2}\,,\ r > 1$$

$$P \rightarrow \alpha\eta\,,\ D \rightarrow \zeta\,,\ \gamma = \kappa b/\sigma\,,\ \Lambda = r - 1$$

Fig. 17 The identical mathematical structure of the Lorenz equations of hydrodynamics and the ones for the laser. (from (Haken, 1975a))

The Lorenz-equations became <u>the</u> paradigm of chaos theory. Haken as well saw the very special meaning of the system of equations: [430]

„The most important result is that spiking occurs randomly though the equations are completely deterministic".

---

[430] (Haken, 1975b)

The Lorenz equations are a typical example of what is called "deterministic chaos."

Completion of this article was such a priority for Hermann Haken that he did not attend the spring 1975 meeting of the DPG in Münster. There again, three topical sub-meetings on phase transitions and one session on non-equilibrium processes were held. Haken also missed the invited talk by Leo P. Kadanoff, speaking on "Renormalization group and critical phenomena." But he knew Kadanoff's ideas from the first *ELMAU Conference* in 1972 and from the 1974 Summer School at Erice. But there was another conference that Haken could not fail to attend: the Budapest conference of the "International Union of Pure and Applied Physics (IUPAP)," held in August of 1975. The reason was that he represented the German physicists in the sub-committee on statistical physics of IUPAP from 1972 until 1978. During the conference, Haken got to know Kenneth G. Wilson who, as an invited speaker, lectured on his theory of renormalisation, which later in 1982 earned him the Nobel Prize. The British physicist Paul C. Martin gave a presentation on experiments and the theory of the onset of turbulence, while Hermann Haken delivered a contribution on the "Statistical theory of self-organising structures."[431]

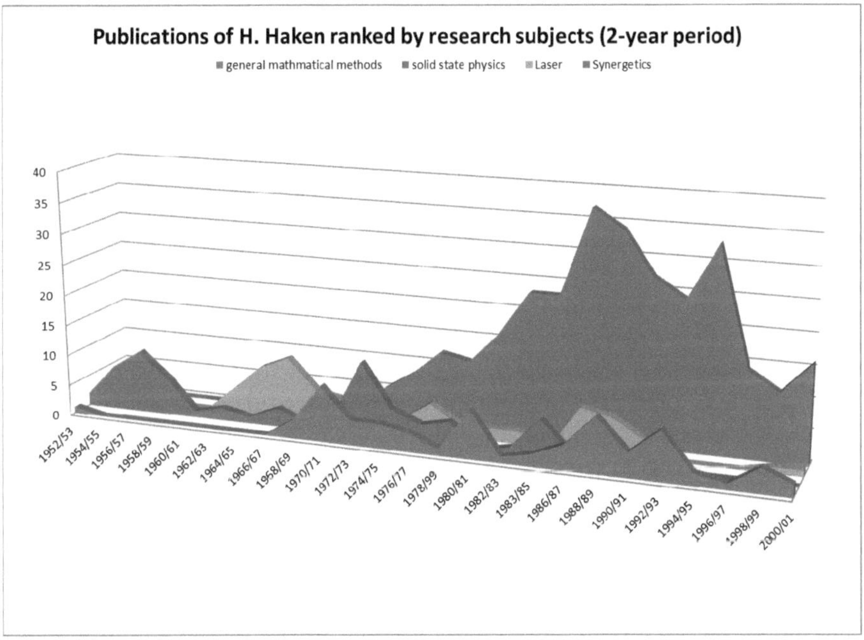

Fig. 18 Number of Haken's publications classified by topic (2 year survey) (after 1980, the articles in the areas of solid-state physics and laser are quite often in cooperation with his doctoral and diploma students)

[431] (Pal, 1975).

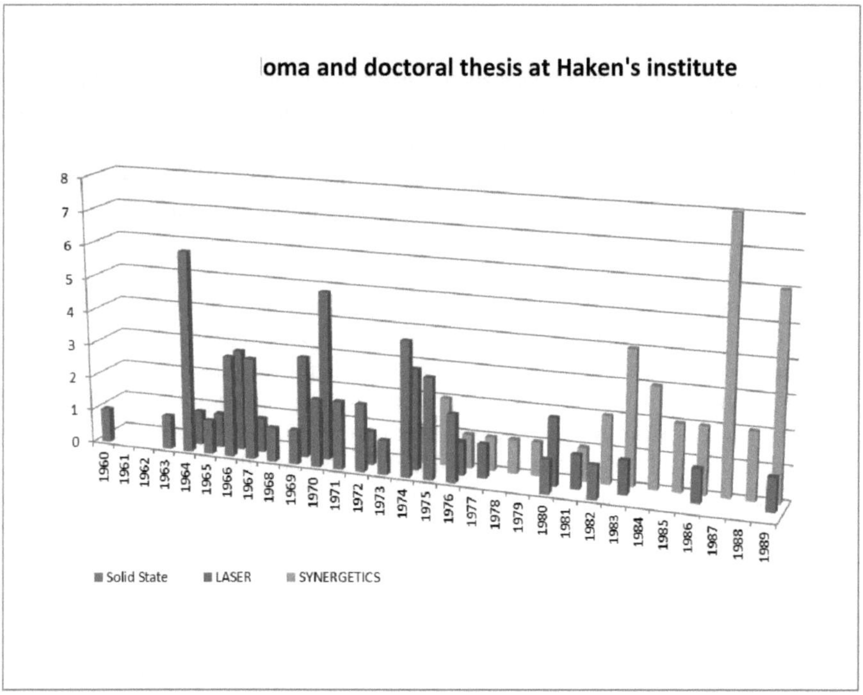

Fig. 19 Topics of diploma and doctoral thesis at Haken's institute ranked by year of publication

It is remarkable that in this international circle of theorists working on statistical physics Haken, in his talk, did not use the word Synergetics; it only shows up as a citation in the bibliography of the printed version in the proceedings.

Nevertheless, 1975 marked a watershed in Haken's research. Induced by the works that we have discussed, Synergetics grew into the focal point of his research interest. Research on solid-state physics and the laser faded into the background. This holds true, even when in the years to come Haken, together with his students, delivered one or another important contributions to these areas. The deflection of his interests towards Synergetics can also be seen by analysing his publication list.

It can be clearly seen that from 1976 until 1977 Haken more and more shifted towards Synergetics. From 1980 on it was the dominant aspect in his work.

Unsurprisingly the same picture shows up if one analyses the topics of the published diploma and doctoral theses of his institute. According to expectations, placing of the subjects of diploma or doctoral theses followed with a certain "delay" of the research interests of the guiding professor. For a short period of one to three years "old" topics dwindled. One should take into account that normally the lead-time before publication of a thesis is 1 – 2 years for diploma work and 3 – 5 years for a doctoral thesis.

Also, this survey shows the thematic change during the 1970s at the institute.

At the end of 1975, Haken rearranged the priorities of his research. Phase transitions in non-linear systems far from thermal equilibrium, renormalisation theory and the theory of dissipative systems were highly topical research subjects. But Haken was in great want of support because his most important co-authors so far in laser theory and quantum optics Wolfgang Weidlich, Robert Graham and Hannes Risken were no longer available due to their own professorships.[432] On top of that, there was a request from the Springer publishing house to write a book on the subject of cooperative phenomena by expanding and bringing up to date the contents of the article that had been published in the *Review of Modern Physics*.

Against this background, in autumn 1975, he directed himself to the *Volkswagenwerk Foundation*, applying for a grant to further develop the research field of Synergetics. The grant was approved in spring 1976, giving his research a definite financial scope for the next four years.[433] He was able to appoint new co-workers and gave himself the chance to organise a second *ELMAU Conference*. But his most important concern was to finish his basic book on Synergetics, which was finally published under the heading "*Synergetics – An Introduction*" by Springer (Berlin), in 1977.

The prevalence of non-linear thermodynamics far from thermal equilibrium was reflected again in the topics discussed at the spring meeting of the German Physical Society held in Freudenstadt (Black forest), April 1976. Within the section "Statistical physics and thermodynamics" there were no less than four sessions on phase transition and three lecture courses on non-equilibrium phenomena. Arne Wunderlin, later on Haken's most important co-worker on the subject of Synergetics, had been sent from the institute, to give a talk on "Symmetriebrechende Instabilitäten in Systemen weg vom thermischen Gleichgewicht."[434]

At this time, Haken received the message that he had been awarded the 1976 Max Born Prize of the *Deutschen Physikalischen Gesellschaf.* This important honour was the first of many awards.[435] The Max Born Prize had been donated in 1972, jointly by the DPG and the London *Institute of Physics,* in honour of the German theoretical physicist Max Born, one of the founders of quantum mechanics.[436] The credentials stated that Haken received the prize for "his

---

[432] Robert Graham in 1974 received a call for the chair of theoretical physics at the *Gesamthochschule/Universität Essen* which he accepted in 1975. From 1972, Hannes Risken already had been appointed professor of theoretical physics at the newly founded *University of Ulm.*

[433] Regarding the *Volkswagenwerk Foundation*, see chapter 7b.

[434] Symmetry breaking instabilities in systems far from thermal equilibrium.

[435] See appendix 6: roster of honors awarded to Hermann Haken.

[436] The prize is awarded alternatingly to a German and a British theoretical physicist every year. Haken was the second German recipient, past Walter Greiner, Professor of theoretical physics at the *University Frankfurt.* (see list of prize winners: Max Born Prize. Deutsche Physikalische Gesellschaft).

outstanding contributions on excitons in solids and for his trend-setting work on quantum optics, especially on laser theory."[437] Synergetics could not be mentioned because at the time, Haken had not officially promoted it.

6.7 The Synergetics Book, 1977

Until 1977 Hermann Haken had often talked about "cooperative phenomena" and sparsely used the word Synergetics. Now, after receiving the grant from the Volkswagenwerk Foundation, he decided to advocate his case more markedly within the scientific community. At the top of the list was writing a basic textbook on Synergetics. However, reworking the *Review of Modern Physics* article from 1975 seemed difficult and Haken chose a different, more pedagogical approach instead.

His book titled *"Synergetics – An Introduction, Nonequilibrium Phase Transitions and Self-Organization in Physics, Chemistry and Biology"* was published by Springer in 1977 and experienced three editions until 1983.[438] In 1981, Arne Wunderlin, co-worker and assistant to Hermann Haken, translated the book into German. Haken took the chance adding a chapter on chaos theory; a topic that in the meantime had become highly topical. This German translation was as well reprinted two times by 1990.[439] Despite the fact that its content was highly mathematical, the book *Synergetics – An Introduction* became the standard reference work of Synergetics. In what follows we sketch its structure.

In the preface of the book Hermann Haken exhibited very clearly his motivation:

> „The spontaneous formation of well-organized structures out of germs or even out of Chaos is one of the most fascinating phenomena and most challenging problems scientists are confronted with. Such phenomena are an experience of our daily life when we observe the growth of plants and animals. Thinking of much larger timescales, scientists are led into the problems of evolution, and, ultimately, of the origin of living matter. When we try to explain or understand in some sense these extremely complex biological phenomena it is a natural

[437] Verhandlungen der DPG 11 (6.Reihe)(1976), p S5 (Ehrungen). The ceremony took place in London on 4 May, 1976. The proposal had come from the British Institutes of Physics. It's safe to say that Herbert Fröhlich has been the driving force.

[438] (Haken, 1977b).

[439] It is remarkable that citations of Haken's book from English speaking countries nearly always refer to the third edition of 1983. That may be due to the growing interest in chaos theory, a topic that Haken had included in the second edition of „*Synergetics – An Introduction*", edited 1978.

> question, whether processes of self-organization may be found in much simpler systems of the unanimated world.
> In recent years it has become more and more evident that there exist numerous examples in physical and chemical systems where well organized spatial, temporal, or spatio-temporal structures arise out of chaotic states. Furthermore, as in living organisms, the functioning of these systems can be maintained only by a flux of energy (and matter) through them. In contrast to man-made machines, [...] these structures develop spontaneously—they are self-organizing."

There is a difference between the book of Glansdorff and Prigogine[440] and the one from Haken in approaching the subject. While the former start from equilibrium thermodynamics Haken developed his ideas *"ab initio,"* according to the methods of statistical mechanics (i.e., without pre-suppositions and limitations). The logical structure of the book was given by this figure:[441]

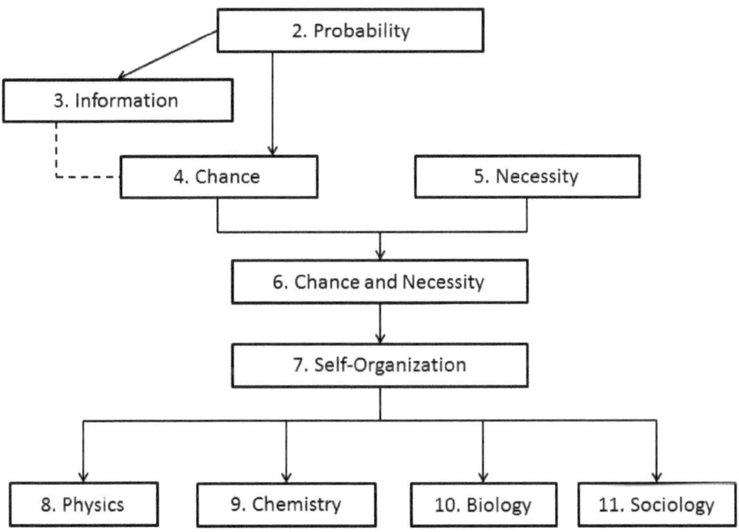

Fig. 20 Logical structure by chapter of the book "Synergetics – an Introduction" (from (Haken, 1977b))

---

440 (Glansdorff & Prigogine, 1971).
441 (Haken, 1977b), p. 16.

The line of thought representing the structure of the book is characteristic for the logical and systematic way of Haken's method. The starting point, when nothing or very little is known about the state of a system, must be probability theory. Applying basic concepts of information theory Haken then moved on to the equations of thermodynamics, deducing the notion of entropy. After that, Haken wrote, "we then pass over to dynamic processes. We begin with simple examples of processes caused by random events. [...] After we have dealt with "chance", we pass over to "necessity", treating completely deterministic "motion".[442] Haken here explicitly gave credit to the book of the French biologist and Nobel Laureate Jacques Monod from 1971,[443] whom he had met at the Versailles Conferences. Monod's considerations were well known to him by his discussions with Manfred Eigen. Haken then connected the term "necessity" with for instance, deterministic forces as described by Newtonian forces. "Chance" is introduced by thermal or quantum-mechanical fluctuations, something that scientists and engineers call "noise" in the case of laser or telecommunications.

Central to the book was the chapter on self-organisation. However, it needed some mathematics to elaborate this phenomenon. Haken introduced equations where „the action (effect), which we describe by a quantity q, changes in a small time interval Δ t by an amount proportional to Δ t and to the size F of the cause."[444] Applying from the outside, a force F, the simplest formula for a damped system reads:

$$dq/dt = - \gamma q + F(t) \tag{1}$$

its solution can be readily written down as

$$q(t) = = \int_0^t e^{-\gamma(t-\tau)} \, F(\tau) d\tau \tag{2}$$

In the case that the system reacts "instantaneously," i.e., q(t) depends only on F(t), we are able to put

$$F(t) = \alpha \, e^{-\delta t} \tag{3}$$

having the solution

$$q(t) = \frac{\alpha}{\gamma - \delta} (e^{-\delta t} - e^{-\gamma t}) \tag{4}$$

The next step is the decisive one: the *adiabatic approximation*. On the premise that $\gamma \gg \delta$ (the time constant of γ is essentially larger than the time constant of δ), the solution of (4) is given by

$$q(t) \approx \frac{\alpha}{\gamma} e^{-\delta t} \equiv \frac{1}{\gamma} F(t) \tag{5}$$

and, Haken wrote, „the time constant $t_0 = 1/\gamma$ inherent to the system must be much shorter than the time constant t' = 1/δ inherent in the orders."

---

442 (Haken, 1977a), p. 15.

443 (Monod, 1971a).

444 (Haken, 1977b), p. 191.

Haken then went on to analyse and present the general problem, something we refrain from discussing in our work. Then, in the next step, Haken introduced a system composed of a single subsystem and an applied external force. He found the pair of equations:

$$dq_1/dt = -\gamma_1 q_1 - \alpha q_1 q_2, \quad (6)$$

$$dq_2/dt = -\gamma_2 q_2 + bq_1^2 \quad (7)$$

Now, if $\gamma_2 \gg \gamma_1$, then (7) can be solved approximately ($dq_2/dt \approx 0$), resulting in

$$q_2(t) \approx \gamma_2^{-1}\, b\, q_1^2(t) \quad (8)$$

Haken explained: „Because (8) tells us that the system (7) follows immediately the system (6) the system (7) is said to be slaved by the system (6). However, the slaved system reacts on the system (6).[445] (8) inserted into (6) results in

$$dq_1/dt = -\gamma_1 q_1 - ab/\gamma_2\, q_1^3 \quad (9)$$

(9) is the fundamental equation describing the behaviour of many synergetic systems. Two completely different solutions exist, depending on the value of γ_1:

If $\gamma_1 > 0$, than $q_1 = 0$, and (because of (8)) also $q_2 = 0$.

A quite different picture occurs for $\gamma_1 < 0$, as is shown in Figure 21:

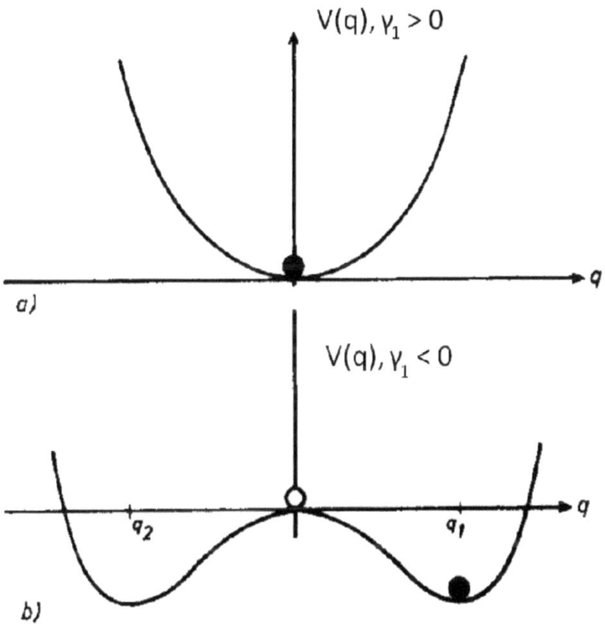

Fig. 21 Representation of the potential representing equation (9) (see text) for different values of γ

[445] (Haken, 1977b), p. 195.

Haken then continues:[446]

„However, if $\gamma_1 < 0$ the steady state solution reads

$$q_1 = \pm (|\gamma_1| \gamma_2 / ab)^{1/2}$$

and consequently $q_2 \neq 0$ according to (8). Thus the system, consisting of the two subsystems (6) and (7) has internally decided to produce a finite quantity q_2, i.e., non-vanishing action occurs. Since $q_1 = 0$ or $q_1 \neq 0$ are a measure if action or if no action occurs, we shall call q_1 an action parameter. For reasons which will become obvious below when dealing with complex systems, q_1 describes the degree of order. This is the reason why we shall refer to q_1 as "order parameter". In general, we shall call variables, or, more physically spoken, modes, "order parameters" if they slave subsystems."

In reality, in most cases there exist many subsystems, not just one. This leads to a reinforcing feedback of the whole system:

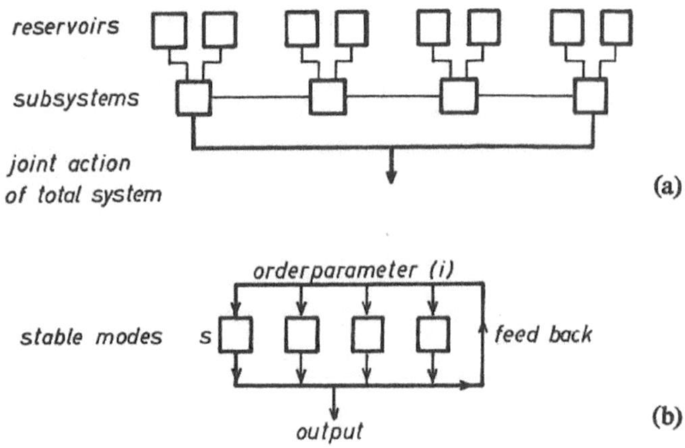

Fig. 22 Self-organising feedback of a synergetic system (from (Haken, 1977b), p. 199)

In such a large system not only a single order parameter exists but many. These different order parameters are competing and define the "output" of the total system.

Fluctuations play an important part. Looking at Figure 21b, if the system is in the unstable state q=0, a fluctuation is needed to reach the left or right stable minimum of the potential. The role of the fluctuations become more visible if one considers different, more elaborate potentials, as are shown in the two following figures:[447]

---

446 (Haken, 1977b), p. 195.

447 (Haken, 1977b), p. 200-201.

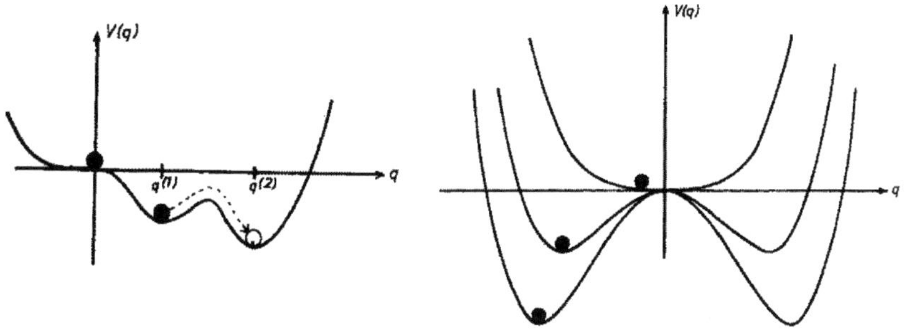

Fig. 23 Shapes of potentials that allow for a change into different stable configurations to perform the change fluctuations as needed. The shape of the configuration on the right hand side traces back to a proposal made by the German-American physicist Rolf Landauer, working in the US with IBM. As early as 1962, he proposed it as a model for an electronic switch by virtue of a deforming potential.[448]

It was only after these seven "introductory" chapters that Hermann Haken finally discussed systems that occur in physics, chemistry and biology in the real world. This shows, how important the sound mathematical foundation of the self-organising phenomena had been to him.

The observed behaviour of the different systems under investigation was based on the mathematical structure of the order parameters and their "slaving" effects – depending on the different time scales of the stable and unstable modes – and thus, much more than just a formal analogy.

In the final chapters, Haken discussed in detail the laser and especially the multiple instabilities of these systems, leading to ultra-short laser pulses. He then reviewed the instabilities in hydrodynamics (Bénard- and Taylor-convection), before mentioning the Gunn instability of electric circuits.

Coming to chemistry, did an in-depth analysis of dynamical systems with diffusion that consisted of two (i.e., three) chemical components, these systems being called the "Brusselator" resp. "Oregenator."[449] The Synergetic systems from biology that Haken mentioned shortly have been population dynamics, predator-prey systems (Lotka-Volterra), evolutionary processes and morphogenesis. The final chapter on self-organisation, comprising only four pages, dealt with a stochastic model of his colleague and friend Wolfgang Weidlich concerning the change of public opinion.

Because of the fact that the examples of synergetic systems in chemistry, biology and sociology rested upon the research of other scientists, Haken only reviewed these results and pointed to the authors articles in previous collective

[448] (Landauer, 1962).

[449] The Oregonator is the name for a chemical model system – analogue to the Belousov-Zhabotinsky reaction - showing oscillations. There are three chemical components in contrast to the Brusselator, consisting of two fluids only. (Field, Körös, & Noyes, 1972)

volumes on Synergetics. The different systems thus, were mere examples of the successful application of his ansatz concerning stochastic many-body problems. They show the transformation of non-ordered unstable systems into stable ordered systems without external ordering forces. The structure of the book mirrors the concept of the lecture course given by Haken and Graham at the University of Stuttgart in 1969 to 1970. Many-body problems are dealt with by means of methods of statistical physics, and solved in a regime far from thermal equilibrium. The analogy between the laser and super-conductivity found earlier was transposed to many other applications. It showed that highly different subjects could be handled mathematically with the same methods in quite disparate scientific areas.

We would like the reader to notice that in the first edition of the book – the manuscript being finished in November, 1976 – no explicit reference was made to the developing "chaos theory." That would change quickly as the first edition sold out in less than a year, an enlarged edition was already edited in 1978.[450] Haken included three new topics that show the dynamics of the new research field: [451]

> „I have added a whole new chapter on the fascinating and rapidly growing field of chaos dealing with irregular motion caused by deterministic forces. This kind of phenomenon is presently found in quite diverse fields ranging from physics to biology. Furthermore I have included a section on the analytical treatment of a morphogenetic model using the order parameter concept developed in this book. Among the further additions, there is now a complete description of the onset of ultra-short laser pulses."

It is instructive to see the rapid development of deterministic chaos theory during this period in the various editions. (See Chapter 7e in this book).

Deterministic chaos was also a topic at the second *ELMAU Conference* that Haken organised shortly after finishing his book in May of 1977.

6.8 The Second *ELMAU Conference* May, 1977

Looking at the speakers and the topics of this meeting one can imagine how the research focus regarding dissipative systems far from thermal equilibrium had changed since the first *ELMAU Conference* in April 1972.

---

[450] The book was translated into different languages: Russian (1980), Chinese (1982, 1984^2) and Hungarian (1984). The German edition (translated by his co-worker Arne Wunderlin) was published in 1982, experiencing further editions in 1983 and 1990.

[451] (Haken, 1977b), preface of the second edition 1978 (finished July, 1978).

Table 8 List of topics at the second *ELMAU Conference* 1977

Conference Topics ELMAU 1977
General Concepts (including Chaos theory)
Bifurcation theory
Instabilities in Hydrodynamics
Solitons
Non-equilibrium phase transitions in chemical reactions
Chemical waves and turbulence (incl. The term chaos in the talk given by O. Rössler)
Morphogenesis
Biological structures
General Structurs (incl. Topics of oeconomy, sociology and linguistic)

Haken included the prevailing mathematical concepts of catastrophe theory and bifurcation theory into the program of the conference, because these methods are important for systems showing symmetry breaking. In particular bifurcation theory turned out to be a starting point for mushrooming chaos theory. Otto Rössler, a biochemist from Tübingen, gave a review titled "Continuous Chaos" on the then known different manifestations of chaos in deterministic systems.

Apart from the highly topical chaos theory, *Synergetics* split into different specialised sub-fields. Particular examples in physics, chemistry and biology (including neurology) were presented and discussed in detail in special sessions.

Another fact seems also noticeable. Compared to the years 1973 and 1974 with its conferences and books on *"Cooperative Phenomena"* and *"Cooperative Effects"* the 2nd *ELMAU Conference* showed "new faces." The list of speakers contains only two participants of the first conference in 1972: Herbert Froehlich and Hans Kuhn. This suggests that the number of scientists and universities being active in the field of non-equilibrium phenomena had increased substantially. It also shows that it was an up-to-date area of research and that Hermann Haken attempted to gain new stimulations for Synergetics by other scientists.

One of the highlights of the conference had been the participation of two eminent French scientists: Alfred Kastler[452] and René Thom. In 1966, Kastler had received the Physics Nobel Prize for his method of "optical pumping" of atoms. This process had been one of the requirements for the laser, a subject vital to Hermann Haken. He also converted René Thom, the "father" of mathematical

[452] Alfred Kastler (1902 - 1984) was a French physicist. Physics Nobel Prize Laureate in 1966. Biographical data see (Perny, 2003).

catastrophe theory to a lecture on this subject.[453] The word catastrophe theory signifies the mathematical theory of differentiable mappings and leads to the classification of discrete, step like changes of certain dynamic systems. Catastrophe theory examines the branching of the solutions of these mappings (bifurcations) when the inherent parameters are changed. It is an important basis for the mathematical treatment of chaos theory. But, contrary to Synergetics, which is a dynamic theory, catastrophe theory is static.

Klaus Kirchgässner, Daniel Joseph (born 1929) and David H. Sattinger (born 1939) presented the new results of bifurcation theory. Kirchgässner (1931 – 2011) was a colleague of Haken at the mathematical institute of the *University of Stuttgart* and made important contributions to bifurcation theory; particularly, he introduced the theorem of "*spatial center-manifold reduction*." Daniel Joseph was professor at the *University of Minnesota* and an expert on stability theory of fluids.[454] The third speaker was the American mathematician David Sattinger from the *University of Minnesota* who had written a fundamental book on stability and bifurcation theory in 1973.[455]

Non-equilibrium phase transitions in chemical reactions and the occurrence of turbulence played an important part at the conference. Haken succeeded in winning four scientists from Poland, Israel, France and Japan giving talks on this subject. Otto E. Rössler, a biochemist from Tübingen (near Stuttgart) delivered a lecture reviewing chemical turbulence. He also discussed deterministic chaos. Despite these important contributors, the absence of members of Prigogine's *Brussels School* working on chemical oscillations is conspicuous. Prigogine should be awarded the Chemistry Nobel Prize for his research on dissipative systems in the same year. Of course, this had not been known in May 1977, the names of the Nobel Laureates being announced only in October of the corresponding year.

At the second *ELMAU Conference* Hans Kuhn (1919 – 2012) from the *MPI für biophysikalische Chemie* at Göttingen and Hans Meinhardt (born 1938) from the *MPI für Entwicklungsbiologie* at Tübingen represented the field of biology. They were longstanding friends of Hermann Haken. That holds true for the American mathematician Jack D. Cowan (born 1933). Haken had met Cowan at the Versailles Conferences, where he, like Haken, had been regularly attending. The topic of his talk was "Neurosynergetics," giving an impetus for the later occupation of Haken with the human brain as a synergetic system.

---

[453] (Thom, 1972). See also (Aubin, 2004).

[454] Later on, he as well wrote a book on the theories of stability and bifurcation. (Joseph, 1980).

[455] (Sattinger, 1973).

Chapter 7
The Propagation of Synergetics: 1978 – 1987

7.1 Synergetics: "Spreading the Word"

The topics discussed during the second ELMAU Conference in 1977 defined Haken's research activities for the next ten years.

Only one month after the second ELMAU Conference, Haken participated at the 4th Rochester Conference in the USA.[456] The subject addressed was quantum optics, his speciality. Haken took the chance to "bifurcate" from his specialty, laser theory, to his new focus on synergetics. He had given his talk the programmatic title "The Laser – Trailblazer of Synergetics".[457] At first glance his contribution must have been confusing to his colleagues, because the meeting was on the highly topical subject of laser physics and non-linear optics. Subtly using this topic, Haken began by outlining the goal of his talk: [458]

> „I want to discuss how a thorough theoretical study of the laser process has led us to understand basic mechanisms of self-organization in physics, chemistry, biology and other sciences".

As examples of self-organising effects outside the framework of physics, he mentioned the Bénard instability, the Belousov-Zhabotinsky reaction and the Gunn effect in semiconductors. In his methodical approach, he first derived the single mode laser equation and then, using the adiabatic approximation technique (slaving principle), demonstrated the equivalence of the laser equation with the Lorenz equations, highlighting:[459]

> „Some time ago Lorenz developed a model of turbulence including 3 variables. We have found that [...] the Lorenz equations are equivalent to those of a single mode laser.

---

456 The 4th. Rochester Conference directed by Leonard Mandel and Emil Wolf took place 8 - 10 June, 1977, at the *University of Rochester*. (Mandel & Wolf, 1978).

457 (Hermann Haken, 1978a).

458 (Hermann Haken, 1978a), p. 49.

459 (Hermann Haken, 1978a), p. 59 (Emphasis added).

B. Kröger, *Hermann Haken: From the Laser to Synergetics*,
DOI: 10.1007/978-3-319-11689-1_7

> Though Lorenz' equations are completely deterministic their solutions show a completely irregular behavior as if caused by a stochastic process. **The laser, when pumped high enough, is the first realistic system obeying Lorenz' equations.**"

This hint towards the Lorenz theory was by no means accidental. Haken drew a connection to the top-notch research field of deterministic chaos that aroused great attraction in the scientific and general audience, thus trying to create interest in the field of synergetics.

In the course of his talk, Haken presented the basic ideas of synergetics. Under certain boundary conditions, open systems can become unstable if external parameters are changed and, taking fluctuations into account, develop into new stable configurations. He showed the following diagram:

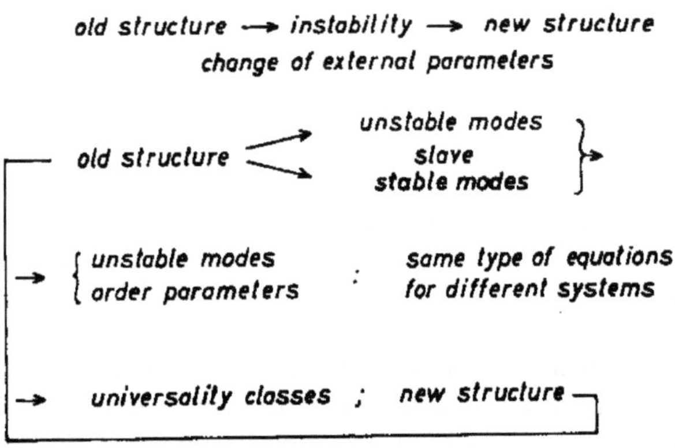

Fig. 24 Diagram representing the passage from an "old" structure to a stable "new" one. (From (Hermann Haken, 1978a))

To us, these descriptions will sound familiar. But it is notable that Haken in this talk presented for the first time the idea of synergetics, as well as the word itself, to a larger audience of physicists in the USA[460], and that he no longer used the more general term "cooperative phenomena".

As we have seen, Haken's emphasis on the Lorenz equations at the 4th Rochester conference was not accidental. For more than two years Haken intensively had been intensively occupied with the connection between phase transitions and bifurcation theory. Additionally he had sent an important article on this subject, co-authored with his assistant Arne Wunderlin, to Physics Letters, where it was received on 9th June 1977. Under the title "New Interpretation and

[460] The conference was attended by more than 300 physicists.

Size of Strange Attractor of the Lorenz model of Turbulence"[461] they presented an illustrative description for the irregular behaviour of the Lorenz attractor.

Bifurcation theory as a path to deterministic chaos was also the central topic of a meeting that was held by the Academy of Sciences in New York in November 1977.[462] The co-organiser was biochemist Otto E. Rössler from Tübingen, who some months earlier in May had delivered a talk on chaos theory at the second ELMAU Conference. The table of contents of the conference proceedings reads like one of the directories of Haken's ELMAU Conferences. Lectures were given under the headings of bifurcation theory in "mathematics, biology, chemistry and chemical engineering, physics, ecology, economics, engineering, experiment and simulation". Among the invited speakers we find no less than six scientists who had also given talks at the second ELMAU Conference six month before.[463] Hermann Haken also visited the New York meeting, where he came upon not only speakers who were well known to him, but also other scientists he had met before.[464]

In his talk, titled "Synergetics and Bifurcation Theory", Haken pointed out the close relation between synergetics and bifurcation theory: [465]

> „Synergetics is a rather new field of interdisciplinary research related to mathematics, physics, astrophysics, electrical and mechanical engineering, chemistry, biology, ecology and possibly to other disciplines. It studies the self-organized behavior of complex systems (composed of many sub-systems) and focuses its attention to those phenomena where dramatic changes of macroscopic patterns or functions occur owing to the cooperation of subsystems. [...] In the course of this research program it more and more transpired that bifurcation theory plays a crucial role."

Taking the laser as his prime example, he discussed the different bifurcation steps of this system, beginning with the onset of laser activity until laser-chaos was reached. For illustrative purposes he used the following picture:

---

[461] (Hermann Haken & Wunderlin, 1977).

[462] *"Bifurcation Theory and its Applications in Scientific Disciplines"*, 31.10.1977 – 4.11.1977. Proceedings of a conference, published in the *Annals of the New York Academy of Sciences 316 (1979).*

[463] D.H. Sattinger, D. Joseph, H. Meinhardt, O.E. Rössler, H. Haken, H. Swinney.

[464] R. Landauer, G. Nicolis, B. Hess, M. Herschkowitz-Kaufman, D. Ruelle, W. Gardiner.

[465] (Gurel & Rössler, 1977) and (Hermann Haken, 1979c), p. 357.

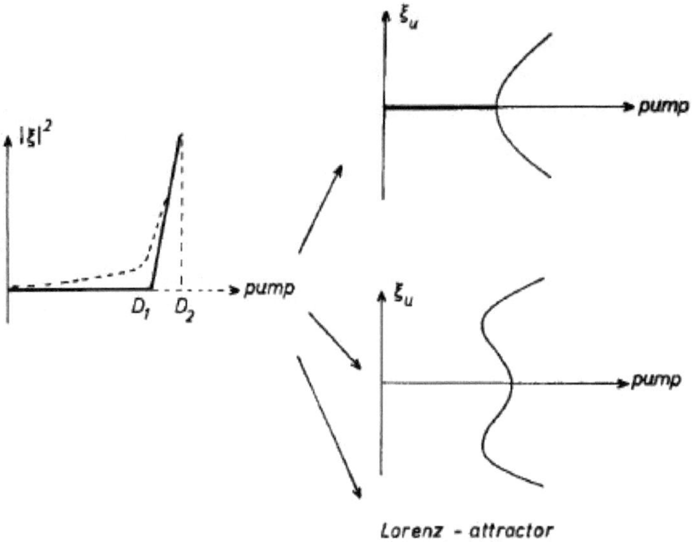

Fig. 25 Bifurcation pattern of the single mode laser depending on the pump strength. The first step denotes the onset of laser activity; the second step gives rise to laser pulses; even higher pump energy leads to chaotic behaviour and the "strange" Lorenz attractor.[466]

The requirements of this bifurcation pattern can be demonstrated via the laser equations, as Haken showed:

$$\frac{\partial E}{\partial t} + c\frac{\partial E}{\partial x} + \kappa E = \alpha P,$$

$$\frac{\partial P}{\partial t} + \gamma P = \beta E D,$$

$$\frac{\partial D}{\partial t} = \gamma''(D_0 - D) + \delta E P,$$

E is the electric field strength, P gives the polarisation of the field, D is the inversion factor of the two-step laser (space-temporal functions), c denotes the velocity of light, γ and γ" are damping constants, and D_o depends on the pump strength. Haken continued[467]:

466 (Hermann Haken, 1979c), p. 369.
467 (Hermann Haken, 1979c), p. 370.

„Whether bifurcation is normal or leads to the Lorenz attractor depends on the relative size of κ, γ and γ″: bifurcation or inverted bifurcation if $\kappa < \gamma + \gamma''$, Lorenz attractor if $\kappa > \gamma + \gamma''$."

Once again the laser had shown its potential as model system, this time as an example for a bifurcation hierarchy in a physical case study that could be exactly calculated.

Of course, in this period Haken was not only involved with synergetics and bifurcation theory. He still had his lecturing duties as a professor, even when his colleagues supported him, taking over many of his obligations. An allusion to his common absences from Stuttgart for conferences and talks was made by his colleague and friend Hans-Christian Wolf, who in a slightly desperate and humorous anecdote noted "that Herrmann Haken is our best-paid guest professor".[468] Nevertheless, Haken's university lectures and courses had been most popular among students, because they were very well structured. Also his contribution on the laser to the Handbuch der Physik had made a great impression. But, being written in English and published within a voluminous and expensive handbook series, it was not very affordable for students. When Heinz Maier-Leibnitz (1911 - 2000), the president of the German Science Organisation (DFG), approached him to write a textbook on "Light and matter", he readily accepted, asking the university and the ministry of education for a research sabbatical.[469] His application was granted; the fact that the Bundesministerium für Bildung und Wissenschaft paid for a substitute made it easy.[470] Therefore we find Haken relieved from teaching and examination duties from 1st April 1978 until March 1979.

While still writing the textbook, Haken made use of the research sabbatical to actively promote his ideas on synergetics. He participated at five big conferences and set about to establish synergetics as a "brand" via lectures and publications.

The first opportunity arose on the occasion of a symposium celebrating the conferment of an honorary doctor's degree to René Thom (1923 - 2002), creator of catastrophe theory, at the University of Tuebingen in May, 1978. Haken gave a talk on "synergetics and a new approach to bifurcation theory", a subject he had talked about recently at the New York conference in autumn 1977. Haken know Thom well, having him invited to lecture at his second ELMAU Conference in May 1977.

Only one month later Haken lectured at the "International School of Statistical Mechanics". The meeting had been organised under the slogan "Stochastic

[468] W. Weidlich in his „Laudatio inofficialis" on the occasion of Hermann Hakens 80s birthday. Unpublished manuscript (Archiv Haken).

[469] Letter from H. Haken to the university from 17.11.1977. (Personal file Haken; University Archive Stuttgart). The book *Licht und Materie Band 1 – Elemente der Quantenoptik* was published in 1979. (Hermann Haken, 1979a)

[470] Letter from the ministry of education from 10.7.1978. (Personal file Haken; University Archive Stuttgart).

Processes in Nonlinear Mechanics" in Sitges, nearby Barcelona, from the 11th to the 23rd June 1978.[471]. Only two weeks later, Haken travelled to Japan where he participated at the Oji Seminar on "Nonlinear Nonequilibrium Statistical Mechanics" (10th - 14th July). This conference marked the end of a research program bearing the same name that had lasted for five years. It therefore offered a kind of "state of the art" review on the topic of non-equilibrium thermodynamics and attracted many renowned scientists working in this field. Paul C. Martin, Günter Ahlers, Harold L. Swinney, R.M. Noyes and R. Zwanzig attended from the United States; Europe was represented by Gregoire Nicolis, David Ruelle, Friedrich Schlögl and Haken, while Japan had sent Ryoku Kubo, H. Mori and T. Shimizu. In his talk, titled "Synergetics: Some Recent Trends and Developments", Haken seized the opportunity to give his colleagues an understanding of the concept of synergetics: [472]

> „Synergetics deals with systems composed of many subsystems such as atoms, molecules, photons, cells, etc. It studies how the cooperation of subsystems can bring about spatial, temporal or functional structures on a macroscopic scale. [...] I first sketch generalized Ginsburg-Landau equations which I derived some years ago and which mainly serve to treat nonequilibrium phase transitions starting from homogeneous quiescent states. I then show how the order parameter concept and slaving principle can be extended so to treat bifurcation and nonequilibrium phase transitions starting from limit cycles or quasiperiodic flows in inhomogeneous media."

He presented his research of the last four years to this circle of mathematicians and physicists in case they had not noticed his latest book "Synergetics – An introduction" that had appeared the year before. Haken emphasised especially the connection between synergetics and statistical physics. He envisioned three distinct stages in its evolution: the first step was due to Ludwig Boltzmann, who had created statistical physics from thermodynamics, which holds true in systems being in thermal equilibrium. The second step, according to Haken, was the theory of irreversible thermodynamics. Here he referred to the works by Kubo and Onsager. The third problematic stage consists of phenomena in systems far from thermal equilibrium, where new qualities arise – one example being the different phase transitions in the laser. At the times, these problems are dealt with by statistical physics. Here synergetics was used to find common mathematical structures in phase transitions where phenomena of a new quality arise. [473]

---

[471] His former student and co-worker Robert Graham, now being professor at the University of Essen had also been invited lecturer. Arne Wunderlin from Haken's institute in Stuttgart was attending too.

[472] (Hermann Haken, 1978b), p. 21.

[473] (Hermann Haken, 1978b), p. 22.

> „This field focuses its attention on those situations where self-organized cooperation of the subsystems brings about qualitative changes of the total system on a macroscopic scale".

Haken emphasised that the sheer size of the subsystems under investigation (atoms, biological cells, photons etc.) make it hopeless to try to calculate the values for each subsystem alone. Like in classical thermodynamics, it seems crucial to work with observable variables, this being the order parameters. Nevertheless, he continued:[474]

> "This is, however, not possible in a straightforward manner and indeed there is nowadays wide-spread agreement between the experts that the method of thermodynamics or irreversible thermodynamics (but not statistical physics) are not sufficient to treat this new field."

By this remark Haken set himself apart from the "Ansatz" of the Brussels School of Glansdorff and Prigogine, who pursued this road of irreversible thermodynamics.

Finally, at the end of his contribution, Haken drew a connection to deterministic chaos, being aware of the fact that Paul C. Martin and David Ruelle – important exponents of this new and active theory – were listening in the audience.[475]

> „More recently we have also considered chaos, i.e. irregular motion produced by deterministic equations. Again there exist classes of phenomena which are governed by the same type of equations but which refer to lasers, models of turbulence, the earth magnetic field, chemical reactions etc."

In this way Haken had shown connections between three important contemporary research areas: the theory of statistical physics, the theory of phase transitions and the developing theory of deterministic chaos.

Back in Europe, he took the road to Bordeaux where he attended a conference on "Far from Equilibrium Instabilities and Structures" in September 1978. Here he again met lecturers and colleagues from the Tokyo meeting earlier in July.[476] The conference was organised by the French chemists Alphonse Pacault[477] and Christian Vidal. Being professor at the University of Bordeaux since 1950, in

---

474 See footnote 402.

475 (Hermann Haken, 1978b), p. 32.

476 R.M. Noyes, G. Nicolis, M. Suzuki.

477 Alphonse Pacault (1918 - 2008), was a French chemist who took his doctorate with Paul Pascal in Paris in 1946. From 1950 he was professor at the *University of Bordeaux*. The Institute Paul Pascal, founded by him in 1965, belonged to the institutes of the French *Centre National des Recherches Scientificques (C.N.R.S.)*. Pacault was member of the French National Order of the Legion of Honor and of the French *Order pour le Mérite*. (`www.crpp-bordeaux.crns.fr/annexes-crpp/images/bis18-pacault.pdf`; recalled 26.1.2012).

1965 Pacault had established the research institute "Paul Pascal" that specialized in the research of chemical oscillations and pattern creation in chemical systems. Pacault's institute was therefore in direct competition with the Brussels School of Ilya Prigogine. Haken was in close contact with the Bordeaux scientists, who until 1982 had always participated in the ELMAU Conferences. Haken, who the year before had already invited Pacault to the second ELMAU synergetics Conference, arranged that the proceedings of the Bordeaux meeting were published as the third volume in the newly edited "Springer Series in Synergetics". Because the conference had a focus on chemical instabilities, the scientists from Brussels were represented prominently. Apart from Paul Glansdorff and Gregoire Nicolis (whom Haken had met in Japan recently) the Belgian fraction was enforced by René Lefever, M. Kaufman-Herrschkowitz and W. Horsthemke. This massive attendance is not surprising, the Brussels School being in the focus of interest after the Nobel Prize in chemistry had been awarded to Ilya Prigogine in 1977. Visitors from Germany were biochemist Benno Hess and physicochemist Otto Rössler. Haken was acquainted with both of them. He knew Hess very well from the annual "Winter Seminars" organised by Manfred Eigen and he supported Otto Rössler with a grant from the Volkswagen Foundation at that time. Japan had sent M. Suzuki and R.M. Noyes had come from the United States, both scientists having been also participants at the recent Oji seminar in Kyoto.

But without doubt, the highlight of Haken's conference marathon in 1978 was the XVII. Solvay Conference that took place at Brussels, between the 20th and the 24th November 1978. Directed by Leon van Hove, the head of the European Nuclear Research Centre CERN in Geneva, a total of seventy-four scientists from sixteen countries assembled to discuss "Order and Fluctuations in Equilibrium and Non-Equilibrium Statistical Mechanics". Subject and location had not been chosen by chance, but rather they were a tribute to Ilya Prigogine and the awarding of the Chemistry Nobel Prize. In the history of 20th century physics the Solvay conferences played an important role. The foundations of quantum mechanics had been discussed and worked out in the twenties and thirties during the earliest of the conference series.[478] Although the number of participants was no longer as strictly limited as it had been in the early days of the conference, being invited to the Solvay conference was still regarded as a high honour. The Stuttgart School was represented by Hermann Haken and Robert Graham.

To quote the organisers, the aim of the conference was:

> „To provide an account on different areas of physics where transitions to ordered behavior occur and to reveal the similarities and the differences in the concepts and techniques involved in the discussion of these problems."

[478] For the history of the Solvay conferences see (Mehra, 1975) and (Marage & Wallenborn, 1999). These essays focus on the early developments at the conferences, until the late thirties. A detailed account on the later Solvay conferences is still missing.

And they continued, saying that[479]

> "[...] the theory of nonequilibrium phenomena has led to the discovery of a new type of nonequilibrium state showing ordered behavior, namely the dissipative structures. Nonequilibrium phase transitions are being intensively studied both from the macroscopic point of view (bifurcation theoretic analyses) and from the point of view of fluctuations".

Ilya Prigogine gave a review talk on "Entropy, Time and Kinetic Description", a topic dealing more with the irreversibility of time and less with the question of arising new order. Robert Graham, a former student of Haken and at that time professor of theoretical physics in Essen, presented a contribution on the "Onset of Cooperative Behaviour in Nonequilibrium Steady States". It is remarkable that the term "synergetics" did not show up in the title of his talk. Within the text of Graham's publication, the term "synergetics" only occurred in the bibliography.[480] This gives some evidence that the word "synergetics" was not familiar in the scientific world at the end of 1978, despite the publication of Haken's book with "synergetics" in the title.[481] The more general term of "cooperative phenomena" was prevailing.

Nevertheless, during the years 1977 and 1978 there had been an intense communication between the scientists working in the field of phase transitions and instabilities far from thermal equilibrium, especially between members of the Stuttgart School of Haken and the Brussels School of Paul Glansdorff and Ilya Prigogine, who met each other bimonthly.

This extremely dense conference activity also hints at an intense research funding by state agencies and other institutions. Since 1976, Haken had received a grant from Volkswagenwerk Foundation that enabled him to organise the second ELMAU Conference in 1977 and the publication of his basic reference textbook "Synergetics – An Introduction" in the same year. At the end of 1978 the funding period drew to a close. Haken therefore applied for a continuation of one year that was subsequently granted.

But to his own great surprise, only some weeks after his application, the Volkswagenwerk Foundation asked him to present an outline of a much broader future research plan on synergetics.

---

479 (Nicolis, Dewel, & Turner, 1981)

480 (Graham, 1981), p. 272.

481 A discrepancy between Graham and his mentor Haken should not be suspected. Graham explicitly thanked Haken in his paper with the words: „I had the privilege of being introduced into this field reviewed in this report about a decade ago by Hermann Haken. Over several years of collaboration he has had a great influence on the development of the views that I hold today in this field. I am very much indebted to him". (Graham, 1981), p. 270.

7.2 Synergetics – Priority Program of the Volkswagenwerk Foundation

The successful approach of synergetics within the framework of phase transition theory in open, non-linear systems far from thermal equilibrium led to the question how to arrange for a wider application basis. In particular, the connections between laser theory and phenomena in chemistry (Prigogine and the "Brusselator", Belousov-Zhabotinsky) and (as it was later called) chaos theory (Lorenz model of strange attractors) required the cooperation of specialists and colleagues in other disciplines. More funds were needed that were not available within a university institute budget. As early as 1975, Haken had had the lucky idea of applying for a grant of the Volkswagen Foundation, knowing that his subject did not belong to the priority programs of the foundation.

Apart from the German Science Association (DFG), the Volkswagenwerk Foundation (from 1988, Volkswagen Foundation) is one of the two large institutions in Germany wherein research is sponsored outside of university or industry institutions. In 1961 the Volkswagen Foundation was created following the transformation of the state owned automotive corporation Volkswagen into an incorporated company and its subsequent privatisation. The proceeds amounted to just over one billion Deutsche Mark.[482] Support is given to all scientific disciplines, not only the natural sciences. Grants were given on the basis of a so-called "positive list" that comprised about twenty research fields. To be able to support other research activities in exceptional cases, the board of trustees in its 49th session from 11th April 1975 confirmed again that "in addition to the priority program also unconventional projects could be supported." Unconventional projects would be research applications "not analysed in full detail, being in an experimental status and having a promising future, as well as applications offering critical ideas that would not fit into existing programs."[483]

In February 1976, during the 52nd board meeting, a research project named synergetics was granted to the Institute for Theoretical Physics at the University of Stuttgart, amounting to 578,200 DM and having a timeframe of four years.[484] The money was spent on three positions for assistant researchers, the organisation of symposia and the subsequent publication of the proceedings.

---

482 For the history of the *Volkswagen-Foundation* see (VolkswagenStiftung, 2002), as well as (Nicolaysen, 2002)

483 (VolkswagenStiftung, 2002), p. 82.

484 Trustees of the *VolkswagenStiftung*: 52. Sitzung, p. 28, Az.: 11 2754. Regrettably, the records of the foundation of that time are incomplete. Switching to EDP-supported administration in the nineties, most of the older applications and correspondence has been destroyed. Only the approval records and the documents on the meetings of the trustees had been kept on file.

The explanatory statement reads:

> "synergetics explores macroscopic phenomena that are created by the collaboration of many subsystems. It tries to reveal the basic and general principles that are valid and responsible for the appearance of spatial, temporal and functional structures in quite different scientific areas. One of the basic assumptions to deal with the structure formation is the concept of order parameters. It could be demonstrated that the transition from one structure to another one, without interference from the outside, is determined by few order parameters only. All other variables are "slaved" by the order parameters and can be eliminated.
>
> The venture has been supported as "unconventional project", outside of the priority programs. During the review process the problem had been named extremely important and future oriented: of central relevance not only for physics but also for all other scientific fields, where cooperative phenomena play a fundamental role. The exceptional qualification of the applicant had been accentuated as well; together with his team he earned a leading role in Germany, receiving international recognition."[485]

In a "tailwind" generated by his research sabbatical, and with the grant from the Volkswagen Foundation, Haken realigned his research activities. He turned away from laser theory and solid state physics and initiated a broad analysis of synergetics at his institute. In the preceding chapter we have shown his various restless activities between 1975 and 1978. During the previous twenty years Haken had acquired a high reputation in the national and international theoretical physics community. He had gained this standing by way of being recognised as one of the "fathers of quantum laser theory", by his extended lecturing activities, and by the positive reception of his book on synergetics in 1977. Nevertheless, he was surprised when the Volkswagen Foundation asked him, in February 1979, to present a research proposal for a priority program named synergetics. This would mean a prolonged timeframe and a much larger scope of research activities than the expiring unconventional program. How did this invitation come about? What could have been the catalyst?

Different effects had come together. Of course, the scientific accomplishments of Haken and his Stuttgart School were at the fore. Apart from laser physics, his research on the theory of phase transitions was highly topical, as we have seen by

---

485 Cited after *Stiftung Volkswagenwerk*: Kuratoriumsunterlage Nr. 1404/62, page 1. At this time, the decision of *Synergetics* as a priority program was pending.

the protocols of the annual meetings of the German Physical Society. Certainly another reason was the holistic "Ansatz" of synergetics. The attempt to discover a common, basal mechanism for the creation of order aroused much interest, because these phenomena were looked at simultaneously in biology. An impressive example is the theory of "hypercycles" in prebiotic evolution, developed by Nobel Prize Laureate Manfred Eigen and his colleague Peter Schuster in the years 1977 / 78,[486] which evoked a strong response nationally and internationally.

But another aspect should not be overlooked: in 1977 the Nobel Prize for Chemistry had been awarded to the Russian born, Belgian physicochemist Ilya Prigogine, who in the same year had published (together with Gregoire Nicolis) a book on the theory of self-organisation in non-equilibrium systems that was thereafter frequently cited.[487] Ennobled by the Nobel Prize, the theory of non-equilibrium systems – also called dissipative systems – had come into focus of attention. This we shall recognise when we come to see the evaluations of the proposal on synergetics as priority program.

In June 1979 H. Plate, the responsible referent, submitted the request on the prolongation of the unconventional program, together with the proposal of synergetics as a priority program to the board of trustees of the Volkswagen Foundation.[488] The office of the foundation recommended acceptance of the prolongation request and asked the trustees for approval to review and detail the priority proposal. This was acceded. Dr. Plate took the chance to present to the trustees the content, status and rating of synergetics. Because this gives us the chance to have a closer look at the status of synergetics in 1979, we cite in more detail from the board documents:[489]

> „Synergetics deals with ensembles (systems) that are composed of mostly equal elements (sub-systems) and does not investigate states but processes that occur far from thermodynamical equilibrium, i.e. near points of instability. Contrary to systems theory that until now had dealt primarily with linear systems, Synergetics differs by including non-linearities, inducing the systems to take on new qualities on a macroscopic scale.
>
> The focus is on transition processes that distinguish themselves by special structures with space-like, temporal and functional features. A characteristic feature of a synergetic system is its self-regulation by cooperative behavior of the

---

486 (M. Eigen & Schuster, 1977), (M. Eigen & Schuster, 1978), (Manfred Eigen & Schuster, 1979).

487 (Nicolis & Prigogine, 1977).

488 *Stiftung Volkswagen*: Kuratoriumsunterlage Nr. 1404/62 from 6 June, 1979.

489 *Stiftung Volkswagen*: Kuratoriumsunterlage Nr. 1404/62 from 6 June, 1979, p. 3.

elements. By this self-regulation the ordered states are stabilized against disturbances.

In contrast to cybernetics these self-stabilizing states are not defined from the very beginning; thus it is not a common feedback control problem with given set value. There are many cases where the system can choose between different states when pushed across the transition point. It is a question of peculiar interest which of these different states will be taken under which conditions. It depends on the initial conditions but especially also on the fluctuations.

Synergetic systems by definition are so-called "open" systems, continuously exchanging energy and, sometimes, matter with the environment. When a system, far from thermal equilibrium, has achieved an ordered state its preservation calls for a constant energy input.

Synergetic systems are described by non-linear differential equations. The elements might be atoms, molecules, photons, cells, animals etc. Accordingly, *Synergetics* relates to disciplines like physics, chemistry and biology (but applications also stretch to fields like sociology, oecology and oeconomy). Working out analogies between phenomena in different scientific fields is a central aspect of *Synergetics*."

The cross-departmental approach of synergetics came into view when the proposal mentioned the different disciplines that were to be addressed by the priority program:[490]

„**Mathematics:**
Bifurcation theory, theory of singularities, theory of stochastic processes
Physics:
Laser, non-linear optics, Hydrodynamics, theory of turbulence, instabilities of electric currents;
Chemistry:
Chemical oscillations, dissipative structures;
Biology:
The cornucopia of problems given it will be advisable to set priorities: population dynamics, morphogenesis, neurol networks

---

[490] *Stiftung Volkswagen*: Kuratoriumsunterlage Nr. 1404/62 from 6 June, 1979, p. 4.

> **Engineering Sciences:**
> Non-linear continuum mechanics, especially problems of deformation of shells, eventually dynamical problems for instance vibrational behavior of surfaces;
> **Informatics:**
> Self-organizing cooperation of computers in computer-networks".

These topics defined a research framework that could be expanded by sociology and economics, if neccessary. The Volkswagen Foundation had arranged for a review process of the proposal by eleven experts. Most of the nine scientists that responded commented positively on this matter.[491] They pointed to the former scientific research done by Haken and his group, the international focus, especially of the ELMAU Conferences, and the interdisciplinary approach. The point that drew some criticism was that the work of the Brussel School of Ilya Prigogine had not been sufficiently appreciated in the proposal. According to the majority of the referees, it should have been included into the concept of the priority program. The following remarks from a university professor of mathematics show the arguments:

> "Anticipating the outcome, I opine that Mister Haken outlines an issue that by its very importance for physics, chemistry, biology and not least for our scientific world view seems worthy of promotion, even as priority program of the Volkswagen Foundation. But the proposal of Mister Haken needs to be modified. On the one hand, an extension is necessary to include approaches of others that are as much as important as the one from the Stuttgart School. [...] In ELMAU we have already spoken about the necessary diversification; that the ideas of Glansdorff and Prigogine (and their subsequent development by the Brussel School) on the "theory of dissipative structures" must be included. Even taking into account reservations concerning the Nobel Prize committees politics of selection, Prigogine, for his seminal work, had been awarded unshared the Chemistry Nobel Prize, in 1977!"[492]

We can see by the evaluation that the referee as well as H. Plate from the Volkswagen Foundation had been participants of the ELMAU Conference in

[491] Four of the nine reviews had been replicated anonymously in the trustees' submission papers. (*Stiftung Volkswagen*: Kuratoriumsunterlage Nr. 1404/62, 6 June, 1979).

[492] *Stiftung Volkswagen*: Kuratoriumsunterlage Nr. 1404/62, 6 June, 1979, Appendix VI. Opinion of a University Professor of Mathematics.

1979. Plate, being the head of division of the foundation, would have had an overview of Haken's research subject and his international connections.[493]

In 1979 / 80, solid state physicist Hans-Joachim Queisser (born 1931) was a member of the board of trustees of the Volkswagen Foundation. He had taken his doctorate with Hilsch at Goettingen University in 1958, at a time when Haken there had been Assistant to Volz. After having post-doctoral positions in the US, among others at the Shockley Transistor Corporation and at Bell Laboratories, and a staging post as professor at the University of Frankfurt, Queisser had been appointed director of the newly founded Max Planck Institute of Solid State Research in Stuttgart in 1971.[494] The suggestion to found such an institute had come from the Stuttgart physicists Pick and Haken.[495] That is why Queisser knew Haken and his work very well. Within the board of trustees, he championed synergetics to be given the status of a priority program.[496] During the spring term meeting in 1980, the new priority program was formally approved and installed. Plate informed Haken immediately and took the chance to ask him about Ilya Prigogine and the Brussels School:

> "I asked Haken about the basic differences in the approach ("ansatz") of the Stuttgart and the Brussels School. He confirmed my opinion that essentially it was about the same subject, he coming from statistics, Prigogine coming from thermodynamics. (I wanted to know which catchwords I shall have to avoid, not precluding the Brussels approach). Today, this would be no problem anymore, the statistical approach being more comprehensive. Prig.[ogine] also had moved in this direction, even not telling so. As early as in his Nobel Prize acceptance speech, one could find a "break" [in the argumentation]. Starting with the Glansdorff-Prigogine-Principle of the minimal entropy increase – hereby it is not possible to deduce further information on the [evolving] structures – he then swings, introducing the "Brusselators".

---

493 The lists of participants at the *ELMAU Conferences* (except the one from 1988) could not be found, neither in the Archive Haken, nor in the papers of the Volkswagen Foundation.

494 See the short CV of Hans-Joachim Queisser, issued on behalf of his presidency in 1976/77 of the DPG: Verhandlungen der Deutschen Physikalischen Gesellschaft 6. Reihe (1976), p. 7, as well as www.de.wikipedia.org/Hans-Joachim_Queisser, recalled 9.2.2012.

495 See interview with Hermann Haken from 21.9.2010, pages 34 and 35. (Archive Haken. University Archive Stuttgart).

496 [...] during the 64th meeting of the board in spring 1980 the priority program [Synergetics] had been added to the priority program list after a detailed explanation by trustee Queisser." Cited after *Volkswagen-Stiftung*: Kuratoriumsunterlage Nr. 2179/95. p.1).

> He would welcome very much, if the Brussels School would be included; to do so, a revision of the priority program proposal would not be necessary."[497]

In the annual report of the foundation, the connection of synergetics with the work of the Brussels School was once again mentioned explicitly: [498]

> „Initiated by a research activity of Professor Dr. H. Haken at the *I. Institute of Theoretical Physics* at the *University of Stuttgart* that had been sponsored as „uncommon project", the Foundation had considered supporting this new research field and decided to include the subject into the priority list. The sponsored program is also to be seen in connection to the work on "dissipative structures", a field of research, Professor I. Prigogine of Brussels had been awarded the Chemistry Nobel Prize, 1977."

Already in the first year of the priority program, four different research activities were supported. Among the scientists involved were the known physicist Hans Otto Peitgen (born 1945) from Bremen, performing research on "pertubations of non-linear differential equations with delay", the molecular biologist Alfred Gierer (born 1929) from the MPI für Virusforschung in Tübingen and the physicist Ernst Dieter Gilles (born 1935) from the Institut für Systemdynamik und Regelungstechnik of the University of Stuttgart. Haken received research grants on "dynamics of synergetic structures" and cooperated in a project with Spanish physicist Manuel G. Velarde (born 1941) and mathematician Ludwig Arnold from Bremen on "deterministic and stochastic approaches, solving instabilities and cooperative phenomena in fluid layers and non-linear reaction-diffusion systems".[499]

The priority program "Synergetics" of the Volkswagen Foundation lasted ten years, from 1980 until 1990. In total, there were one hundred and fifteen projects from forty-eight research teams. It is not the aim in this work to review these activities in detail. A list of projects and the researchers involved is given in Appendix 3. We will just present a brief survey to show the wide scope of the activities. The work of Hermann Haken and his group will be shown in the following chapter.

An important role was played by the development of the mathematical foundation of synergetics and the dynamics of non-linear systems. Apart from Haken and Weidlich in Stuttgart, their colleagues Güttinger from Tübingen as well

---

[497] Transcription of a handwritten note by Plate, from 19. 3. (presumably 1980). (*Stiftung-Volkswagen*, Ordner Synergetik, Az.: 4150.9).

[498] (Volkswagenwerk, 1981) p. 124.

[499] (Volkswagenwerk, 1981) p. 126/127.

as Arnold, Peitgen and Richter from Bremen took part in this endeavour. The latter produced large interest from the public, creating fascinating computer generated pictures (which were also published as a book[500]), popularising terms like "fractals" and "Mandelbrot-Männchen" (Mandelbrot set). In the physical sciences, convection phenomena and their multiple phase transitions had been the focus. To name just a few, we mention the groups of Schulz-Dubois from Kiel, Lücke from Saarbrücken, Stierstadt from München and Busse from Bayreuth. Structure formation in solids was investigated by Hübener from Freiburg, Klingshirn from Kaiserslautern and Purwins and Jäger from Münster.

It is hardly surprising that chemistry of dissipative systems had been another focal point of the research activities. These systems, as well as oscillating chemical systems and autocatalytic phenoma at surfaces, were researched by not less than nine groups, to be found in the Appendix.

The biological sciences had been interested in the space and time structure of cells and cellclusters. We like to make reference to the activities of Benno Hess from the MPI für Ernährungsphysiologie from Dortmund, a long time acquaintance of Haken. Following up the recent work on the theory of hypercycles by Manfred Eigen and Peter Schuster, Dress from Bielefeld performed research on stochastic analysis of molecular evolution. Some money also flowed to Vienna, supporting Peter Schuster and his students in their research on biological evolution in sequence spaces.

Halfway through the financial supportive term, research proposals on subjects like neurology and the sciences of the brain were granted more frequently. There was now a common understanding that the brain with its myriad neurons could show synergetic behavior and attempts to model this mathematically were made. The theory of neural networks became a central research field, not only in synergetics, but in computer science as well. In the framework of the Volkswagen Foundation priority program, von der Malsburg and von Seelen from Bochum, Kinzel from Gießen and Reitböck and Eckhorn from Marburg devoted themselves to this matter. Single projects from the engineering sciences, sociology, linguistics and the humanities complete the picture.

In 1990, at the end of the priority program, the head of division from the Volkswagen Foundation made up the following balance:

> "synergetics, as a transdiciplinary research activity, has found its way into many areas of the natural sciences. [...] There are young scientists versed in synergetics, continuing research at some places and established communication structures across the borders of disciplines. [...] Furthermore, – scientifically particularly interesting – synergetics is part of the theory of non-linear dynamics (some people would call it chaos theory), that – run ahead and cross-fertilised by synergetics – had

[500] (Peitgen & Richter, 1986).

> developed in the [university] institutes after the availability of powerful computers during the Eighties".[501]

From 1989, Haken and his group were supported by budget resources from the state Baden-Wuerttemberg and his institute renamed "Institut für Theoretische Physik und Synergetik" of the University of Stuttgart. Therefore, the Volkswagen Foundation considered the discipline of synergetics as established, and finished the priority program at the end of 1990.[502] Altogether, grants amounting to twenty-three million DM had been awarded.

7.3 Research Activities of the Stuttgart Synergetics School

From 1980, the financial resources of the Volkswagen Foundation priority program enabled Hermann Haken to pursue three strategies for promoting synergetics:

1. Application of the synergetic methods to leading research fields in physics and beyond: synopses, surveys and stimulation was given by the bi-annual ELMAU Conferences.
2. Promoting the concept and the term "synergetics": expedient here were the books in the series "Springer Series in Synergetics"; regular talks and presentations at conferences by Haken and his students, a second textbook addressed to specialists titled "Advanced Synergetics"[503], popularisation of the subject by his bestselling book "Erfolgsgeheimnisse der Natur – Synergetik, die Lehre vom Zusammenwirken"[504] and increasing numbers of articles in popular science magazines.
3. Theoretical research at his Stuttgart institute: wide-ranging activities concerning the mathematical theory and application of synergetics by Haken, assisted by a growing number of diploma and doctoral students.

As we have shown in the preceding chapter, the research field of synergetics took shape only from 1977 / 1978. But at that time, Haken had been largely on his own, because Robert Graham and Hannes Risken had been called to theoretical physics chairs at other universities. It took some time before Haken could attract fresh co-workers to the field of synergetics from his group of doctoral students.

Taking a look at the university calendars from the University of Stuttgart for the years 1975 to 1985, one can see that synergetics manifested itself only slowly in the curriculum. Even though the word "synergetics" had been used in the 1970 summer term, the first course devoted to synergetics only showed up four years

---

501 *Volkswagen-Stiftung*: Kuratoriumsunterlage Nr. 2179/95. p. 8.

502 (Volkswagen-Stiftung, 1991), p. 147.

503 (Hermann Haken, 1983a).

504 (Hermann Haken, 1981).

later, in 1974. Another two and a half years later, Hermann Haken established an advanced seminar. This graduate seminar than constituted an on-going basis for the activities on synergetics in the institute.

Table 9 Lectures and seminars held by Hermann Haken at the University of Stuttgart during the years 1975 to 1985, according to the university calendars. (SS denotes summer term; WS denotes winter term).

1974 SS	Introduction to Synergetics 1 and tutorials
1974/75 WS	Introduction to Synergetics and tutorials
1975 SS	
1975/76 WS	Selected problems of non-linear optics Special questions of statistical physics
976 SS	Laser and non-linear optics Selected problems of statistical physics
1976/77 WS	Introduction to theoretical Physics Advanced seminar Synergetics
1977 SS	Laser and non-linear optics 2 Kuramoto (guest professor): Instabilities in systems far from thermal equilibrium Advanced seminar Synergetics
1977/78 WS	Theoretical Physics III: Quantum Theory Advanced seminar Synergetics Advanced seminar non-linear optics
1978 SS	Theoretical Physics IV: Thermodynamics Quantum field Theory of Solids Advanced seminar Synergetics
1978/79 WS	W. Dieterich: Synergetics
1979 SS	Laser and non-linear optics Advanced seminar Quantum Optics Advanced seminar Synergetics
1979/80 WS	Laser and non-linear optics 2 Advanced seminar Synergetics
1980 SS	Synergetics – statistical physics out of thermal equilibrium
1980/81 WS	Quantum Optics
1981 SS	Synergetics IIStatistical Physics Quantum Optics
1981/82 WS	Theoretical Physics III: Quantum Theory Advanced seminar Synergetics Advanced seminar Laser Theory
1982 SS	Electrodynamics and tutorial Advanced seminar Synergetics Advanced seminar Quantum Optics

Table 9 (*continued*)

1982/83 WS	Laser and non-linear optics Advanced seminar Synergetics Advanced seminar Laser and non-linear optics
1983 SS	Non-linear Optics and Laser Theory II and tutorial Advanced seminar Synergetics Advanced seminar Laser and non-linear optics
1983/84 WS	Synergetics I Advanced seminar Synergetics Advanced seminar Laser Theory
1984 SS	Synergetics II: Cooperative phenomena Advanced seminar Synergetics Advanced seminar Laser and non-linear optics
1984/85 WS	Deterministic Chaos Advanced seminar Synergetics Advanced seminar Laser Theory
1985 SS	Selected problems of Synergetics Advanced seminar Synergetics

Arne Wunderlin became the most important collaborator of Haken on the subject of synergetics. In 1971, Wunderlin had written his diploma thesis at Haken's institute on a subject of solid state physics[505], taking his doctorate on non-equilibrium statistics in 1975.[506] The essential role for Hermann Haken, which Robert Graham played in laser theory, was played in synergetics by Arne Wunderlin. He supported him in the mathematical consolidation of the slaving principle in synergetics, did lot of work supervising Haken's diploma and doctoral students and, once in a while, took on lecturing duties when Haken was participating at the numerous conferences he attended.

The first clearly recognisable work concerning synergetics by one of his students originated from Harald Pleiner who, working on the subject of the Bénard-instability, took his diploma with Robert Graham in 1974. At that time, Johannes Zorell (who received his diploma in 1976) had also been occupied with the question of non-equilibrium phase transitions in chemical reactions.[507] Supervised by Haken, he explored in detail the reaction model of the Brussels School of Prigogine, the so-called Brusselator. But Zorell still called it the "Reaction model of Prigogine, Lefever and Nicolis PLN". In his thesis, Zorell called attention to an essential difference between Haken's work of 1975[508] and the calculations of the Brussels School: treating the chemical model, Haken applied stochastic equations that included fluctuations, whereas Prigogine and his collaborators only worked with rate equations that did not allow for fluctuations.

[505] (Wunderlin, 1971).
[506] (Wunderlin, 1975).
[507] (Zorell, 1976).
[508] (Hermann Haken, 1975b) and (Hermann Haken, 1975a)·

However, in non-linear systems, fluctuations play an essential role at the critical transition point. For our argument it is also relevant that in his bibliography Zorell cited statements from Haken from the second and third Versailles Conferences on "From Theoretical Physics to Biology"[509] that had been reprinted in the conference proceedings. This shows that the topics of these conferences were read and discussed at the institute at least as long ago as 1976.

In the period that followed, no other papers on chemical space-time ordered models like the Brusselator and the related Belousov-Zhabotinsky reaction could be found. Haken concentrated on physical issues like the Bénard-phenomena, or on hydrodynamic problems, especially the Taylor-Couette effect.

After having the solution of the space-like form of the Bénard effect in 1973, Haken advised his doctoral student Herbert Klenk to look into the phenomenon in the plasma regime. Plasma – a strongly ionised gas – may be considered as an extremely diluted fluid. The motivation for this work probably came from a talk given by E. Lothar Koschmieder (born 1940) at the Second ELMAU Conference in May 1977.[510] Klenk took his doctorate in 1979, theoretically analysing so-called MHD Plasmas (Magneto-Hydrodynamical Plasma). These are characterised by the fact that the mean free path between two collisions of the atoms is much shorter than the characteristic length at which the distribution function changes its values. In such a dense plasma, one is therefore allowed to assign mean values like temperature and velocity, analogous to a normal gas. Nevertheless, the calculations are more difficult than in a gas because feedback mechanisms of electromagnetic fields have to be taken into account. MHD Plasmas were a matter of peculiar interest to Haken because they can be found in nature quite often. Examples are convections in the Earth's mantle, fusion plasmas, the sun, interstellar gases and earth's ionosphere.

In solving his difficult equations, Klenk took advantage of computer calculations. He detected stable solutions of the Bénard effect in MHD Plasmas under certain boundary conditions. Figure 26 shows one nice example that Haken used sometimes in his later publications.[511]

Even later on, at the beginning of the Eighties, another student – Michael Bestehorn – occupied himself with pattern formation in the Bénard problem,[512] investigating the transitions between rolls and hexagons under certain boundary conditions (round boundary, angular boundary, two-dimensional case). In addition he calculated numerical solutions of the general Ginzburg-Landau equations with the assistance of computer programs.

---

[509] (Marois, 1971) and (Marois, 1973).

[510] (Koschmieder, 1977).

[511] For instance in his best selling popular book „Erfolgsgeheimnisse der Natur" (Hermann Haken, 1981).

[512] (Bestehorn, 1983).

Fig. 26 Flow pattern of a plasma in a vertical magnetic field that is heated from beneath. (From (Klenk, 1979)).

Another important research topic at Haken's institute had been the theoretical examination of the different ordered states in hydrodynamic phenomena, especially in the so-called Raleigh-Taylor effect. In this phenomenon, the flow characteristic of a fluid that is encased between two rotating concentric cylinders is investigated. Haken assigned the work to his diploma student Rudolf Friedrich (1957 - 2012), who calculated the different ordered states that occur consecutively, until the fluid shows chaotic motion. The idea behind this work had probably come from the experimental work of Fenstermacher, Swinney and Gollub, presented at the second and third ELMAU Conferences in 1977 and 1979.[513] It had been an important insight from these investigations that, by increasing the rotational velocity of the inner cylinder, only a few stable patterns occur before chaos persists. Friedrich commented on this subject:[514]

> „If the inner cylinder rotates only, by increasing the rotational velocity, the following flow patterns show up: for small angular velocities, the fluid flows laminar between the two cylinder walls. This pattern is called Couette-flow.
> If the angular velocity exceeds a critical value, the Couette flow gets unstable and a timelike stable vortex flow occurs in the longitudinal direction of the cylinder, the so-called Taylor vortex flow.

---

[513] (Swinney, Fenstermacher, & Gollup, 1977). Also (Fenstermacher, Swinney, & Gollub, 1979).

[514] (Friedrich, 1982).

> Still increasing the angular velocity, a timedependent periodical flow of the fluid can be observed, the Taylor rolls start to oscillate. The oscilation is superimposed on the stationary vortices and the waves move in azimuthal direction around the cylinder. This flow is called wave-vortices. (Wellenwirbel) [...] The time-like behavior of a weakly turbulent flow is reminiscent of the so-called chaotic behavior of finite dimensional systems. Therefore quite often the weak turbulence phase is named chaos."

In the following period these approaches were deepened by two doctoral theses from Klaus Marx and Michael Bestehorn, focusing on numerical solutions by means of computer programs.[515]

But not only topics from chemistry and physics attracted the interest of Haken in his synergetics research program that had been analysed by him and his co-workers. Remembering the motto of the bi-annual Versailles Conferences "From theoretical Physics to biology" and the fact that Haken attended the annual "Winter Seminars" organised by Manfred Eigen from the Max Planck Institute of biophysical Chemistry in Göttingen, it is unsurprising that Haken also took up biological subjects. At the aforementioned conferences and gatherings he became acquainted with Alfred Gierer and Hans Meinhardt, inviting them to his ELMAU Conferences.[516] Both scientists were engaged in the topic of cell differentiation, resorting to an approach by Alan Turing (1912 - 1954), dating from 1952.[517] In his seminal publication, Turing had shown that coupled diffusion-reaction equations were able to explain the formation of space-like patterns in biological systems. Turing assumed concentration gradients of substances creating "pre-patterns", thus inducing cell differentiation. In the early Seventies, Gierer and Meinhardt had developed a theory explaining this process by means of two substances that acted as activators and inhibitors to the system.[518] The Gierer-Meinhardt equation became an important and often-discussed model system on cell differentiation. The formulas on activator-concentration a and inhibitor-concentration h are given by

$$\dot{a} = \rho + d\frac{a^2}{h} - \mu a + d_a \Delta a \quad \text{und}$$
$$\dot{h} = c\, a^2 - \nu\, h + D_h \Delta h$$

ρ, d, μ, c and ν being constants; D_a and D_h are the diffusion constants of the activator (resp. inhibitor). Δ denotes the Laplace-Operator.

---

515 (Marx, 1987), (Bestehorn, 1988).

516 (Meinhardt, 1977).

517 (Turing, 1952).

518 (Gierer & Meinhardt, 1974). See also (Gierer & Meinhardt, 1972).

Gierer and Meinhardt had solved these equations numerically. It was Haken's ambition to solve the equations analytically and, additionally, to look for a two-dimensional solution instead of a linear one. As early as 1977 he succeeded, working in collaboration with his student Herbert Olbrich.[519] They explained the significance of their model in a subsequent publication:[520]

> "Since most tissues or organs are two-or three-dimensional it appeared highly desirable to explore the properties of the Gierer-Meinhardt model in more than one dimension. The most striking result of our above treatment is the great variety of different possible patterns (leaving aside the possibility of oscillatory phenomena).
>
> The evolving patterns are determined by several factors: a) the form of the original equations, b) boundary conditions, c) initial conditions and d) fluctuations".

Also emphasised explicitly were the advantages of their approach and its connection to synergetics:

> „It allows us to study time-dependent processes during which the pattern evolves, it permits a classification of patterns by a maximum probability principle and it allows us to put pre-pattern formation in complete analogy to nonequilibrium phase transitions in physics and chemistry. There are indeed profound analogies between quite different systems with respect to the formation of patterns, and it is a main objective of synergetics to elaborate these analogies and to unearth their common roots."

To give an example: the method of Gierer and Meinhardt allows for an explanation of polarity (head-foot) in the development of cell differentiation of the fertilised egg or the occurrence of patterns in animal skin or snails.

After these first steps in applying synergetics to biology, Haken had to wait for some while until he came upon another student that was educated in biology.

In 1981, he finally recruited Christoph Berding, who had studied biology and physics at the University of Hamburg and wanted to write his diploma thesis with Haken on the subject of space- and time-like structures in morphogenesis.[521] Berding then deepened this topic in his doctoral thesis that he finished in 1985.[522]

---

[519] (Olbricht, 1977).

[520] (H. Haken & Olbrich, 1978). „Note added in proof".

[521] (Berding, 1981).

[522] (Berding, 1985).

Fig. 27 Seashell pattern compared to a computer generated pattern according to the Gierer-Meinhardt model (© Hans Meinhardt)

This procedure, dealing with the same subject in both theses, was a normal one when working with Hermann Haken. Having the right candidate, he always tried to get a quick access path towards a theoretical problem, followed by an early publication. In the doctoral thesis an intensified research then followed that could take up to five years. Haken remarked:

> "Often it has been the case that the main work was done in the diploma thesis. Here the most important ideas were found that were detailed later on in the dissertations. That is why it is very interesting to read the diploma thesis as well.[523]

Halfway through the Eighties the subjects of the students' work changed, now concentrating on two new topics that Haken had taken up after publication of his second textbook "Advanced Synergetics" (1983): the theory of information (informatics) including the "synergetic" computer and applications of synergetics to the coordination of movement and brain research.[524]

---

[523] Interview with Hermann Haken from 16.10.2010, p. 17. (University Archive Stuttgart; Archiv Haken).

[524] These later works of Hermann Haken do not constitute the focus of this analysis. Nevertheless, a short survey is given in chapter 8.

7.4 The ELMAU Conferences of 1979 and 1980

The scientific conventions in the secluded and scenic beauty of ELMAU at the foot of the Wetterstein Mountains in the Bavarian Alps have been especially important to the thinking of Hermann Haken. Unmolested by his university duties, he could enter into most intensive discussions with leading experts from different scientific areas. The participants were personally hand-picked by him and allowed for inclusion of interdisciplinary professionals. Because of the fact that at the ELMAU Conferences there existed no program-committees, as with other international meetings, no consideration on the different partialities of its members concerning the selection of the lecturers had to be taken into account. In the end, Haken invited to ELMAU only those researchers he felt to be interesting for the subject under discussion. That is why the list of speakers and the particular topic of the conferences reflect the scientific research interests of Haken during the Eighties.

The first two ELMAU conferences of 1972 and 1977 dealt with the first steps of synergetics, coming from phase transitions in systems far from thermodynamic equilibrium and instability theory. (See Chapter 6d and 6h). In the following nine conferences a single research topic took centre stage each time. Table 10 gives a survey of the subjects dealt with at the different meetings from 1972 until 1990. After 1990, with no grants from the Volkswagen Foundation at hand anymore, Haken was not able to organize separate conferences at ELMAU.

Table 10 Topical list of subjects of the 11 ELMAU Conferences, 1972 till 1990

Date	Subject
30.4.-6.5. 1972	Synergetics: Cooperative Phenomena in Multi-Component Systems
2.5.-7.5. 1977	Synergetics - A Workshop
30.4.-5.5. 1979	Pattern Formation by Dynamic Systems and Pattern Recognition
27.4. - 2.5. 1981	Chaos and Order in Nature
26.4. - 1.5. 1982	Evolution of Order and Chaos
2.5. - 7.5. 1983	Synergetics of the Brain
6.5. - 11.5. 1985	Complex Systems - Operational Approaches in Neurobiology, Physics and Computers
4.5. - 9.5. 1987	Computational Systems - Natural and Artificial
13.6. - 17.6. 1988	Neural and Synergetic Computers
4.6. - 8.6. 1989	Synergetics of Cognition
22.10. – 25.10. 1990	Rhythms in Physiological Systems

Already the third ELMAU Conference in1979 had moved beyond the scope of phase transitions and focused on pattern creation and pattern recognition. The space-like and time-like stability of the macro-state after phase transition was the focal point of research interest. Haken was not only concerned with pattern formation. He also wanted to discuss the question of how such states can be recognised, i.e. the question of pattern recognition. Even in this early stage he seeded reflections, taken up later on, on how the human consciousness (or a technical system) might detect patterns. In his introductory talk of the meeting Haken presented his assumption that pattern formation and pattern detection constitute two sides of the same coin that are connected by dynamical processes.[525] With the help of Figure 27, representing the analogies between pattern formation and pattern recognition at different levels, he illustrated his thoughts.

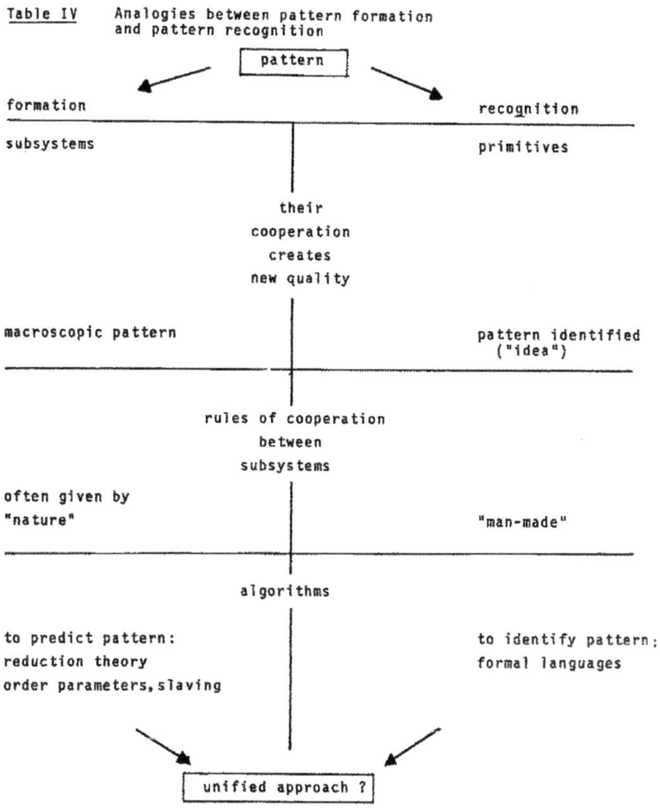

Fig. 28 Analogy between pattern formation and pattern recognition. (from (Haken, 1979a), p. 11)

Ten years later, the questions raised on this subject led Haken to his research on pattern recognition by means of his "synergetic" computer. An essential condition

[525] (Hermann Haken, 1979b), p. 2.

for that endeavour had been the advent of powerful (and affordable) computers that only came in the late Eighties.

The third ELMAU Conference assembled specialists working on pattern formation and pattern recognition. One focus of the meeting had been presentations on time-like patterns in lasers and quantum optics, given by Hermann Haken's Italian friends Rudolfo Bonifacio and Fortunato Tito Arecchi. Pattern formation in fluids, up to the level of fully developed turbulence and deterministic chaos, as well as questions about pattern formation and pattern recognition in biology, constituted the other focal point of the conference.

In addition to some newly recruited speakers[526] who had not yet participated at the first two ELMAU Conferences, other well-known scientists from the Versailles Conferences attended: Werner Reichardt, Hans Meinhardt, Alfred Gierer and Jack Cowan. This is not very surprising, because one of the focal points of the meeting had been pattern creation and recognition in biology, a topic central to the early Versailles Conferences. Haken had met some of the participants to the conference quite often at the "Winter seminars" organized by Manfred Eigen in Klosters.

7.5 The ELMAU Conferences on Chaos Theory, 1981 and 1982

The two ELMAU Conferences in 1981 and 1982 were devoted to the then-highly topical research on the theory of deterministic chaos.[527] The history of modern chaos theory has been written up to now only to some extent.[528] Whereas Gleick and Waldrop have written typical American "heroic stories", centered around single researchers and institutions like Mitchell Feigenbaum, Steven Smale and the Santa Fe Institute – the book from Gleick being an international bestseller – the historical lines of development are presented in great detail in the article of Aubin and Dalmedico.

According to them, one of the roots of deterministic chaos was dynamical systems theory, which has a long tradition. Without doubt, one of its originators was the French physicist Henri Poincaré. As early as 1890 he wrote an article, "Sur le problème des trois corps et les équations de la dynamique", describing the instability problem of the three-body system. He showed that the long-term stability of such a system cannot be calculated, raising questions about the stability of Earth's orbit around the sun. In 1908, in his monumental book "Science et méthode", Poincaré made another important statement:[529]

> „il peut arriver que des petites differences dans les conditions initiales en engendrent de très grandes dans les phénomènes finaux; [...] La prédiction devient impossible [...]".

---

526 L. Glass, R. Larter, P.H. Richter, E.E. Sel'kov.

527 (Lewin, 1992).

528 (Aubin & Delmedico, 2002), (Leiber, 1998) (Kellert, 1992) (Gleick, 1988 (engl. Original 1987)) (Waldrop, 1992).

529 (Poincaré, 1908), p. 68 – 69.

Poincaré's remark addresses a crucial element of deterministic chaos: although the equations of a system are completely determined, the long-term prediction of the behaviour of the system is not possible, because of the influence of noise and fluctuations and the fact that very often the initial values are not known exactly. The American meteorologist Edward Lorenz had rediscovered this phenomenon in 1963.[530] The American physicist Okan Gurel[531] also pointed out that Poincaré was the founder of bifurcation theory, the central elements of which were described as early as 1885. In this theory, by varying certain parameters a dynamical system "branches" into new, stable states. Contemporaneous with Poincaré, the Russian mathematician A. M. Liapunov occupied himself with the stability of dynamical motion and derived certain stability criteria.[532] It is hard to state why, but the works of Poincaré and Liapunov were not followed up for many years. The works of the so-called Russian Andronov School[533] only became known in the US during the Sixties of the last century. In 1967, it was a publication by the American mathematician Steven Smale (born 1930) on differentiable dynamical systems and their topological features that became the starting point of modern mathematical systems research.[534] Another mathematical field contributed to research on stability: in 1968, French mathematician René Thom published his seminal work on "catastrophe theory", dealing with discontinuous changes in systems.[535] A third road to deterministic chaos theory was hydrodynamics, this theory being not completely understood up to the present day. In addition to the work of Edward Lorenz, who demonstrated that even few and relatively simple deterministic equations lead to strongly divergent solutions[536], the research of David Ruelle and Floris Takens became important. In 1971, they published a fundamental paper about turbulent motion of a fluid (described by classic deterministic Navier-Stokes equations).[537] Therein they stated, without being able to prove it exactly, that the motion of such a fluid can be described more closely by generic "strange attractors" (in the meaning of E. Lorenz) than by the Landau theory, the latter always assuming a superposition of infinite modes at full turbulence. The assumption of Ruelle and Takens had been that the appearance of strange attractors could be synonymous for the appearance of (deterministic) chaos.[538] In the words of Aubin und Dalmedico:[539]

530 (E. N. Lorenz, 1963).

531 (Gurel & Rössler, 1977) Preface p. 1.

532 (Liapunov, 1907).

533 (Andronov, 1973). English translation of a work by Andronov (1901-1952) and co-workers, first published in Russian.

534 (Smale, 1967), see also (Gurel & Rössler, 1977), p. 2.

535 (Thom, 1972). However, catastrophe theory is static, not resulting in any dynamic prediction.

536 (E. Lorenz, 1979).

537 (Ruelle & Takens, 1971).

538 See also: „Period-doubling route to chaos shows universality". Physics Today, March 1981, p. 17 – 19.

539 (Aubin & Delmedico, 2002), p. 33 – 34.

> „In this explanation of turbulence, disorder stemmed from the topological character of the system of equations governing fluid flow, rather than external noise; it was a dynamical, not statistical, property of the system. [...] From this viewpoint, the confluence of two disciplines, the mathematical theory of dynamical systems and the theory on nonlinear hydrodynamic stability, constituted a major turning point".

Looking at the problem from the physical and chemical perspective, the roots of chaos theory originated from phase transition theory, the research of Haken and Ilya Prigogine being of special interest.

An understanding of the different approaches was provided by several international conferences in the Seventies. The most important were the "Conference on Instability and Dissipative Structures" in Brussels (1973)[540], the Budapest IUAPAP Conference (1975)[541], the ELMAU Conference in May 1977, "Synergetics – a Workshop"[542] and the conference "Bifurcation Theory and Applications in Scientific Disciplines"[543] held in November 1977 in New York.[544] Paul C. Martin (born 1931), a US physicist, in his talk at the Budapest meeting displayed the different theoretical approaches of the "road to chaos" in the following picture

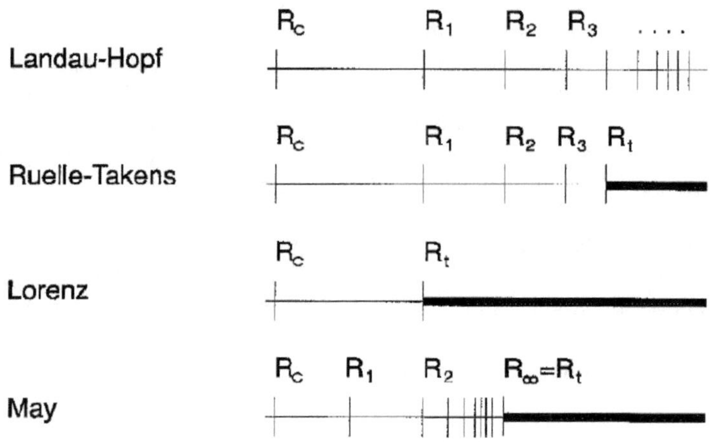

Figure 15: Landau's Picture for the Onset of Turbulence, and Three Alternatives. Redrawn from P. C. Martin, "Instabilities, Oscillations, and Chaos," Fig. 4, C1-60.

Fig. 29 Four different "roads to chaos", assumed by different authors. At point R_n phase transition takes place and the system adopts a new stable state. (From (Martin, 1976))

---

540 (Prigogine & Rice, 1973).

541 (Pal, 1975).

542 (Hermann Haken, 1977).

543 (Gurel & Rössler, 1977). See the detailed account of this conference and Haken's contribution on pp. 155-157.

544 (Aubin & Delmedico, 2002), p. 35ff.

Whereas, according to the theory of Landau, there should be infinite bifurcations (and no chaos should appear), the theory of Lorenz predicted only one transition. Ruelle and Takens predicted only a few bifurcations before onset of chaos. Only some month later, at the ELMAU Conference, 1977, Gollub and Swinney[545] reported their experimental results. They concluded that the theory of Ruelle and Takens seemed to be the most probable one.

Against this background of scientific discussions the two ELMAU Conferences "Chaos and Order in Nature" (1981) and "Evolution of Chaos and Order" (1982) took place, having chaos and order in hydrodynamics as focal point. Hermann Haken had succeeded in inviting two of the most prominent experimentalists in this field – Pierre Bergé, coming from the Institut des Haut Études Scientifiques (IÉHS), and Albert Libchaber from the École Normale Superieur, both in Paris. The two scientists were colleagues of David Ruelle, who had initiated the new experimental research on hydrodynamics with his article (co-authored by Floris Takens) in 1971.[546]

Bergé had made a name of himself working on the Raleigh-Bénard phenomenon, while Libchaber had experimentally proven quite recently the so-called period-doubling on the road to chaos.[547] In his experiment, the dimensionless Reynolds number R_c takes on the role of the order parameter. Increasing the value of the Reynolds number, the system undergoes several bifurcations, whereby the distances between the bifurcations halve themselves. Figure 30 shows this effect (a-c; always with increasing Reynolds number):

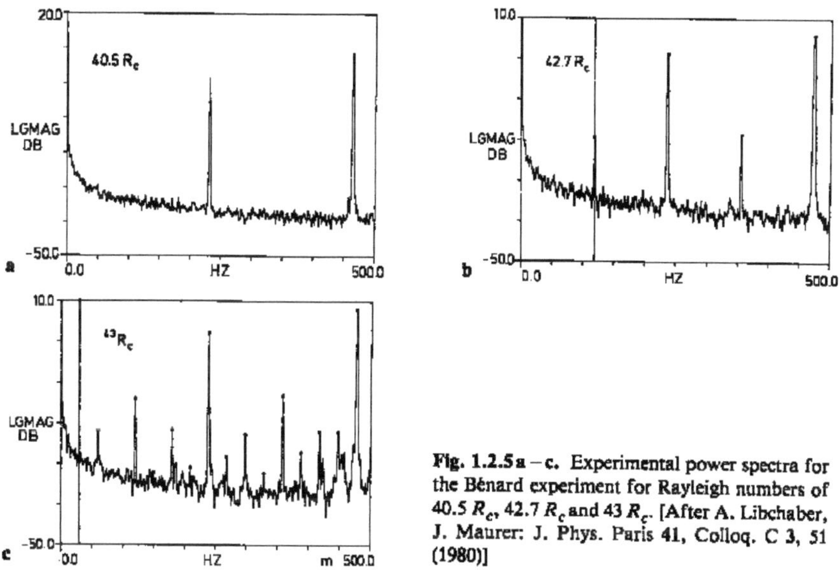

Fig. 1.2.5a – c. Experimental power spectra for the Bénard experiment for Rayleigh numbers of 40.5 R_c, 42.7 R_c and 43 R_c. [After A. Libchaber, J. Maurer: J. Phys. Paris **41**, Colloq. C 3, 51 (1980)]

Fig. 30 Period-doubling in the Bénard experiment – experimental confirmation by Libchaber and Maurer

545 (Gollub & Swinney, 1975). (Swinney et al., 1977).

546 (Aubin, 1998).

547 (Maurer & Libchaber, 1979). (Libchaber & Maurer, 1980).

Period-doubling played an important role in the so-called universality discussion. Grossmann and Thomae[548] as well as Feigenbaum[549] (later on) had proved that for a certain class of one-dimensional mappings the ratio of the difference between two successive transformations leads to a limit value. For instance, if we look at the quadratic mapping

$$x_{j+1} = \lambda x_j (1 - x_j)$$

the difference

$$\frac{\lambda_{n+1} - \lambda_n}{\lambda_{n+2} - \lambda_{n+1}} \lim_{n \to \infty} n = 4{,}6692016 \ldots$$

leads to the value 4,669..., later known as the famous Feigenbaum constant.[550] Siegfried Grossmann, Haken's colleague from Marburg, well known to him from many meetings, was one of the leading German mathematicians on chaos theory. Grossmann attended both ELMAU Conferences on chaos theory, delivering a talk at the second meeting in 1982.[551] Haken had invited a bunch of young American scientists to the second conference: James Crutchfield, Norman H. Packard, Joyne D. Farmer, R. Shaw and Stuart Kauffman, who afterwards became famous by their activity at the Santa Fe Institute and by the book of Gleick. Haken had met Stuart Kauffman at the 6th Versailles Conference the year before. Crutchfield, Packard and Farmer who at the time were still students at the University of California in Santa Cruz had aroused Haken's interest by an article published in Physical Review Letters in 1980.[552] In this work they described a method by which the dimension of an attractor could be derived by single measurements. This in turn was of great interest to Haken, because it opened up the possibility of drawing inferences from brain waves about underlying dynamic processes.

Having brought together the above-mentioned scientists, with the addition of Günther Ahlers, F. H. Busse, O. E. Rössler and Christian Vidal, Haken had succeeded in assembling at the ELMAU Conferences a bigger part of the leading researchers in turbulence and chaos theory.

Haken performed his own research on chaos theory together with his doctoral student Gottfried Mayer-Kress (1954 - 2009), who had joined his institute in 1979. Afterwards Mayer-Kress took on a post-doctoral position at the Center for Nonlinear Studies (1984) und then moved to the newly-founded Santa Fe

---

548 (Grossmann & Thomae, 1977).

549 (Feigenbaum, 1978).

550 In September, Haken met Feigenbaum at the Bordeaux conference on "*Nonlinear Phenomena in Chemical Dynamics*", organized by his longtime acquaintance Vidal and Pacault. (see (Vidal & Pacault, 1981).

551 S. Grossmann: „Diversity and Universality. Spectral Structure of discrete Time evolution".

552 (Packard, Crutchfield, Shaw, & Farmer, 1980).

Institute[553]. From 1981 until 1984 Haken and Mayer-Kress jointly published three articles on the so-called logistic equation.

At the ELMAU-Conference of 1981 the appearance of chaos was the focus of interest, as can be seen by the headlines of the different conference sections: "*Order and Chaos in Fluid Dynamics*", "*Chaos in Fluids, Solid State Physics and Chemical Reactions*", "*Instabilities and Bifurcations*", "*Once again Chaos: Theoretical Approaches*", the next conference, only one year later, centered on order in chaos. This corresponded much better to Hakens research that concentrated on order creating ("Gestaltbildend") phenomena and was less interested in fully developed chaos.[554] Topics at the meeting like "*Order in Chaos*", "*Emergence of Order or Chaos in Complex Systems*" and "*Coherence in Biology*" are a manifestation of it.

A special highlight of the conference had been talk by the Nobel Prize Laureate Manfred Eigen on "Ursprung und Evolution des Lebens auf molekularer Ebene"[555]. After participating in the Winter Seminars organised by Eigen, Haken had succeeded in inviting him to his own conference. Eigen, being a Nobel prize recipient, played a prominent role in the discussion about the origin of life, not least due to his publication of the concept of "hypercycles" in 1977 (see chapter 9c).

7.6 The Textbook "Advanced Synergetics", 1978

In 1982, Haken felt that the time had come to summarise the mathematical and physical results on synergetics of the last five years in another textbook. Looking back, one might say that with the book "Advanced synergetics" a certain completion of the first phase of synergetics had been achieved.[556] The book is highly "technical" in most of its parts. It contains the mathematical results from 1975 onwards, especially the generalised Ginzburg-Landau equations and the detailed elaboration of the so-called "slaving principle". The book created a solid mathematical basis for synergetics, giving a methodical framework for the years to come.

---

[553] Gottfried J. Mayer (born 1954 in Germany) died as early as 2009 in Tübingen, where Haken visited him a few times during his illness. (Private communication from Hermann Haken). In 2005 he became US-American citizen and played an important role within the growing chaos theory community of the USA. He aroused some attention with two publications, researching the consequences of the so-called "Strategic Defense Initiative (SDI)" by means of chaos theory. This led to a consultancy contract with the US-American administration. Mayer-Kress also founded the "Complexity Digest", an internet information channel that grew to an important communication tool of the members of the "complexity community". (For a CV see `www.gottfriedjmayer.de/GMayerCVTAS06.htm`, recalled 7.8.2010)

[554] Interview with Hermann Haken from 20.4.2011, p. 7ff. (Archive Haken (University Archive Stuttgart)).

[555] "Origin and evolution of life at the molecular level".

[556] This book too has been very successful. It has been translated into three languages: Russian (1985), Japanese (1986) and Chinese (1989). In 1987, a second English edition appeared.

As to content, Haken built upon his basic textbook of 1977: "Synergetics – An Introduction". During the last five years this work had seen three new editions and had been translated into four more languages: Russian (1980), Chinese (1982), Hungarian (1984) and German (1982). His new book concentrated on the phenomenon of the instability of a system, dealt in detail with the concept of order parameters and described to great accuracy the mathematical methods of the "slaving" of fast relaxing modes. In Haken's own words:[557]

> „These concepts represent the "hard core" of synergetics in its present form and enable us to cope with large classes of complex systems ranging from those of the "hard" to those of the "soft" sciences".

By the "hard" sciences were meant the natural sciences, especially physics and chemistry, whereas the so-called "soft sciences" comprised economics, ecology and sociology at that time. Medicine, psychology and neurology only came later.

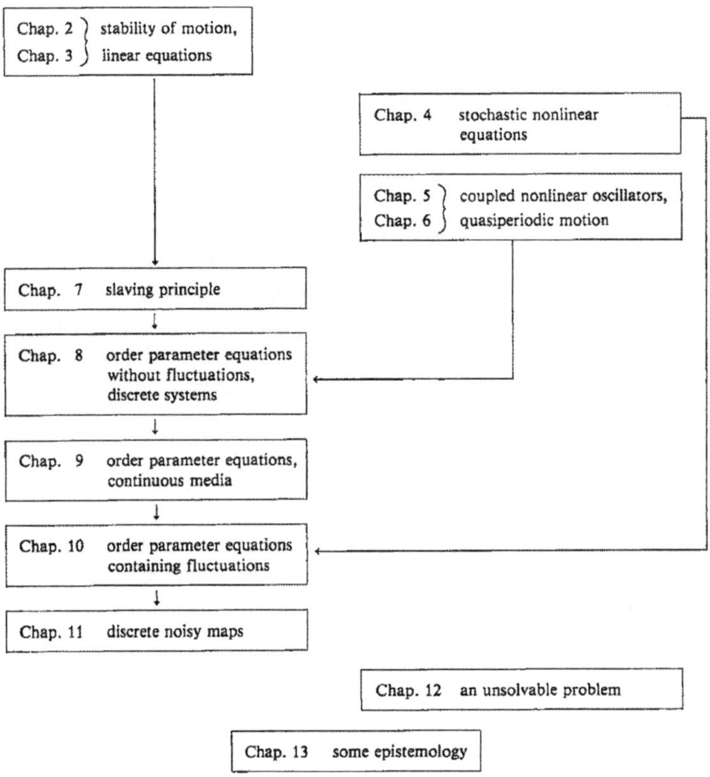

Fig. 31 Logical structure of Haken's textbook "Advanced Synergetics" from 1983, p. 59

[557] (Hermann Haken, 1983b), p. VIII.

Being a skilled and well-trained university professor, Haken structured his book very systematically. First of all, it was necessary to ask when and why large systems lose their stability, driven by internal and external forces. This was the prerequisite for a transformation into a new and stable state. For this transitional behaviour the action of the so-called "slaving principle" is fundamental, because it then becomes understandable why a large system (consisting of numerous components) can be described mathematically by few order parameters only in an ideal case. But looking into the details, one encounters several different cases: discrete (discontinuous) systems without fluctuations, continuous systems without fluctuations, and – mathematically the most difficult – systems with fluctuations. Figure 31 illustrates the structure of Haken's way of proceeding.

If one compares this survey from 1983 with that of of his first book from 1977, the much more mathematically detailed approach becomes apparent. After the initial years, it was essential for Haken to free synergetics from its (potential) odium of a speculative theory and raise it to the rank of a serious and mathematically well-founded theory. Comprehension of the "Advanced Synergetics" textbook required detailed mathematical knowledge from its readers. Haken was well aware of this fact, writing:[558]

> „The basic concepts of synergetics can be explained rather simply, but the application of these concepts to real systems call for considerable technical (i.e. mathematical) know-how".

Table 11 Examples of synergetic systems: comparison from three different books of Hermann Haken (years 1972, 1977 and 1983)

Synergetics – Cooperative Phenomena[559]	Synergetics – An Introduction[560]	Advanced Synergetics[561]
Physics: ▪ Laser ▪ Superconductivity ▪ Ferromagnetism	Physics: ▪ Laser ▪ Bénard-effect ▪ Hydrodynamics ▪ Gunn-effect	Physics: ▪ Laser ▪ Bénard-effect ▪ Hydrodynamics ▪ Plasmaphysics ▪ Solid State Physics
Chemistry: ▪ Bénard-effect	Chemistry: ▪ Belousov-Zhabotinsky-reaction ▪ Brusselator ▪ Oregenator	Chemistry: ▪ Belousov-Zhabotinsky ▪ Brusselator ▪ Oregenator

558 (Hermann Haken, 1983b), p. VIII.
559 (Hermann Haken, 1973c).
560 (Hermann Haken, 1977).
561 (Hermann Haken, 1983b).

Table 11 *(continued)*

Biology: ▪ Hypercycle (Eigen)	Biology: ▪ Cooperative systems (oscillations) ▪ Morphogenesis ▪ Evolution	Biology: ▪ Biological clocks ▪ Coordinated muscle motions ▪ Morphogenesis ▪ Evolution ▪ Immunological system
Sociology: ▪ Opinion forming	Oecology: ▪ Population dynamics (Lotka-Volterra)	Computer: ▪ Pattern formation ▪ Self-Organization (parallel computing) ▪ Reliable systems from unreliable components
		Oeconomy
		Engineering Sciences
		Oecology: ▪ Phase transitions in natural systems
	Sociology: ▪ Opinion forming	Sociology: Opinion forming

How far had the applications of synergetics evolved in the period from 1971 to 1983? We can receive an impression by comparing the examples that Haken mentioned in his ground-breaking works from 1972, 1977 and finally 1983. Whereas in the beginning applications in physics and chemistry dominated, later on synergetics captured parts of biology and, being a model system for neural networks, information theory.

It is striking that many topics and subjects had been in place early on, "in a nutshell", without being dealt with in detail. This was done subsequently in the following years. In the early Eighties, the applications of synergetics in physics stayed more and more on the side-line, while synergetic processes in biology, medicine and information theory (pattern recognition by computers) came to the fore.

The different ways in which order could arise from non-ordered states has been of special interest for Hermann Haken. The most important way is the change of one or several control parameters, the so-called order parameters, inducing a system at a certain critical value (and triggered by fluctuations) to switch to a new and stable state. A second route appears when the number of sub-systems is raised. This too can lead to qualitatively new states.[562] Haken then demonstrated a third method to achieve new self-organised order: the "transient" transition, which is only possible due to the existence of inherent fluctuations (within the system).[563]

According to Haken, the textbook "Advanced Synergetics" provided the "building blocks of self-organisation".[564] Sometime later in a lecture he gave a graphic presentation of this structure (figure 32 and 33).

562 (Hermann Haken, 1980).

563 (Hermann Haken, 1983b), p. 58.

564 (Hermann Haken, 1983b), p. 56.

If synergetics is viewed being a living tree, processes by chance and deterministic processes are the most vital roots. This reminds us at the chapter "Chance and Necessity" of his basic textbook from 1977.[565] Deterministic processes are described by the mathematical methods and fields of control theory, differential equations and bifurcation theory. The processes by chance are dealt with by probability theory, information theory, statistical mechanics and thermodynamics.

Fig. 32 The structure of synergetics according to Herrmann Haken (same as Figure 33)[566]

---

565 (Hermann Haken, 1977). See also figure. 20 in this book.

566 Drawing by Hermann Haken's own hand. (Archive Haken).

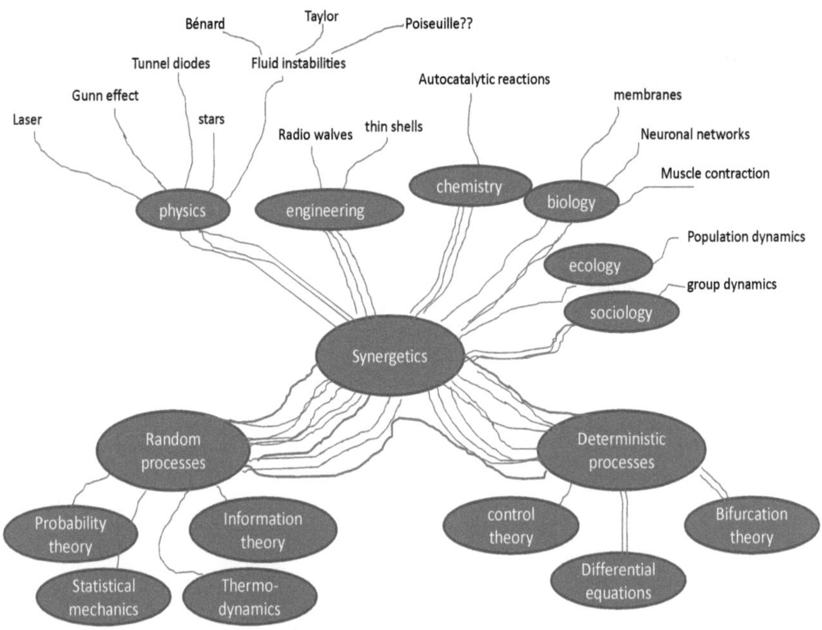

Fig. 33 Transcription of Figure 32

Taking into account the size and complexity of the "roots of synergetics", it doesn't take wonder that its application to real processes demanded for significant "technical" (i.e. mathematical) know how, as Haken mentions above.[567] And to use the same metaphor: the main branches of synergetics are in the fields of physics, chemistry, biology, ecology, economics, sociology and the engineering sciences, the leaves being single phenomena such as the laser etc.

7.7 The Book Series: Springer Series in Synergetics

In addition to the bi-annual ELMAU Conferences and the extensive national and international lecturing of Hermann Haken, the book series "Springer Series in Synergetics" figured prominently in the dissemination of synergetics. Under the aegis of Haken, who acted as author, book editor as well as series editor, the publication series ran to seventy-seven volumes during the twenty years from 1977 till 1997. In an interview he referred to the series as "one of three pillars he

[567] See footnote 522.

could count on".[568] A detailed list of the single volumes and their respective titles, as well as the year of publication can be found in Appendix 4.

It seems interesting that the first proceedings concerning the subject of synergetics – published on occasion of the first ELMAU Conference in 1972 – did not appear within the "Springer Series in Synergetics", but was published by Teubner of Stuttgart. [569] In the same year, Haken had published his first textbook "Quantentheorie des Festkörpers"[570] with Teubner and the publisher agreed to issue the ELMAU volume as well. A book from Haken had been published before in 1970 by Springer, the famous volume "Laser Theory", but this had been a volume of the renowned "Handbook of Physics" under the editorship of Siegfried Flügge. Therefore Haken had no influence in choosing the publisher. Because the response to the "Laser Theory" was highly positive, Springer took on the publication of the "Festschrift" on the occasion of the retirement from Herbert Froehlich, the editors of this volume being Hermann Haken and Max Wagner. As Haken remembered, the editor Helmut Lotsch from Springer came up with the idea of publishing a comprehensive book on the subject of synergetics. This idea could then be realised with the help of the grant from the Volkswagen Foundation, and the single volume developed in time into a large series of titles on synergetics:

> "He [Lotsch] had read my article in the Review of Modern Physics and asked if I couldn't write a book on this subject, in an extended version. That had been the impetus".[571]

This idea gave birth to the basic textbook "Synergetics – An Introduction" (1977) and constituted the start-up volume of the "Springer Series in Synergetics".[572] From the very beginning, Haken aimed for a wider international audience and the volumes were published exclusively in English. During the early years from 1977 to 1981, the funds from the Volkswagen Foundation allowed for the publication of international conferences on synergetics that were held in quick succession. This is shown by seven conference volumes: ELMAU (1977 and 1979), Bordeaux (1978), Tübingen (1978), Bielefeld (1979 and 1980) and Carry-le-Rouet (1980).

Having started with conference proceedings, it was of the greatest concern for Haken to stress the mathematical foundations of synergetics and self-organisation. He therefore decided to include monographs of this subject into the series. Reading through the list we find publications like the "Handbook of Stochastic Methods" written by C.W. Gardiner, "Concepts and Models of a Quantitative Sociology" from W. Weidlich and G. Haag, "Noise Induced Transitions" from the members of the

---

568 Interview with Hermann Haken from 21.09.2010, p. 34.

569 (Hermann Haken, 1973a).

570 "Quantum theory of the solid state". (Hermann Haken, 1973b).

571 Interview with Hermann Haken from 21.09.2010, p. 34 (Archiv Haken (University Archive Stuttgart)). See also the preface of the first edition of "Synergetics – An Introduction".

572 See Appendix 4: list of volumes of the "Springer Series in Synergetics".

Brussels school W. Horsthemke and R. Lefever, as we well as the fundamental book of his first assistant Hannes Risken "The Fokker-Planck Equation". Many of these books experienced several editions, such as the monographs by Haken himself. Altogether a balanced pattern emerged between proceedings of synergetics conferences and fundamental monographs, as is shown by Table 12:

Table 12 Thematic and editorial classification of the seventy-seven volumes of "Springer Series in Synergetics" that were published while Haken was series editor

Subject Synergetics of ...	Edited book (multiple authors)	Monograph
interdisciplinary	11	4
Physics	4	6
Chemistry	5	3
Mathematics	3	9
Biology	-	3
Chaos theory	5	1
Sociology	1	2
Medicine/Psychology	6	6
Informatics/Computer	2	4
Other	1	1

The transformation in the applications of synergetics from the "hard" sciences to the "soft" sciences like medicine and psychology that we will sketch later on, also shows up in the publication history of the series. While in the beginning subjects and examples from physics, chemistry, biology and mathematics had been the focus, in the Eighties the interest shifted towards the human sciences. At that time, Haken's research activities concentrated on pattern recognition by the human brain and by computer, neural networks and the question of how mental processes could be modelled. Therefore the editorial emphasis was on topics from information theory (informatics) and medicine.

After 1995, when Haken had retired, Springer gradually ceased to promote the word synergetics. The emphasis was now on "Nonlinear Dynamics" and especially "Complexity". This annoyed Hermann Haken, because he felt that many volumes that now appeared under the heading "Complexity" could have been published under the heading synergetics as well.[573] In 2013, the Springer book catalogue showed diverse series concerning this subject: for instance "Springer Briefs in Complexity", "Nonlinear Systems and Complexity", "Nonlinear Systems and Complex Systems" and – most prominently – "Understanding Complex Systems", of which the series editor is J. A. Kelso, the US colleague of Hermann Haken.[574]

Many volumes of the "Springer Series in Synergetics" are still available (2013).

573 Interview with Hermann Haken from 21.09.2010, p. 33.

574 See also chapter 8b.

Chapter 8
Synergetics 1987 – 2010: Applications in Medicine, Cognition and Psychology – A Survey

8.1 The Sixth ELMAU Conference 1983: Synergetics of the Brain

Having written the textbooks *Synergetics – An Introduction* in 1977 and *Advanced Synergetics*, finalized at the end of 1982,[575] Hermann Haken felt that the mathematical foundation of the subject was laid. Increasingly his interests shifted to experimental applications of the mathematical formulas. Biological processes came to the fore. The human brain, in particular, became a focal point as the most complex system on earth, with retarding and inhibiting feedback systems and diverse thresholds. Looking at the human brain, Haken was reminded of the laser. To promote the new field he organized the Sixth ELMAU Conference in 1983, choosing as his subject "Synergetics of the brain". For the first time, Haken made use of the help of other specialists in selecting the invited lecturers. Until then, he had always made the selection on his own.

> „[I] asked the experts whom I should invite. I was able to win over Başar, Flohr and Mandell to act as co-organizers and co-authors of the proceedings".[576]

Haken had met Arnold J. Mandell[577] two years earlier, when the latter gave a lecture on "Strange Stability in Hierarchically Coupled Neuropsychobiological Systems" at the ELMAU Conference of 1981. He had no former connections with

[575] The preface was written in January 1983.

[576] Haken (2005b), p. 67.

[577] Arnold J. Mandell (born 1934) is an US-American psychiatrist and neuroscientist. From 1969 he was professor of psychiatry at the *University of California in San Diego*. For a survey of his scientific work see his autobiography (Mandell, 1977), as well as the detailed biographical interview by David Healy (recalled 16.3.2013): http:d.plnk.co/ACNP/50th/Transcripts/Arnold%2520J.%2520Mandell%2520by%2520David%2520Healy.doc

B. Kröger, *Hermann Haken: From the Laser to Synergetics*,
DOI: 10.1007/978-3-319-11689-1_8

the other co-organizers Erol Başar[578] and Hans Flohr.[579] It seems that thanks to his good relations with Mandell Haken was able to attract a larger group of US scientists to Elmau, the most prominent being Nobel Prize Laureate Donald A. Glaser. The creator of the bubble chamber (a detector of electrically charged particles) subsequently abandoned elementary particle physics, switching his interest to brain research.[580] The conference had 35 invited lecturers, thirteen from overseas and another ten from other European countries. Leading researchers on the human brain were invited from Germany, among them Wolf Singer, Gerhard Roth and Christoph von der Malsburg. In 2005, Haken's pride was evident when he recollected this meeting:

> "I just want to mention some scientists who also later on stood out consistently: Walter Freeman, with his highly interesting research on the olfactory system, H. Reitböck, who early on developed a multiple-electrode system to measure simultaneously the electric activity of several neurons, Erol Başar at that time - as well as today - performing pioneering work measuring electric fields of the brain, Christoph von der Malsburg, today one of the leading scientists on the recognition of facial expressions, and Kohonen, who developed basic ideas how in the computer or maybe in the brain, objects with similar characteristics are stored in adjacent segments".[581]

For the attending physicians and biologists, Haken, in his opening talk, moved via synergetics to physics and chemistry. Following an introduction to the methods of synergetics he addressed the problem of how to deal with complex systems in medicine. He pointed to the fact that the brain itself had found ways of reducing complexity, for example by reducing the abundant air molecules, interacting with the skin to reach a mean value of a "felt" temperature.[582] Such a compression of information could be described by the order parameter concept of

---

578 Erol Başar (born 1938), is a Turkish-German physicist who took a doctoral degree in physiology. From 1980 till 2000 he was professor of neurophysiology at the *University Lübeck*. Since 2000 he is professor at the *Dokuz Eylül University* in Izmir (Turkey).

579 Hans Flohr (born 1936) is a German neurobiologist. From 1975 he is professor of neurobiology at the *University Bremen*.

580 Donald Glaser (1926 - 2013), US-American physicist, creator of the bubble chamber. At the time of the *ELMAU Conference*, 1983, Glaser held the post as professor of physics and molecular-biology at the *University of Berkeley* (California). For his life and scientific achievements see the interview: Donald Glaser, "The Bubble Chamber, Bioengineering, Business Consulting, and Neurobiology," an oral history conducted in 2003-2004 by Eric Vettel, Regional Oral History Office, The Bancroft Library, University of California, Berkeley, 2006.

581 (Haken, 2005b), p. 67.

582 Hermann Haken in (Basar, Flohr, Mandell, & Haken, 1983), p. 7.

synergetics. On top of that, the brain with its special "architecture" of interacting neurons is a splendid example of a synergetic system. He reminded his audience not to be frightened by the large number of neurons (ca. 100 billion neurons, each with 10.000 connections), because similar information compression existed in the case of the laser, which had ca. 10^{18} photons. Answering the obvious question of what would be the equivalent to the laser order parameter "light wave", he speculated that in the brain thoughts could take on the role of order parameters: [583]

> „Order parameters can be identified with our thoughts while the subsystems are electrochemical processes. It is suggested that in the way order parameters and subsystems condition each other, thoughts and electrochemical processes condition each other."

As regards the brain, Haken saw the synergetic method as particularly promising because synergetics deals with systems at critical points. And, as he remarked, the brain inevitably had to work at such instability points, because only there could something new originate:[584]

> „[...] when a certain set of subsystems [neurons] is governed by a single or few order parameters, we have a compression of information and an increase of reliability of the total system. On the other hand, when a system must be adaptable, the original sets of subsystems must be decoupled and must then be governed by a new order parameter (or some news). [...] in the brain there must be a delicate balance between reliability and adaptability and we shall suggest that the brain operates close to instability points when we compare the brain with a dynamic system."

Looking for analogies between dynamic systems, Haken then drew a comparison between the laser and its different phase transition hierarchies and the appearance of certain electric patterns in EEG brain measurements. He gave the following analogy:

It should be noted that the coherent phase in the laser activity corresponds to the epileptic seizure phenomenon in the brain, and the "normal" brain activity corresponds to the chaotic radiation of laser light. Table 13 summarizes Haken's ideas on the brain as a dynamic self-organizing system:

---

[583] (Basar et al., 1983), p. 9.

[584] (Basar et al., 1983), p. 10.

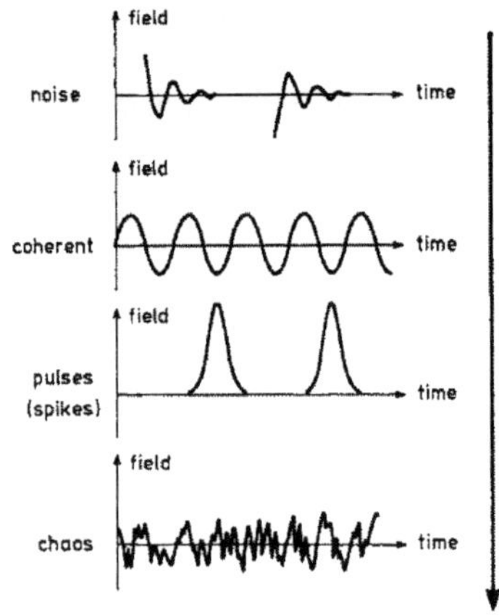

Fig. 34 Different dynamic radiation characteristics of the laser (from(Basar et al., 1983), p. 12 figure 4)

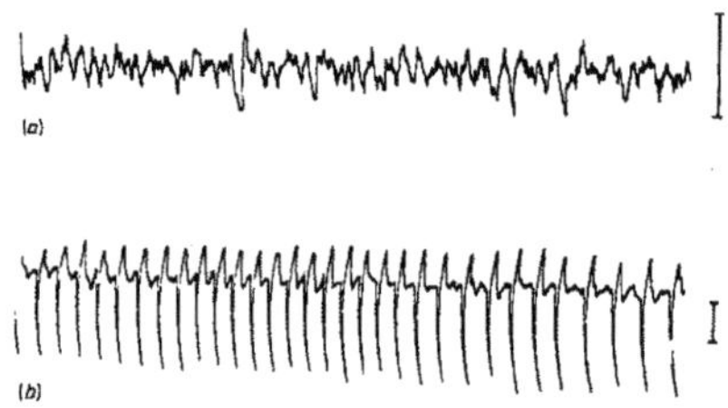

Fig. 35 Two EEG-measurements Sane person in process of "normal" thinking Epileptic seizure (from (Haken, 1989a), p. 529)

Table 13 Formal analogy between the laser and the brain (adapted after table I (Basar et al., 1983), p. 12)

Laser	**Brain**
Sub-systems (showing complex intrinsic dynamics) Atoms and molecules	Sub-systems (showing complex intrinsic dynamics) Neurons
Modes of collective action	
Non-correlated emission of light by atoms	Non-correlated "firing" of neurons
Coherent wave trains	Coherent wave trains (e.g. α-waves)
Pulses	Collective spikes at epileptic seizures
Chaotic light	„normal" brain function

Haken's introductory remarks at the opening of the Sixth ELMAU Conference should be considered as the first stage of looking at the brain as a synergetic system. He was fully aware of the difficulties of such a physical and materialist approach. Again and again he stressed how little was known experimentally about the functioning of the brain.[585] Looking back to 1983, we should recall that from today's perspective neither powerful computers nor evaluation software were at hand. Also lacking were fine-grained experimental non-invasive measurement methods, later made available by computer tomography and multi-electrode electroencephalography, of course.[586] According to Haken, even if such a "flood of data" existed, they could only be dealt with if basic ideas and the corresponding mathematical modelling existed.[587] Synergetics, of course, was such a basic idea. That was Haken's conviction.

In the course of 1983, unexpectedly and unplanned, a colleague from the fields of medicine and physiology arrived who was to strongly influence the research of Hermann Haken in the years to come.

8.2 Cooperation with Scott Kelso

Slowly but surely, Haken's research focus shifted towards medical topics. Not least, this is attributable to the many discussions at the winter seminars in Klosters, organized by Manfred Eigen, where these topics were abidingly popular. The human brain, built and structured by billions of neurons, is so similar to other synergetic systems (in relation to the many interacting sub-systems) that it was natural to attempt an analogue perspective: [588]

> „Once in a while I myself had been interested in brain research, albeit marginal. I listened to the interesting lectures of Wilson and Cowan explaining their model of excitation patterns in the brain or the interesting experiments of Julesz on stereoscopic vision that impressed me very much. [...] Even then I realized that here a broad field of activity for *Synergetics* was to be found. And early on I made bold claims about thoughts (Gedanken) being nothing else but order parameters, whereas the activity of the neurons must be the slaved parts."

As we have shown, in 1983 this interest brought about the 6th ELMAU Conference on "Synergetics of the Brain".[589]Some weeks later, at the end of

[585] (Basar et al., 1983), especially p. 17ff.

[586] On the history of these important medical diagnostic tools see: on computer-tomography the Nobel Prize lectures (Cormack, 1992) and (Hounsfield, 1992). On electro-encephalography see (Brazier, 1961) and (Borck, 2005). See also the article of Klaus Hentschel in (Hentschel, 2012), p. 113ff.

[587] (Basar et al., 1983), preface, page V.

[588] (Haken, 2005a), p. 66.

[589] See the preceding chapter.

July, Haken received a letter from the young Irish-American neuroscientist Scott Kelso.[590] It reported his research on movement coordination in animals and humans and explained his great interest in phase transitions. After receiving a positive answer from Haken, Kelso visited him at his home in Sindelfingen near Stuttgart as early as September, accompanied by his wife, while on a research visit to France. This meeting was the start of close collaboration during the years to come. Kelso explained his experiments on coordinated finger movements: starting with a symmetric (parallel) movement, the finger spontaneously switches to an anti-symmetric movement (anti-parallel), when the frequency is enhanced. These seemingly "simple" finger movements require the coordination of many muscles. Kelso was eager to learn whether Haken's synergetic method could explain the coordinated behaviour of these muscle sub-groups.

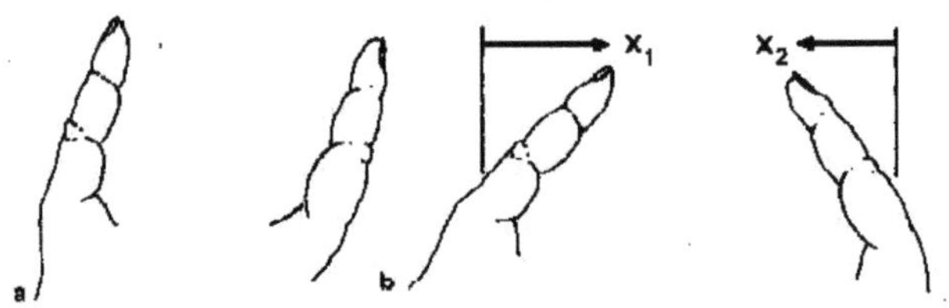

Fig. 36 The finger movement coordination experiment by Kelso. a) anti-symmetric b) symmetric (from (Haken, 1988c), p. 219)

Haken immediately invited Kelso to his institute as guest professor for the summer term of the following year.[591] Kelso accepted and, in June 1984, stayed for four weeks in Stuttgart, before leaving for the *Wissenschaftszentrum* (Science Centre) at Bielefeld, where he became a temporary fellow. Haken solved the problem quite quickly by using an *Ansatz* that had been successfully applied to the laser problem before. He conceived a "potential landscape" that was deformed by an order parameter; in this case the order parameter was the relative phase of the fingers. In this way he succeeded in describing the experimental results mathematically.

[590] J. A. Scott Kelso (born in 1947) is an Irish-American neuroscientist. He studied neuropsychology in Belfast and Calgary, taking his PhD in Madison (USA), in 1975. In 1983, while getting into contact with Hermann Haken, he was a researcher at the *Haskins Laboratory* of the *Yale University* in New Haven, being simultaneously professor at the *University of Connecticut*. (`http://en.wikipedia.org/wiki/J._A._Scott_Kelso`, recalled 12.12.2012).

[591] Letter from Hermann Haken to Scott Kelso from 12.1.1984. (File Kelso, Archive Haken).

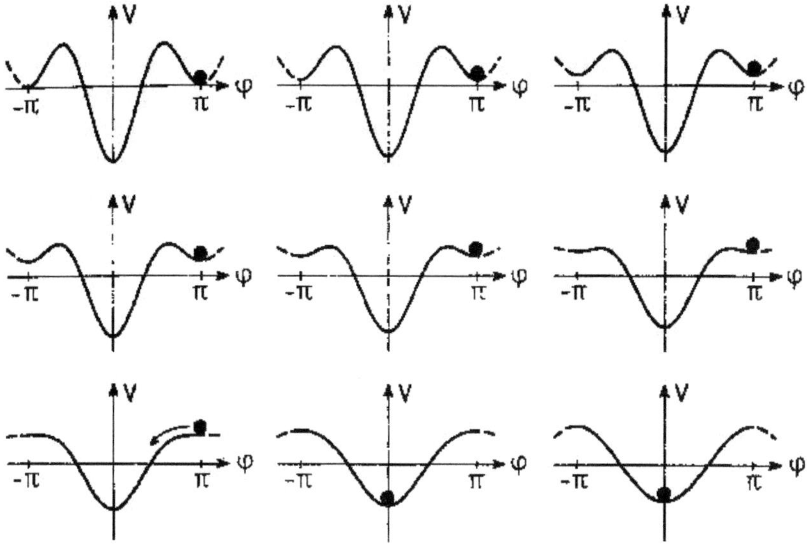

Fig. 37 Potential describing the transition of the finger movement (from (Haken, Kelso, & Bunz, 1985), p. 350)

The potential is deformed by increasing the frequency of the finger movement. The dip at point π levels out and finally the system (symbolized by the ball) takes on a new stable configuration. Haken immediately informed Kelso of his finding:[592]

> „[…]In the meantime I could formulate two coupled equations for the two hands which exhibit a number of features we want, namely the decrease of the amplitude with increasing frequency, the coupling of their phases, and the switching behavior of the relative phase including hysteresis. The equations are slightly more complicated than we originally thought."

They published their results as soon as possible. A doctoral student of Haken, Herbert Bunz, who had numerically solved the equations with the help of a computer, featured as co-author. The article appeared in Biological Cybernetics[593] under the heading "A theoretical Model of Phase-Transitions in Human Hand Movements" , on 17 November. With 850 citations, this publication is the most

---

[592] Letter from Hermann Haken to Scott Kelso, 24.7.1984. (File Kelso, Archive Haken).
[593] (Haken et al., 1985).

cited of Haken's articles.[594] To establish a stronger connection to the synergetic method, Haken then proposed further experiments that Kelso subsequently performed, confirming Haken's theoretical predictions. The experiments showed the occurrence of hysteresis (i.e. when the finger movement had switched from parallel to anti-parallel, the symmetric phase rested stable, even though the frequency slowed down), the observation of "critical fluctuations" and the retardation of disturbances of the system in the transition zone. These were effects that also showed up in other synergetic systems.[595] According to Haken, it was a crucial insight:[596]

> „Until then scientists of the brain and researchers on the movement dynamics of limbs always had assumed that the brain operates like a computer. But, - and that is the crucial insight – with computer the observed phenomena like hysteresis, critical fluctuations and critical slowing-down would never appear. In consequence, the brain does not operate like a computer, but rather like a self-organizing system."

Haken now felt confident about applying the synergetic methodology to problems of brain research, working closely with Scott Kelso.[597] Therefore Kelso quite often visited Haken at his Stuttgart institute as guest professor and attended the 7th ELMAU Conference on "Complex Systems – Operational Approaches in Neurobiology, Physics and Computers" in May 1985. In the course of this work its focus changed from movement physiology to the dynamics of the brain. Apart from Haken and Kelso, many of Haken's students were involved, including Gregor Schöner, Armin Fuchs and Viktor Jirsa.

---

594 Web of Science Database, recalled 17.12.2012. The seminal work on the quantum theory of the laser *"Nonlinear Theory of Laser Noise and Coherence I"* (Haken, 1964) has been cited 115 times "only". Of course, one has to take into account the different size of the respective scientific community.

595 See (Kelso, Schöner, Scholz, & Haken, 1987). (Haken, 1988b). (Haken, 2005b).

596 (Haken, 2005b), p. 70 (Archive Haken).

597 In summer 1985, Scott Kelso received a call to the neurophysiology chair at the *Florida Atlantic University* in Boca Raton (Florida). In the following years he created a large research institute and he suggested that Haken should be "Co-Chair". Haken accepted, but was unable to carry out this task due to health problems. Details can be found in the correspondence of H. Haken and S. Kelso. (Archive Haken). The first visit to Florida by Haken only took place in March 1989.

8.3 Application of Synergetics to Brain Research

In addition to the research performed in cooperation with Scott Kelso, in the 1980s Hermann Haken devoted himself to information and computer theory. His motivation was the insight that in nature many systems exist where it is not possible, as in the case of the laser, to establish the mathematical equations of the sub-components ab initio, reducing them subsequently to a few order parameter equations by means of his "slaving" methodology. Often it is only possible, by measuring brain waves for example, to obtain noisy macroscopic data (i.e. an EEG). That is why Haken searched for a method to reason from these macroscopic data to the underlying order parameter equations. He found a solution in the maximum information entropy principle, introduced by Edwin T. Jaynes to statistical physics in 1957.[598] In simplified terms the principle states that if little is known about a system, from all possible states of the physical system one has to choose the one that maximizes the entropy. Finally, in 1988, Haken published a textbook on information and self-organization that summarized his research results of the last six years.[599] He himself declared it to be „the second foundation of synergetics".[600]

> „We wish now to develop an approach which can be put in analogy with that of thermodynamics. Namely, we wish to treat complex Systems by means of macroscopically observed quantities. Then we shall try to guess the microscopic structure of the processes which give rise to the macroscopic structure or the macroscopic behavior."

Haken restricted the application of his method to open systems and found a complete conformity between the microscopic approach (bottom-up) and the macroscopic approach (top-down). Using this method, it is crucial to find the restricting boundary conditions of the system under consideration. The path taken by Haken allowed for this possibility:[601]

> „Using the results of the microscopic theory as a guide, we are then able to do much more; namely, we can do without the order parameters from the outset. Instead our approach will start from correlation functions, i.e. moments of observed

598 (Jaynes, 1957).
599 (Haken, 1988d).
600 (Haken, 1988d), p. 33.
601 (Haken, 1988d), p. 35.

variables from which we may then reconstruct the order parameters and the enslaved modes. Incidentally, we can also construct the macroscopic pattern, or in other words we may automatize the recognition of the evolving patterns which are produced in a non-equilibrium phase transition."

With the aid of this *Ansatz*, Haken and his co-workers – we must mention especially his old assistant Rudolf Friedrich – could explain some of the basic phenomena of neural activity. In a publication dating from 1992, Friedrich showed that the so-called *petit mal* epilepsy creates a coherent firing of neurons ("neuron thunderstorm") the basic pattern of which could be reduced to only a few order parameters.[602] Kelso's research group also worked on related problems. They analysed the patterns of magnetic field variations (MME) taken from a test participant while performing a special cognitive task. Also a space-time pattern occurred that showed the dynamics of a few underlying order parameters.[603] It is noteworthy that Rudolf Friedrich abstracted the mathematical tools for this analysis from the treatment of the spatial Taylor phenomenon of hydrodynamics.[604] For Hermann Haken, this was clear evidence that these two scientific fields, hydrodynamics and dynamics of neurons, could be described by the same mathematical method of synergetics.

8.4 Pattern Recognition in the Visual System and by Synergetic Computer

Having discovered this new macroscopic approach from information theory, Haken, by means of synergetics, could tackle the problem of pattern recognition. For many years he had been especially interested in how the visual system recognizes gestalt. His reasoning followed these lines: if his *Ansatz* was correct that the human brain operates close to instability points, then from his experiences with the order parameter concepts in physical systems – especially the laser – phenomena like bi-stability (i.e. symmetry breaking), hysteresis and fluctuations should occur. Bistability in visual perception had been familiar for a long time, as can be seen in the famous Necker cube, a "bi-stable figure". Subsequently, Haken used another well-known figure, the so-called Rubin vase, where the perception oscillates between the vase and the heads, to illustrate this effect.

---

602 (Friedrich & Uhl, 1992).

603 (Friedrich, Fuchs, & Haken, 1992).

604 (Friedrich & Haken, 1986). (Friedrich & Haken, 1987).

Fig. 38 An illustration of bi-stable figures used by Hermann Haken. (from (Haken, 1990a)). This figure, also called Rubin-vase, was first shown by the Danish psychologist (1886 - 1951), in 1915. (Rubin, 1921). Other famous examples of this effect are for instance the Necker cube invented by the Swiss geologist Louis Necker (see (Forbes, 1863)) or the ambiguity of rabbit-duck head discussed by Ludwig Wittgenstein. (see (Wittgenstein, 1953).

An example of visual hysteresis is demonstrated by the following sketch:

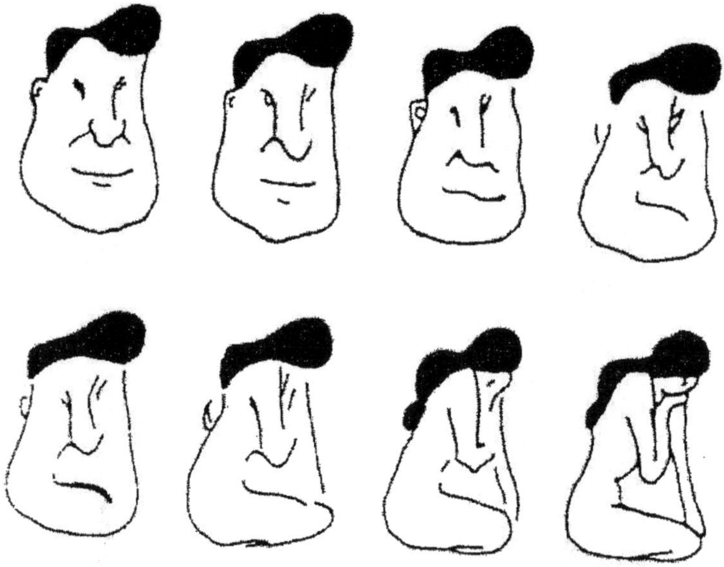

Fig. 39 Example of visual hysteresis (see text) (from (Haken, 2005b)))

Starting from the upper left-hand side, if the eye follows the sequence of figures from left to right, the perception of the woman mostly appears with the antepenultimate picture of the lower row. If the starting-point is the picture of the woman, the perception of the man's face does not appear with the third picture in the sequence, but again only with the antepenultimate picture of the upper row. Haken explained this feature using the following potential:

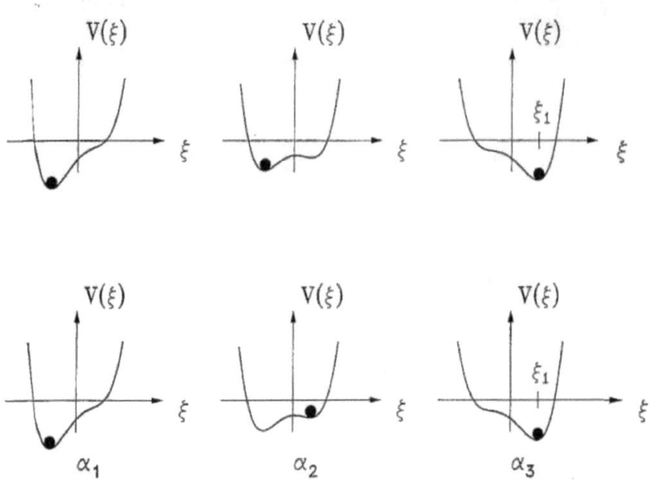

Fig. 40 Potential applying to figure 39 (from (Haken, 2005a))

The different position α_2 of perception (denoted by the ball) is crucial, despite the fact that starting-point and endpoint are the same. This corresponds to the different position where the observer recognizes the emergence (=transition to) of the new figure. Haken and his doctoral student Thomas Ditzinger (born in 1960) managed to obtain order parameter equations for the difference in visual perception that in this special case included an additional "attention" parameter. Two order parameters were in oscillating competition.[605] Afterwards Haken tried to translate these analytical results into a computer model (end of the 1980s).

As we have seen, Hermann Haken assumed neurons were sub-systems in the brain whereas "thoughts" (*Gedanken*) represent the order parameters. According to the concepts of synergetics both parts have to react on different timescales in order for the sub-systems to be "enslaved". That holds true for the brain: while the neurons fire on a timescale of milliseconds the perception of new content (*Gedanken*) varies by tenths of a second.[606] We should remember that in the early 1980s – without non-invasive imaging techniques – little was known about the functioning of neurons and their interactions. The concept of the so-called Hebb's synapses was central.[607] The idea behind it was that, by simultaneous activation of neurons, physical enhancement of the connection between them would occur, leading to a learning process. Models of this kind were applied to develop so-called neural networks, important concepts for the development of artificial intelligence by means of computers.

From a physicist's point of view an important sub-class of neural nets are spin-glass models. They assume that neurons can take on only two different states (spin up or spin down) and that the triggering of neurons is achieved by the input of signals from other neurons.[608] As early as 1943, two US scientists, Warren McCulloch (a neurologist) and Walter Pitts (a logician), had proved that elements of this kind could perform every logical function desired.[609] An oft-discussed spin-glass model is that of John Hopfield (1982).[610] He used the analogy of a potential landscape, the symmetry of which could be broken by additional information. The "valleys" occurring in the potential acted as attractors. Haken was immediately reminded of the symmetry-breaking model that he had developed for the laser threshold in 1964.[611] But there is a twist: Hopfield's concept is ambiguous because many potential landscapes correspond to a stored state.[612]

At this point Haken's approach with the "synergetic computer" came into play. It utilized the fact that it is a feature of some open systems to draw the whole system into a well-defined final state (to "enslave the system") by specification of

---

605 (Ditzinger & Haken, 1989).
606 (Pöppel, 1987).
607 (Hebb, 1949).
608 (Haken, 1989b), p. 168
609 (McCulloch & Pitts, 1943).
610 (Hopfield, 1982). See also (Little, 1974).
611 See page 66.
612 (Haken, 1989b), p. 169.

some starting values. Then the (difficult) task remains "to define the potential V in such a way that the valleys of V correspond exactly to the stored patterns".[613] According to Haken, the main advantage of this synergetic approach is the fact that no undesired attractors – no ambiguity – crops up to mar the approach. The synergetic computer is a "top-down *Ansatz*", starting with the desired features and afterwards constructing the corresponding neural net. To put it another way, this computer can only be used for one specialized task, for instance pattern recognition, contrary to a common serial computer that can perform any task.

During the period from 1985 to 1989 Hermann Haken and his co-workers published no fewer than thirty articles on the subject of pattern recognition and the synergetic computer. In addition, the Eighth and Ninth ELMAU Conferences were dedicated to this subject, running under the headings "Computational Systems - Natural and Artificial" and "Neural and Synergetic Computers".[614]

8.5 Synergetics in Psychology

After years and years of work on the problems of movement coordination, neural networks and pattern recognition it is not surprising that Hermann Haken turned his attention to the difficult problem of the possible application of synergetics to psychology and research on the human conscious mind. Not least, it was clear to him that he was entering the long-standing philosophical field of the so-called body-mind problem, denoting the assumed dualism of the brain as the material substrate and the mind as an autonomous medium.[615] In his own words, thinking about order parameters led him "necessarily to the discussion of psychic processes". He continued: "in fact that is what also happens to me; and in the 1980s I gave talks to psychotherapists and psychologists hinting at such concepts".[616] In 1985, he seized such an opportunity when he was invited to lecture at the Heidelberg Congress on Family Therapy, talking about "Synergetics and its application to psychosocial problems".[617] He was fully aware of the fact that he would encounter strong reservations. A statement normal to him like "a central consideration of synergetics is looking for general principles of systems, no matter if the different parts of the system consist of atoms, molecules, cells or people" would lead to the accusation of blatant materialism. He rebutted the charge and explained: [618]

---

613 (Haken, 1989b), p. 170.

614 (Haken, 1987a) and (Haken, 1988e).

615 Literature on the body-mind problem seems to be boundless, starting with Plato (see (Ebert, 2004) and (Descartes, 1987)). For a today's perspective see for instance (Edelman & Tononi, 2002), as well as (Seifert, 1979).

616 (Haken, 2005a), p. 83.

617 (Haken, 1987b).

618 (Haken, 1987b), p. 36-37.

> "... we are only looking for typical relations – in a highly abstract fashion – between the system and its parts. The result is that we do not put the word for reductionism, where the features of the system as a whole are derived by the features of its parts alone. Quite the contrary, we emphasize that at the level of the total system we need new conceptualizations".

He was also aware of a different linguistic usage in psychology. It was not by chance that he explained "that synergetics introduces notions that are impartial. The [different] value judgement can only be made in the context of the single discipline". And, of course, the word "enslaving" had to be explained: "despite protestations to the contrary being a technical expression – it appeared a decidedly inflammatory term to sociologists".[619] In the course of his lecture Haken described the methods of synergetics, indicating some analogies that resulted if one moved from the abstract relation of synergetics to applications in psychology and psychiatry.

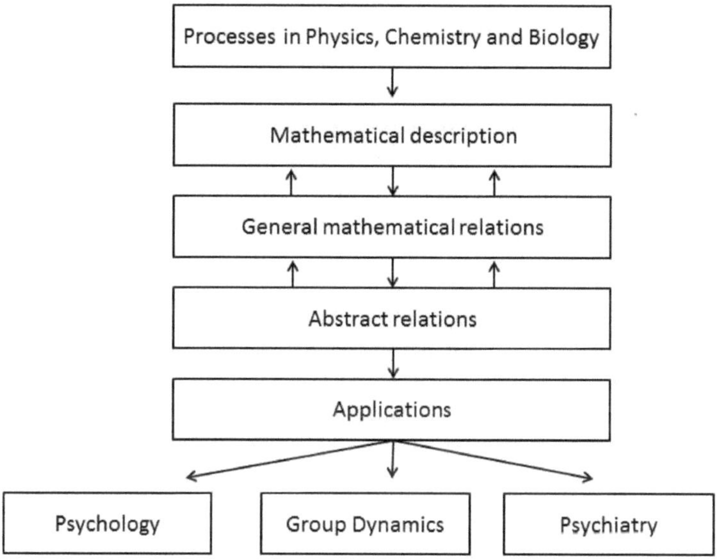

Fig. 41 Proceeding from physics to psychology (adapted from (Haken, 1987b))

In psychology, synergetics is seen as a theory of self-organization.

His talk had no immediate consequences. What is more, Haken did not mention explicitly the extension into the field of psychology in his comprehensive review article of 1988.[620] The situation changed in the following year when Haken invited

---

619 (Haken, 1987b), p. 40.
620 (Haken, 1988a, 1988b).

participants to the penultimate ELMAU Conference "Synergetics and Cognition", in 1989. He had prevailed upon Michael Stadler[621] to act as co-organizer, a psychologist from Bremen who specialized in cognitive research. The conference roster comprised network models, oscillatory processes of the brain, perception and action, language processes, psycho-emotional development of the human being and social cognition.

In his introductory talk Haken explained the basic principles of synergetics, especially the order parameter concept. He explicitly stressed that self-organizing processes are characterized by symmetry-breaking, critical fluctuations and critical "slowing down", nearing the phase transition point. In terms of applications he mentioned research on the self-organized change in movement types which he had performed with Scott Kelso as well as his work on pattern recognition. Of special interest had been the Necker cube and the Rubin vase, phenomena well-known to the psychology community. Self-critically he asked how far the synergetic approach could go and emphasized the operational approach of his research:[622]

> „I have tried to present an operational approach to the theoretical study of brain functions. [...] Our approach is based on the interdisciplinary field of synergetics where we now exploit a pronounced analogy between pattern formation via self-organization and pattern recognition."

Finally, he ventured the suggestion that computers might mimic the human brain but could never show the same "enormous integrative power". In relation to psychology he concluded:[623]

> „In the realm auf psychology there are deep problems for which I can hardly imagine a satisfactory solution. For instance, how can we model emotions? Of course, we can look at them at the behavioral level, and for instance we may construct computers that can read off emotions of persons from their gestures, faces, etc. But this will not give us any access to the internal world of humans."

Haken's co-organizer, Michael Stadler, had studied psychology at the University of Münster and afterwards obtained an assistant position with

[621] Michael Stadler (born in 1941), he studied psychology at the University of Münster. From 1980, he was professor of psychology at the University of Bremen where he founded the Institute of psychology and cognitive research.

[622] (Haken, 1990b), p. 25.

[623] (Haken, 1990b), p. 29.

Wolfgang Metzger, a *Gestalt* psychologist. *Gestalt* theory, most notably founded in the 1920s by Wolfgang Köhler, makes use of notions that are also relevant to synergetics. To give an example: the formation of order and structure finds its expression in the term "tendency of *gestalt* pregnancy (*Prägnanz*)". Analogue uses of notions are shown by the phenomena of multi-stability in perception (Necker cube), the appearance of hysteresis or symmetry-breaking.[624] Therefore, synergetics with its detailed mathematical and operational approach opened a new way to tackle problems in psychology. Yet checking the different concepts by experimentation was of utmost importance to Hermann Haken.

Hermann Haken developed personal contacts with various German psychologists who became important for his later research. First and foremost we must mention Wolfgang Tschacher,[625] Günter Schiepek,[626] Ewald Brunner[627] and Michael Stadler.[628] It was Günter Schiepek who became the driving force in the application of *synergetics* to psychology. In 1990, co-operating with Tschacher and Brunner he organized the first *Herbstakademie* (autumn meeting) under the heading "Self-organization and clinical psychology", that had the programmatic sub-title "Empirical Approaches to a Psychological Synergetics". To quote from the proceedings "The Herbstakademie is an outcome of cooperation between psychologists from the Universities of Bamberg and Tübingen who seek an empirical and theoretical foundation of systemic thinking in psychology."[629] The realization of the meeting was made possible by a grant from the Volkswagen Foundation under the priority programme "Synergetics" that Günter Schiepek had applied for. The role of Hermann Haken was emphasized: "… in cooperation with the physicist Hermann Haken, originator of synergetics". It is fair to say that Hermann Haken was the "spiritus rector" of the congress. The *Herbstakademie* enjoyed great popularity and is repeated almost annually.[630]

---

[624] (Stadler & Kruse, 1990).

[625] Wolfgang Tschacher (born in 1958) took his doctorate in psychology at the University of Tübingen in 1990. From 2002 he is professor of psychology in Bern (Suisse), director of the department of psychotherapie at the University Hospital of Psychiatry. His main research activities center on psychotherapy and psychopathology seen from the perspective of systems theory.

[626] Günter Schiepek (born in 1958), studied psychology in Salzburg and habilitated in Bamberg, 1990. He is professor of psychology at the private Paracelsus Medical University Salzburg and holds tenure at the Ludwig-Maximilians-University Munich.

[627] Ewald J. Brunner (born in 1941) studied theology, psychology and sociology in Heidelberg, Bielefeld and Tübingen. From 1982 till 1992 he was professor for pedagogical psychology in Tübingen. Since 1992 he is professor for pedagogical psychology in Jena.

[628] http://de.pluspedia.org/wiki/Psychosynergetik (recalled 21.01.2013).

[629] (Tschacher, 1991).

[630] In 2012, the 17th Herbstakademie took place. A survey of all meetings, including a short description of the subjects can be looked up at http://www.upd.unibe.ch/research/symposien.html (recalled 21.1.2013).

Because in 1990/91 the priority programme "Synergetics" of the Volkswagen Foundation had been terminated, Haken was not able to organize further ELMAU Conferences.[631] In some respects the annual autumn academies continued the tradition, although the thematic focus was on research in psychology and psychiatry. Nevertheless, subjects ranging from sociology to scientific theory were also covered. In 2006, the relentless activities of Günter Schiepek as well as Haken's interest in the applications of synergetics to psychology led to the publication of a voluminous book, *Synergetik in der Psychologie – Selbstorganisation verstehen und gestalten.*[632] In the preface the authors stated: "During the last years, thanks to detailed investigations, synergetics has become a paradigm for psychology. It connects different problems and issues of psychology with a common kernel".

The book is also of great interest to those readers more interested in synergetics relating to physics, biology and chemistry, because it contains an extensive chapter on the philosophical questions of synergetics comprising no fewer than sixty pages.

---

[631] In October 1990, the final *ELMAU Conference* was held having the title "Rhythms in Physiological Systems" (Haken & Koepchen, 1991).

[632] (Haken & Schiepek, 2006).

Chapter 9
Theories of Self-organization: The Role of Synergetics

Synergetics is the science of systems that shows the dynamic and cooperative behaviour of sub-systems. Therefore, it is a systems-theory and belongs to so-called theories of self-organization.[633] The detailed history of self-organization – a research field that has flourished since the mid 20th century – has only been rudimentarily investigated.[634] In the first place, we have to mention the members of the "core theme science research" group at the University of Bielefeld (Germany), Wolfgang Krohn, Günter Küppers and Rainer Paslack. Even as the different sub-disciplines developed at the end of 1980s and the beginning of the 1990s, these sub-disciplines reached for a systematic classification and valuation of the different theories of self-organization.[635] Nevertheless, up to now there is no common agreement about a consistent and comprehensive definition of self-organization. In '*Enzyklopädie Philosophie und Wissenschaftstheorie*' self-organization is defined as: [636]

> „Term to designate the spontaneous formation of ordered macroscopic structures that result from self-amplifying microscopic fluctuations and their selection by boundary conditions or given constraints. The state of a self-organizing system essentially depends on intra-system factors and is not totally determined by outside circumstances. S.[elforganization] is seen as the basis of many patterns of order

---

[633] The term „Self-Organization" was introduced by Ashby in 1947 (Ashby, 1947). Ashby was a British psychiatrist and is one of the founders of cybernetics. For some time he worked at the Biology Computer Laboratory, founded by Heinz Foerster (as explained below). Biographical details may be found in the "W. Ross Ashby Digital Archive" (`http://www.rossashby.info`, recalled 18.3.2013).

[634] More recent work on the theory of self-organization are (Skar, 2003), (Corning, 1995), (Kauffman, 1995). A historical account is offered by (Keller, 2008), (Keller, 2009).

[635] This research of several years was supported by grants from the *Volkswagen foundation* in the framework of the *Synergetics* priority program. The most important contributions are (Wolfgang Krohn, Krug, & Küppers, 1992), (W. Krohn & Küppers, 1987), (Wolfgang Krohn, Küppers, & Paslack, 1987), (Dress, Hendrichs, & Küppers, 1986), (Küppers, 1987), (Paslack, 1991), (Paslack & Knost, 1990). An account that is more mathematical is given by (Bushev, 1994).

[636] (Mittelstraß, 2005ff.), p. 761.

B. Kröger, *Hermann Haken: From the Laser to Synergetics*,
DOI: 10.1007/978-3-319-11689-1_9

(Ordnungsmuster) and coherent behavior in the animated and in-animated world and in civilization."

Contrary to the definition above, the *Encyclopedia of Complexity and System Science* emphasizes the systems and processes aspects of self-organization: [637]

„The term Self-organizing Systems refers to a class of systems that are able to change their internal structure and their function in response to external circumstances. By self-organization is is understood that elements of a system are able to manipulate or organize other elements of the same system in a way that stabilizes either structure or function of the whole against external fluctuations. [...]
Self-organizing systems have been discovered in nature, both in the non-living (galaxies, stars) and in the living world (cells, organisms, ecosystems), they have been found in man-made systems (societies, economies), and they have been identified in the world of ideas (world views, scientific believes, norm systems)."

The following remarks are guided by the works „*Urgeschichte der Selbstorganisation: zur Archäologie eines wissenschaftlichen Paradigmas*" by Rainer Paslack[638] and „*Die Entdeckung der Komplexität- Skizzen einer strukturwissenschaftlichen Revolution*" by Reiner Hedrich.[639] Concerning synergetics, our evaluation sometimes differs from the evaluations of these two authors due to the facts presented in this work. Furthermore, we also present statements from Hermann Haken that isolate synergetics from other approaches to self-organization.

As one can read, especially in the book by Paslack, the theory of self-organization of the animate world and of matter has many predecessors reaching back to ancient Greek times. Time and again, Aristotle is cited as stating, "The whole is more than the sum of its parts".[640] The German philosophers Immanuel Kant[641] and Friedrich Wilhelm Schelling[642] are also mentioned.

---

637 (Meyers, 2009ff.).

638 (Paslack, 1991).

639 (Hedrich, 1994).

640 An abridged citation following Aristotle: Metaphysics VII: "Since that which is compounded out of something so that the whole is one, not like a heap but like a syllable-now the syllable is not its elements, ba is not the same as b and a, nor is flesh fire and earth."

641 „In such a natural product as this every part is thought as *owing* its presence to the agency of all the remaining parts, and also as existing *for the sake of the others* and of the whole, that is as an instrument, or organ... The part must be an organ *producing* the other parts—each, consequently, reciprocally producing the others... Only under these conditions and upon these terms can such a product be an *organized* and *self-organized* being, and, as such, be called a *physical end*.." (Kant, 1790) 65 B 291 (cited after (Paslack, 1991), p. 20-21).

642 (Heuser-Keßler, 1986).

The new self-organization concepts, emerging originally during the 1960s, have quite different roots. According to Paslack, in addition to synergetics, one can distinguish at least five separate approaches:

1. The systems-theoretical or cybernetical approach (Bertalanffy);
2. Autopoiesis and self-referentiality (Maturana, Varela);
3. The theory of auso-catalytic hypercycles (Eigen, Schuster);
4. The theory of dissipative structures (Prigogine); and
5. Chaos-theory (Lorenz, Peitgen and Richter, Santa Fe Institute).

To describe the somehow parallel developments and to show the problem of demarcation between these approaches and synergetics in what follows, we are going to present the "Ansatz" of the different theories of self-organization, their main actors and their central contents. We limit ourselves to the first four theories because chaos theory has already been dealt with in Chapter 7e.

9.1 The Systems-Theoretical or Cybernetic Approach

Systems-theory and cybernetics emerged in the 1940s. Whereas systems theory mainly was a brainchild of Ludwig von Bertalanffy,[643] cybernetics was coined by Norbert Wiener.[644] During the 1950s and 1960s, Heinz von Foerster, working at the *Biological Computer Laboratory* at the *University of Illinois,* then extended this "Ansatz" into a so-called cybernetics of second order.[645]

Bertalanffy[646] had studied philosophy and biology in Vienna. Opposing vitalism theory in biology, he introduced the notion of "flux equilibrium" (Fließgleichgewicht) to denote biological organisms as open and dynamic systems that could be described analytically. In doing so, he found basic similarities and analogies in different scientific fields that led him to develop an "Allgemeinen Systemtheorie" (general systems theory).[647] According to him, the reason for these analogies is to be seen in the limited number of simple differential equations that are needed to describe phenomena.[648] In 1940 he wrote: [649]

---

643 (Bertalanffy, 1972) and (Davidson, 1983).

644 (Wiener, 1948). There exists a vast secondary literature to Norbert Wiener, the "Father of Cybernetics". Some recent works are (Bluma, 2005), (Conway & Siegelman, 2005), (Heims, 1993).

645 (Müller, 2000).

646 Ludwig von Bertalanffy (1901 - 1972), was an Austrian biologist and systems theoretician. After the Second World War he emigrated to Canada and the United States. Biographical data can be found in (Pouvreau, 2006).

647 (Bertalanffy, 1949), (Bertalanffy, 1950a).

648 (Bertalanffy, 1950b).

649 (Bertalanffy, 1950b), p. 28.

„Conceptions and systems of equations similar to those of open systems in physicochemistry and physiology appear in biooecology, demography and sociology. The formal correspondence of general principles, irrespective of the kind of relation or forces between the components, lead to the conception of a "General System Theory" as a new scientific doctrine, concerned with the principles which apply to systems in general. Thus, the Theory of open systems opens a new field in physics ..."

These words show a close resemblance to the verbalization that Hermann Haken used for synergetics some years later on, i.e.: [650]

„Synergetics [...] deals with the spontaneous formation of structures in completely different systems. The systems may belong to physics, chemistry, biology, computer science or economy. As synergetics has shown, a large class of systems exists, in which the formation of spatial, temporal or functional structures is governed by the same principles."

In the article from which the citation above is taken, Haken also deals with the general systems theory of von Bertalanffy: [651]

„It might be seen as an indication of my ignorance or as a sign of the fragmentation of the sciences that only later on, i.e. after the foundation of Synergetics, I learned about the "Allgemeine Systemtheorie" of the biologist von Bertalanffy. Probably his most important thesis had been to search for fundamental isomorphism (in the mathematical sense of the word) between the elements of the different systems. It is interesting to note that the far reaching analogies in the behavior of quite different systems, revealed by Synergetics, are not based on isomorphism of the elements that, by the way, are most often not found. On the contrary, they are based on abstract isomorphism in the behavior of order parameters, the collective variables that govern the system, when qualitative macroscopic changes in the behavior of the system occur. The latter is a constraint to the problem, but on the other hand allows for a breakthrough finding general principles".

---

[650] (Haken, 1988a), p. 163.
[651] (Haken, 1988a), p. 172.

In the same article, Haken refers to Bertalanffy's book, "*Biophysik des Fließgleichgewichts*", that was published in German in 1953.[652] As we have seen, Wolfgang Weidlich, his co-researcher and friend, had drawn Haken's attention to von Bertalanffy as early as 1971.[653] But, at that time, Haken's work was centred mainly around the subject of stationary states of open systems – the prerequisite for phase transitions far from thermal equilibrium – and not as much on systems theory. Haken's way of seeing Bertalanffy's systems theory might have been influenced by his acquaintance with Mihajlo D. Mesarovich (born in 1928), whom he had invited as guest speaker to the first ELMAU Conference, 1972. Mesarovich defined the study of systems theory as "a formal relationship between observed features or attributes".[654] On the other hand, in his "historical account", a sub-chapter of the same publication, he indicated that the aim of Bertalanffy was less mathematical:[655]

> „While von Bertalanffy proposed a *theory-of-general-systems* meaning of systems, which will reflect the universal laws or principles valid for biological, social, physical, and any other phenomena, we are interested in a *general-theory-of-systems*."

Continuing, Mesarovich then argued that such a theory is not needed for connections to specific scientific disciplines because – being a formal theory – systems theory is, by its very nature, interdisciplinary (one may say a-disciplinary). His characterization of Bertalanffy's systems theory moves it much closer to synergetics as Hermann Haken had been aware of even in 1988.

But, it was not only Haken who, at that time, did not notice Bertalanffy's work. The reverse is also true. Hedrich states, "On the side of the system theorists, there is not even the act of mentioning *Synergetics* that is oriented likewise. [...] Synergetics, whereby there readily had been overlappings in time and content."[656]

It was the Austrian-American physicist Heinz von Foerster who opened up a second way for a systems-theoretical and cybernetic approach.[657] Von Foerster had been influenced by the ideas of the "Vienna circle". In 1948, he published an article on brain research "Das Gedächtnis. Eine quantenphysikalische Untersuchung", that caught the attention of the US-American neurologist

---

[652] (Bertalanffy, 1953).

[653] See p. 113.

[654] (Mesarovich & Takahara, 1975). Mesarovich is also the author of the *Second Report to the Club of Rome* running under the title "Mankind at trhe turning point" (Mesarovic & Pestel, 1975). Biographical data can be found at `http://systemsbiology.case.edu/participants/faculty/Mesarovic.shtml`, recalled 18.3.2013.

[655] (Mesarovich & Takahara, 1975), p. 247. (Italics in the original).

[656] (Hedrich, 1994), p. 123.

[657] Heinz von Foerster (1911 - 2002), was an Austrian born physicist who emigrated, after the Second World War, to the US. Biographical data as well as information on the history of the BCL ist given by (Müller & Müller, 2007).

Warren McCulloch.[658] The latter helped to provide to him with a physics professorship (in telecommunications) at the University of Illinois and appointed von Foerster as secretary of the influential *Macy conferences* that were held under the title "Circular causal and feedback mechanisms in biological and social systems" from 1945 until 1953. It had been the aim of these meetings "to create principles for a general science of human brain functioning". [659] At the suggestion of von Foerster, the title of the conferences was changed to "Cybernetics" in 1949. Two years earlier, this term had been coined by Norbert Wiener,[660] defining cybernetics as the science of feedback systems in technical and biological systems. The *Macy conferences* became the nucleus of cybernetics. Among others, the meetings were attended by John von Neumann, Norbert Wiener, Warren McCulloch and Walter Pitts, all-important exponents of the new line of research. Starting from problems in communication systems and electronics, cybernetics initially dealt with maintaining a desired condition by means of negative feed-back. This work is also the nucleus of automata theory and of the theory of neural networks that gained importance in artificial intelligence research in the 1970s and 1980s. But, von Foerster went beyond the concept of negative feedback, emphasizing the aspect of self-organizing systems being open systems (dissipative systems). With the help of the *Biological Computer Laboratory (BCL)*, von Foerster then created an institutional framework in 1958 where research on self-organization could be practiced interdisciplinarily.[661] The members of the BCL and their guest scientists, among others W. R. Ashby, G. Pask, L. Löfgren, G. Günther, H. Maturana and F. Varela, worked on a broad scope of subjects, all having some connection to machine learning and cognitive human processes. Von Foerster's "meta-approach", called the theory of "cybernetics of cybernetics" or "second order cybernetics", influenced the work of Maturana and Varela on autopoiesis very much.

However, while von Foerster and the *Biological Computer Laboratory* became the first nucleus of self-organization research, it is quite remarkable that there existed no points of contact with the theories of self-organization of Prigogine, Eigen and Haken. These theories, developed from natural sciences, had no impact on the reflections of the somehow epistemological and sociological oriented American tradition, even though one could have expected such influence from the research of Hermann Haken on cognition, undertaken from the late 1980s.

---

[658] Warren S. McCulloch (1898 - 1969), was an US-American neuroscientist and one of the founders of cybernetics. From 1941 till 1952 he was at the *Department of Psychiatry* of the *University of Illinois*, changing in 1952 to the *Research Laboratory of Electronics* of the *MIT* (Mass.). McCulloch (working jointly with Pitts) developed important brain theories. Most notably, they could show that neurons are capable to perform all logical operations of the so-called Boolean algebra. For his works see (Husbands & Holland, 2012).

[659] (Pias, 2003).

[660] (Wiener, 1948).

[661] For biographical data on Heinz von Foerster and on the history of the BCL see especially (Müller, 2000). On the philosophical implications of his work see (Hedrich, 1994), p. 121ff and (Paslack, 1991), p. 133ff.

9.2 Autopoiesis and Self-referentiality

The theory of autopoiesis was defined by the Chilean neurophysiologist Humberto Maturana. His research centred on the question: what is the essence of an organism? Maturana studied biology and afterwards became scientific assistant to Warren McCulloch at MIT, where he did neural network research. In 1970, during a later appointment at von Foerster's *Biological Computer Laboratory,* he published his famous article "Biology of cognition", considered to mark the birth of the theory of autopoiesis.[662] Starting from his research on visual perception, Maturana finally came to the conclusion that the following three attributes characterize the essence of a living organism: the ability of *self*-creation, the ability of continuous *self*-regeneration and the ability of *self*-containment. The term autopoiesis then was introduced in an article written jointly with his doctoral student F. Varela:[663]

> „An autopoietic system arises spontaneously from the interaction of otherwise independent elements when these interactions constitute a spatially contiguous network of production which manifests itself as a unity in the space of its elements. The properties of the components of an autopoietic system do not determine the property as a unity."

According to Maturana, autopoietic systems are self-contained in their organization and, with regard to this aspect, are autonomous. All the information necessary for its circular organization is situated inside the system. It is in this respect that autopoietic systems are *self*-referential. The functional organization of these systems is explained by feedback and self-referential ties of self-organizing processes. The system itself is "structurally coupled" to the environment and hence "open". It follows that the human nervous system is also functionally closed and its single purpose is the synthesis of behaviour.[664] In this respect, the theory of autopoiesis interconnects inextricably the organization of living beings and their cognitive abilities that are produced and mediated by the nervous system. Maturana's theory became the origin of a new biological oriented epistemology, the theory of "radical constructivism".[665]

Despite the fact that autopoiesis is very influential in biology and especially in sociology, its relation to self-organization is mainly to be seen in the circular action of existing order in organizations. With respect to the emergence of order and the transition between different order hierarchies, autopoiesis gives no definite statements. Autopoiesis presupposes order. That is the reason why Hermann Haken felt no call to argue intensely about autopoiesis. It was widely discussed in the framework of the theory of radical constructivism.

---

662 (Maturana, 1970).

663 (Varela, Maturana, & Uribe, 1974).

664 (Schmidt, 1987).

665 (Paslack, 1991), p. 165.

9.3 The Theory of Autocatalytic Hypercycles

In the years 1970 and 1971, Nobel Prize laureate Manfred Eigen developed a highly specialized self-organization theory on the prebiotic evolution of life.[666] During the following years, he and his colleague, Peter Schuster, then extended his approach leading to a theory of "autocatalytic hypercycles".[667] Eigen directly referred to the question: can physics – in its current form – explain biology? And, his answer was:

> Biological phenomena and processes, sufficiently examined so far, show no signs at all that physics – in its current form known to us – should not be able to do so. Even though – as in the macroscopic phenomena of the inanimate world – there are limits in describing it in all details. There are no general [limits], but the limits are inherent in the complexity of the processes and phenomena.[668]

The theory of Manfred Eigen dealt with the prebiotic development of live, the evolutionary stage before cell creation. It is of critical importance how information is stored within a biological structure and – if or how – the structure is then subjected to natural selection. Eigen's earlier work on hyper-fast reaction kinetics, that gained him the Nobel Prize in 1967, led to the conclusion that "the genesis and stabilization of biological information must be understood as the result of self-organizing interactions between nucleic acids and proteins (enzymes).[669] Eigen gave details in an interview with Günter Küppers in 1985:

> Formerly we had done much more research on the kinetics of super-fast reactions and developed these processes methodically. Next came different phases [...] and then we examined enzymatic reactions. Finally, in one of our Winter Seminars, the following question sprang up: how does it come about that enzymes always optimally function; that these complex mechanisms are geared so perfectly to each other? One of the participants said: that must be Darwin's principle, that's selection [...] Therefore we tried to develop a kind of selection theory that could be applied to molecules.[670]

---

666 (Eigen, 1971).

667 (Eigen & Schuster, 1977), (Eigen & Schuster, 1978), (Eigen & Schuster, 1979).

668 (Eigen, 1971), p. 520.

669 (Paslack, 1991), p. 110.

670 Cited from (Paslack, 1991), p. 110. Neither the original audio record nor the transcription of the interview of Günter Küppers with Manfred Eigen from July 1985, still exist. (Private information Günter Küppers).

Eigen summarized the results of his research in an article published in 1971:[671]

> „1. A detailed analysis of the reproduction mechanism of nucleic acids and proteins show no indication for the assumption of special forces or interactions only inherent to living organism. Selection behavior, characteristic for the evolution of living systems, occurs already at this [evolutionary] step as a material property, inherent to special reaction systems.
> 2. Every system that is preserved by mutation and selection is undetermined with respect to its individual structure, nevertheless the resulting process of evolution is inevitable – hence law. The occurrence of a mutation possessing a selective advantage corresponds to an instability that, as such, can be explained by means of the principle of Prigogine and Glansdorff for stationary, irreversible thermodynamic processes. Therefore, the process of the optimization of evolution is in principle unavoidable, but not determined with respect to the individual route it will take.
> 3. Finally, it becomes apparent that the creation of life is tied to a number of properties, all of which can be explained conclusively by physics.[...]"

Mathematically, the hypercycle is described by a set of non-linear differential equations. In the words of Shneior Lifson, Eigen, with his findings, gave "the final death blow to vitalistic theories of the functioning of animate matter, in whatever disguise they may appear".[672]

Eigen especially pointed out that self-organization and Darwinistic selection occur already at the molecular level. Jointly working with Peter Schuster, he then showed how sensitive the process is with respect to modification and stability – the equivalence of Monod's "chance and necessity". Their "formula" for an "ur-gene" reads that the length of the molecule must have comprised between 50 to 100 nucleotides because a shorter molecule would have been too unstable and a longer one would have shown to many replication errors.[673] In the 1980s, the US-American scientists Sidney Altman and Thomas Cech could present a realistic *hypercycle*, earning them the chemistry Nobel Prize in 1989.[674]

---

671 (Eigen, 1971), p. 521-522.
672 (Lifson, 1987), p. 303.
673 (De Duve, 1994), p. 197. The original article from von Eigen and Winkler-Oswatitsch had been „Transfer-RNA, an Early Gene?" (Eigen & Winkler-Oswatitsch, 1982).
674 See (Cech, 1986).

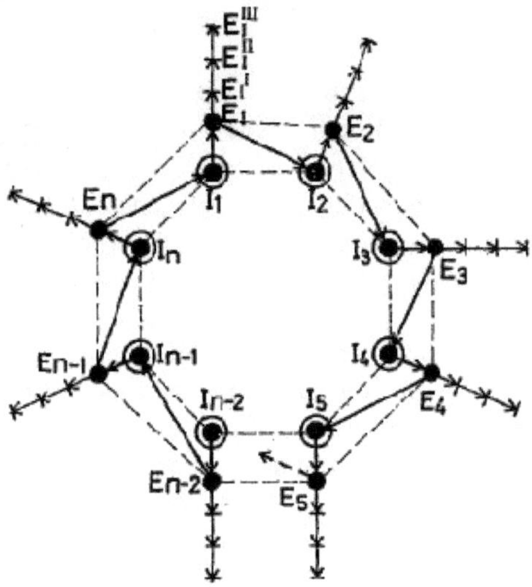

Fig. 42 The autocatalytic hypercycle. The I_i are single-thread information carriers (RNA), the E_i represents the catalytic functions (enzymes) (from from (Eigen, 1971), p. 504)

Due to the high specificity of his biological model, Manfred Eigen never attempted to extend his hypercycle model of prebiotic self-organization to other domains or disciplines.[675]

It is very difficult to judge how these ideas of Manfred Eigen influenced the work of Hermann Haken. Haken had met Eigen during the second *Versailles conference*, "*From Theoretical Physics to Biology*" and later on had been regularly invited to (and participated at) the famous winter seminars organized by Eigen in the alpine ski-resort of Klosters (Austria).[676] On one side, Haken denied a direct influence of the *hypercycle* theory concerning his work: [677]

> „The Hypercycle, while being important for the development of evolution theory, is not so relevant for Synergetics. That [the process consists of] complex auto-catalytic reactions that, as far as I can see, do not fit into Synergetics – but, maybe I was only too lazy about doing enough on this subject."

---

[675] (Hedrich, 1994), p. 130.

[676] Interview with Manfred Eigen and Ruthild Winkler-Oswatitsch in Göttingen, 24.5.2011. (Transcript of the interview in Archive Haken (University Archive Stuttgart).

[677] Interview with Hermann Haken from 16.11.2010, p. 19 (Archive Haken (University Archive Stuttgart)).

On the other side, Haken was a regular and well-received participant at the annual winter seminars of Manfred Eigen, which were characterized by a special atmosphere leading to intensive scientific discussions. Eigen's long-time colleague and partner Ruth Winkler-Oswatitsch once described this special atmosphere:[678]

> „The name of Manfred Eigen is associated by friends and colleagues with the Winter Seminar. His idea was simple: freed from the distractions of the daily scientific nitty-gritty, in a tranquil place in winterly surroundings, he and his co-researchers could "take on" one or two topical scientific themes in a concentrated manner otherwise impossible. During the light hours, the programme consisted of ski-ing: when darkness fell, the science would start. In the course of the evening discussions, current problems would be purpose-fully worked through, projects concluded and new ones conceived. [...] The first Winter Seminar in January 1966 was a great success, and everyone was clear that the institution would become permanent.
>
> Naturally, the Winter Seminar was also subject to evolution in the course of time. It began in the narrowest circle of co-workers, and soon other scientists, from home and abroad, began to come too. Under the general theme "Molecules, Information and Memory" the Winter Seminar became and established tradition.

While synergetics was never the direct subject of the winter seminars, Haken had been a "pleasant and very lively [participant], both in socializing as well as in his lectures". He often "presented his thoughts" and, according to Eigen, "his talks [were always] outstanding".[679] Because there had been close analogies between the laser theory of Haken and the *hypercycle* model of Eigen, these discussions in Klosters should be seen as important "background noise" for Haken's ideas.

---

[678] (Winkler-Oswatitsch, 1987), p. 112 – 113.

[679] Interview with Manfred Eigen and Ruthild Winkler-Oswatitsch in Göttingen, 24.5.2011. p. 13-14. (Transcript of the interview in Archive Haken (University Archive Stuttgart).

9.4 The Theory of Dissipative Systems

In 1967, Russian born Belgian physio-chemist Ilya Prigogine[680] coined the term "d*issipative systems*" in an article jointly published with his long-time colleague Grégoire Nicolis, "On symmetry-breaking in dissipative systems".[681] Dissipative systems were not new phenomena. They denote systems in physics and chemistry that are "open" to the environment, i.e. they exchange energy and/or matter with their surroundings. The work of Prigogine and Nicolis had been stimulated by an earlier article of A. Turing on morphogenesis.[682] Their calculation showed a close analogy to the Bénard-phenomenon and the authors pointed to a possible application in biological processes, especially in the prebiotic phase: [683]

> "Even on a broader scale, it is difficult to avoid feeling that such instabilities should play an essential role in biological processes and especially in the first biogenetic steps".

Prigogine, taking his doctoral degree in 1945, had studied with the Belgian physicist Théophile De Donder. From the very beginning, Prigogine was highly interested in the problem of the creation of life and the associated question of time's role in natural laws.[684] He wondered about the discrepancy between the reversible laws of mechanics and thermodynamics if compared to the irreversibility of animated processes in nature. In the 1940s and 1950s, he occupied himself by performing research on irreversible thermodynamics, developing his "principle of minimal entropy production" regarding the stability of thermodynamic systems near equilibrium states. During the 1960s he then concentrated on systems far from thermal equilibrium and in 1971 published the influential work (together with Paul Glansdorff), *Thermodynamic Theory of*

---

680 Even though Ilya Prigogine received the Nobel Prize in Chemistry for his work on dissipative systems in 1971, no comprehensive biography of his life and work exists. One reason might be the sheer size of his oeuvre: He published more than 1000 articles and wrote 21 books, encompassing such diverse fields as thermodynamics, the theory of dissipative systems – especially the question of the direction of time -, self-organization and complex systems, cosmology. In his later years, he was heavily involved in the discussion of scientific-philosophical questions about nature. He received not less than 54 honorary doctorates. During the twenty years after having been awarded the Nobel Prize he was one of the most apprehended and discussed scientists worldwide. (see for instance (Altner, 1986). His autobiography, written for and published by the Nobel Prize Committee, is often referred to. (Prigogine, 1993). An informative survey of the scientific achievements of Prigogine is given by his former student and colleague Balescu. (Balescu, 2007).

681 (Prigogine & Nicolis, 1967).

682 (Turing, 1952).

683 (Prigogine & Nicolis, 1967), p. 3550.

684 (Prigogine, 1993).

Structure, Stability and Fluctuations.[685] According to Paslack, this book "rather marks the breakthrough of dissipative systems".[686] Reaching a critical value, a new structure might show up, triggered by fluctuations. The system shows self-organization, not influenced by outside forces but as an inherent feature. Initially, chemical processes had been the focus of Prigogine's research. The interest in these phenomena rose dramatically when chemical oscillations of the so-called Belousov-Zhabotinsky reaction became known.[687] In his autobiography, Prigogine recalled how he pondered about the question, whether his principle of minimal entropy production still held true in systems far from thermal equilibrium (in the so-called non-linear regime):

> „Since the formulation of the minimum entropy production theorem, the study of non-equilibrium fluctuation had attracted all my attention. It was thus only natural that I resumed this work in order to propose an extension of the case of far-from-equilibrium chemical reactions".
> This subject I proposed to G. Nicolis and A. Babloyantz. [...] Nicolis and Babloyantz developed a detailed analysis of linear chemical reactions. [...] They added some qualitative remarks which suggested the validity of such results for any chemical reaction.
> Considering again the computations for the example of a non-linear bio-molecular reaction, I noticed that this extension was not valid".

In this statement from Prigogine, two developments and results were consolidated that happened at different times. Encouraged by the work of his co-workers Gregoire Nicolis and Agnessa Babloyantz, at first he proceeded on the assumption that the theorem of minimal entropy production would remain valid also for processes far from thermal equilibrium. This assumption is an important component of his book *Self-Organization in Nonequilibrium Systems,*[688] published jointly with G. Nicolis in 1977. But, four years earlier in July 1973, the US-American scientists J. Keizer and R. Fox, in an article published in the *Proceedings of the National Academy of Sciences (USA)*, had already doubted whether the theorem holds true in these cases.[689] Because this article and the argumentation are central to the criticism Haken expresses on Prigogine's approach, we cite the argument in more detail:[690]

685 (Glansdorff & Prigogine, 1971).
686 (Paslack, 1991) S. 93.
687 (Prigogine, 1993).
688 (Nicolis & Prigogine, 1977).
689 (J. Keizer & Fox, 1974) and (Joel Keizer, 1976).
690 Emphasis by BK, (J. Keizer & Fox, 1974), p. 192.

„When another example, of the type considered by Glansdorff, Prigogine, and Nicolis, is examined we find a regime of stable steady states on the basis of "normal mode" analysis, which their stability criterion, based upon the so called excess entropy production, cannot demonstrate is stable. These results strongly emphasize that **the Glansdorff- Prigogine criterion for stability is at best only a sufficient condition for stability**, a fact recognized, but not sufficiently stressed by Glansdorff and Prigogine, and so the violation of this criterion does not necessarily imply the lack of stability. We believe that a stability criterion based upon the second differential of the entropy provides a useful condition only in a neighborhood of full equilibrium, and it does so there because of its initimate connection with the second law of thermodynamics. For general non-equilibrium states, the examples treated in this paper show that the time derivative of the second differential of the entropy may be either positive or negative and that a negative sign does not imply instability".

Ten years later, this opinion was acknowledged even by Grégoire Nicolis himself.[691] The above statement by Prigogine in his autobiography, "Considering again the computations for the example of a non-linear bio-molecular reaction, I noticed that this extension was not valid", was only expressed at a later point in time than the point in time represented by the publication of his book in 1977.

Even though in chemistry the term "dissipative systems" was fairly new, the phenomenon of open systems was rather old. As we have seen in Chapter 6b, in 1964, Hermann Haken had already described the laser system mathematically as a "dissipative system", recognizing in 1969 that it resembled a phase transition. Haken was aware of Prigogine's approach but considered it to be false. In an interview, he explained his position:[692]

„Although I noticed the work of Prigogine, where he emphasized the Bénard instability as well as the Belousov-Zhabotinski reaction, I never understood the idea of his entropy principle. [...]

He looked at the problem from the point of thermodynamics, at best of thermodynamics of irreversible processes, but no step further. Then he formulated two theorems, the principle of minimal entropy production and another one, on excess

691 (Jiu-li, Broeck, & Nicolis, 1984).

692 Interview with Hermann Haken from 21.9.2010, p. 26. (Archive Haken (University Archive Stuttgart)).

entropy production. But afterwards other people, among others his co-worker Nicolis had proved that these principles are not sufficient to explain the formation of structures. Personally, I came from statistical quantum physics and always dealt with the problem from the microscopic side. That was the fundamental difference.

Another difference: whereas I tried to reveal general principles, Prigogine – in his work with Nicolis – dealt with a special model, the Brusselator."

In his book *Die Selbststrukturierung der Materie*, published with Arne Wunderlin in 1991, Haken once again explained the "fundamental difference" of his approach and the one of Prigogine.[693] Haken's criticism centred on two major points. For one thing, strictly speaking, the notions of temperature and entropy are only defined in classical (equilibrium) thermodynamics. For another thing, only the microscopic and stochastic approaches detailed further down allow for the inclusion of fluctuations that are central at the critical bifurcation points.[694]

„We may construct a mathematical network of relationships from the objects that we gained from statistics, like partition function, information etc. that is consistent with phenomenological thermodynamics.

$$S' = -k_B \sum_K p_k \ln p_k = -k_B < \ln p_k >$$

[S' = Entropy ; k_B =Boltzmanns constant ; p_k = Probability of states for particle k]

S′ can be seen as a mean value. Therefore it is to be expected that fluctuations will occur around this mean value S′. But entropy fluctuations are not known to classical thermodynamics".

„If we would like to speak of the entropy in a system far from thermal equilibrium we only can make use of the statistical definition of entropy. ... There have been famous scientists that claimed: In systems far from thermal equilibrium holds

$$dS < 0$$

having the meaning: the entropy decreases if the system transforms from a non-ordered state to an ordered one.

693 (Haken & Wunderlin, 1991).

694 (Haken & Wunderlin, 1991), p. 122-123 and p. 445.

> Actually one can calculate the statistically defined entropy with the help of the laser distribution function. As a result it arises that in the ordered state the entropy decreases not at all but rather increases. This is contradictory to the common way of thinking and against intuition nourished by statistical physics since many years. Traced back to Boltzmann, in statistical physics the opinion is held that an increase in entropy is necessarily connected to an increase in disorder."

With the methods utilized by Prigogine, the behaviour of a system at the critical points cannot be predicted. Only the application of statistical methods allows for prediction of such behaviour, later also used by Prigogine's students and co-workers G. Nicolis and A. Babloyantz.

Against this background, it becomes readily understandable that Haken saw his approach, with order parameters and the "slaving principle", as superior to Prigogine's Ansatz.[695]

> „Glansdorff and Prigogine developed a second principle based on "excess entropy production" [...] If the principle is fulfilled a new structure may appear but not neccissarily. In addition to it this principle cannot predict or calculate the new structures and says nothing about the dynamics in structure formation. Alltogether, from the very beginning, the methods we developed for and applied to the laser have been superior to this principle."

Haken wondered quite often why, in public perception, Prigogine "was so much valued".[696] At one occasion, citing the statement of a third person, he made use of a drastic choice of words – without personally naming Prigogine: "Some years ago a physic-chemist had been awarded a high distinction in chemistry. A well-known theoretical physicist described these works as 'non-sense', whereas a bio-mathematician reckoned, the person in question received the honor in the field of linguistics, hinting at [sic] his word coining activity".[697]

In conclusion, it remains to be said that the approach of Hermann Haken was much more fundamental than the chemical oriented approach of Prigogine and bore more fruit in the scope of its applications.

---

695 (Haken, 1988b), p. 214.

696 Interview with Hermann Haken from 21.9.2010, p. 26. (Archive Haken (University Archive Stuttgart)).

697 (Haken, 1991), p. 189.

Nevertheless, in the public discussion of each theory's philosophical consequences, people were much more aware of Prigogine's self-organization, "Ansatz", than they were of the synergetic approach. This can be traced back to the fact that, after having received the Nobel Prize in 1977, Prigogine worked closely with Isabelle Stengers, jointly publishing two books, *La Nouvelle Alliance* ('Order out of chaos') and *The End of Certainty: Time, Chaos and the New Laws of Nature*, that aroused widespread attention from philosophers and sociologists.[698] The statements on "man's new dialogue with nature" aroused interest because of notions of holism, "everything is bound to anything". The idea of self-organisation and emergence – from chaos to order – synchronized in many ways with the ideas of the New Age movement of that time.[699]

While Hermann Haken published a successful book, *Erfolgsgeheimnisse der Natur: Synergetik – Die Lehre vom Zusammenwirken*,[700] that has been translated into many languages, his other popular books remained restricted to German speaking consumption.[701] And, in contrast to Prigogine, he took a backseat concerning philosophical and epistemological reflections. He always remained the fact-oriented scientist. The characteristics of a preacher or "guru" did not show up in him.

---

[698] (Prigogine, 1980), (Prigogine & Stengers, 1979), (Prigogine & Stengers, 1993) (Prigogine, 1979).

[699] The "Age of Aquarius", an esoteric movement, greatly influenced the perception of natural sciences by works from authors like James Lovelock, Fritjof Capra and Prigogine/Stengers as well. See for instance (Capra, 1977), (Lovelock, 1979), (Ruß, 1993), (Hemminger, 1987), (Mutschler, 1990).

[700] (Haken, 1981).

[701] (Haken & Haken-Krell, 1989), (Haken & Haken-Krell, 1992), (Haken & Haken-Krell, 1997).

Chapter 10
Summary

In retrospect of Hermann Haken's scientific achievements, we are able to answer the questions posed at the outset of this work. Relatively clearly defined periods of his life emerged that were crucial for the development of his research. The first two periods comprised his beginnings in Halle and Erlangen, followed by his "laser period" during the first ten years of his Stuttgart career. From 1970 until 1985, the third stage encompassed the development and promulgation of *Synergetics*. This was followed by a fourth period, dating from 1985 until the new millennium, where he concentrated his work on applications of *Synergetics* in the fields of medicine, brain research and psychology.

Halle and Erlangen, the period after the Second World War until his appointment at Stuttgart in 1960, might be described - loosely based on Goethe – as his "Years of Learning and Wandering." It was a fortunate coincidence, caused by his mother's wish, that Haken did not choose physics as the main subject of his studies but rather mathematics, in which he also obtained his doctoral degree. The analytical approach, inherent to this science, and the attention to boundary conditions and its rigorous stringent execution were formative for his scientific methodology. But Hermann Haken was not interested in pure mathematics; he was (and still is) interested in finding out if the theoretical calculations can be found and verified in nature using experiments. Therefore, it is no wonder that he turned towards physics after his later wife ceded her "Hilfskraftstelle" (teacher's aid position) to him.

In the early 1950s, the *University of Erlangen* was a centre of solid-state physics and crystal optics research. Its experimental physicists, Hilsch and Mollwo, as well as professor Pick – later being important in his appointment at the *University of Stuttgart* – all originated from the famous Goettingen School of Robert Wichard Pohl.[702] Questions of the movements of electrons and atoms in solids constituted the research focus. It was Hilsch who performed experiments in the phenomenon of super-conductivity where the flow of electric currents shows no resistance at very low temperatures. Therefore, it is not astonishing that young Hermann Haken occupied himself with the theory of excitons, describing the dynamic behaviour of coupled electron-"hole" pairs in solids. This theory was

[702] See especially (Teichmann, 1988).

B. Kröger, *Hermann Haken: From the Laser to Synergetics*,
DOI: 10.1007/978-3-319-11689-1_10

seen as particularly promising for solving the question of the nature of superconductivity.

It was another mere coincidence that Hermann Haken, at the intercession of Helmut Volz, came into contact with Walter Schottky and his co-worker Eberhard Spenke, both working on semi-conductor research at the Siemens laboratory at Pretzfeld.[703] A book project by Spenke lead young theoretical physicist Haken to thoroughly study the features of the so-called second quantization, that later on, was crucial for his appointment to chair of theoretical physics at Stuttgart, as well as for the formulation of his quantum-mechanical laser theory.[704]

Haken's "Years of Wandering" were less marked. In addition to a short-term professorship replacement at the University of Munich, he was very much influenced by an invitation from Herbert Fröhlich[705], who invited him for a research stay at Cambridge (GB). At the end of the 1960s, the evolving long-time friendship paved the way for Hermann Haken to participate in the *Versailles Conferences*, an important interface on his way to *Synergetics*. (See especially Chapter 6a). Another, also unintended direction of his studies was prompted by his research stay in the USA at end of 1959 and early 1960. During his time at the famous *Bell Laboratories*, his friend Wolfgang Kaiser peaked his interest in the laser experiments, a research field totally unknown in Germany at that time.

Returning to Germany shortly afterwards, Haken was awarded the chair of theoretical physics at the (then) *Technische Hochschule Stuttgart*, initialising his "laser period" of about ten years. (See Chapter 5). We have to keep in mind that this is only an arbitrary time limitation because Haken continued publishing on laser theory in the years to come, especially in co-operation with his doctoral students. Without a doubt, the formative years of laser theory were the years from 1960 until the publication of his seminal book "Laser Theory" in the *Handbuch der Physik,* in 1970. It was the second time that Haken entered into a scientific race. While working on the theory of excitons and trying to explain superconductivity, he did not sense this race[706]; but this time he was fully aware of its existence. Not least because of his meeting with Nobel Prize winner Willis Lamb Jr. on the occasion of the Heidelberg Conference on "Optical Pumping" in 1962, where Lamb informed him about his latest results, thus making the competitive situation obvious to Haken. (With respect to Lamb, see especially Chapter 5c). In the race to the quantum-mechanical formulation of laser theory, detailed in Chapter 5, Hermann Haken benefitted from several exceptional circumstances. On the one hand, he was able to attract Hannes Risken and Wolfgang Weidlich as co-workers, two extraordinary able scientists. It was Wolfgang Weidlich especially, who was thoroughly trained in quantum field theory by Günther Ludwig in Berlin, who helped pave the way from the semi-classical theory to the fully quantized laser theory. In addition, Haken was inspired by the fact that Maiman

703 See (Handel, 1999).

704 More details can be found in (Schweber, 2012).

705 For biographical data of Herbert Froehlich see (Hyland, 2006).

706 See citation on page 35.

demonstrated the first laser effect in a ruby crystal – a solid. Therefore, Haken immediately could make use of the quantum-mechanical mathematical tools, tools familiar to him because of his longstanding work on the theory of excitons. Even though some new mathematical tools had to be subsequently developed, as gas- and semi-conductor lasers quickly showed up, but the work done on exciton theory was a firm basis upon which to build. In 1964, Haken experienced a climax of his scientific achievements with the publication of his article "A Nonlinear Theory of Laser Noise and Coherence I" in the journal *Zeitschrift für Physik*. Fifty years later, he rated this work in retrospect:

> "Decisive for me and qualitatively new – still – was the quantum theory of the laser. The article of 1964. There, in a model like manner, I introduced features like noise and fluctuations. Previously noise had been calculated for the [electric] field, but noise with respect to the atoms arose for the first time. The next big thing was the operator calculus in the context of Heisenberg's so-called second quantization. For one thing the elimination procedure: the adiabatical procedure had been known, but that it was possible to do it with quantum-mechanical operators – was new. And the "Ansatz" for the field operator: That it was possible to split the operator into a classical and a quantum-mechanical part, all of this was new. To my mind, the article of 1964 contained a bundle of features, all of them totally new."[707]

Furthermore in this article, Haken realised that the laser experiences symmetry breaking at the threshold, leading to a totally different radiative behaviour below and above the threshold. Below the threshold, the laser radiates like a common light bulb, above the threshold, the radiation shows the quality of quasi-monochromatic coherent light; this was not due to an ever narrowing of the so-called line-width, as people had always assumed. Responsible was a completely new phenomenon, later on named "slaving principle" by Hermann Haken. One single electromagnetic mode wins the selection competition and forces all other modes – via a self-organised feedback behaviour – to a coherent radiation. The laser behaviour was out of all reason as the energy input (the "optical pumping process") was entirely inhomogeneous.

In the following three years, the basic articles presenting the fully quantized laser theory appeared in quick succession. Three different scientific groups were engaged in a fierce competition: Hermann Haken and his "Stuttgart School" including Risken and Weidlich, Willis Lamb and his doctoral student Marlan Scully, as well as Melvin Lax together with William Louisell. A fundamental

[707] Interview with Hermann Haken from 9.10.2012, p. 8. (Archive Haken (University Archive Stuttgart)).

aspect was the insight that when describing the laser phenomenon, one has to apply non-linear equations and take into account fluctuations and/or noise.

At the end of 1968/69, the collaboration with his student and, later on, his most important colleague, Robert Graham started a new period in the scientific life of Hermann Haken. (See especially Chapter 6b and c). In his doctoral thesis, Graham, while analysing an extended quantum-mechanical system, the so-called quantum-mechanical oscillator, had discovered that the resulting mathematical equations showed an astonishing analogy with the equations of the Ginzburg-Landau theory of super-conductivity. Haken in particular recognised this fact, remembering his former studies on super-conductivity and exciton theory during his early career in the 1950s. In super-conductivity, when cooled to a few degrees above absolute zero, the closed system experiences a phase transition. Due to their research, Graham and Haken were now able to demonstrate that – in complete analogy – when energy is supplied the open laser system shows the phenomenon of phase transition as well. They concluded that phase transitions are of much higher relevance than previously thought. "The concept of a phase transition is much more general than usually thought of."[708] Through the concepts of phase transition, the laser and "open" systems, Hermann Haken had methodical tools at his disposal to look for transition phenomena from disorder to order in physics, but also in other sciences. Some of these phenomena were familiar to him for a long time, for instance the Taylor-Couette effect in hydrodynamics, the convection cells named after Bénard; other effects, like the famous Belousov-Zhabotinsky reaction in chemistry or the theory of hypercycles by Manfred Eigen in biology, developed nearly about the same time as Haken's reflections. The fact of the laser experiencing a phase transition was crucial to Haken. Thereby, Haken gained a heuristic analytic prototype that comprised concepts like symmetry breaking, order parameter, critical "slowing down," growing of fluctuations and hysteresis.

In the period from 1967 to 1972, new ideas and stimuli came about while Haken was attending the *Versailles Conferences* entitled "From Theoretical Physics to Biology." These meetings discussed questions that influence how physics concepts might influence biological phenomena. It was here that Hermann Haken became acquainted with Nobel Prize laureate Manfred Eigen from Göttingen and Belgian physicochemist Ilya Prigogine. Both scientists occupied themselves intensely with the aspect of order out of disorder (or chaos): Eigen investigating the theory of the (prebiotic) hyper-cycle, published in 1971, and Prigogine with a modification of the Belousov-Zhabotinsky reaction, later called the "Brusselator." Haken developed a long-time friendship with Manfred Eigen and his circle of biologists and chemists, which led to his continuous participation in the annual "Winter Seminars," organised by Eigen and his team in the Alpine ski-resort of Klosters. These annual discussions, talks and information at the *Versailles Conferences* and the simultaneous intensive work on the concept of the laser showing phase transitions were the fertile ground upon which the idea of *Synergetics* was aroused.

---

[708] (R. Graham & Haken, 1970).

After the first publication on the subject of *Synergetics* in the popular scientific journal "Umschau," in 1971, showed no noticeable resonance, in 1972, Hermann Haken laid the foundation for his interdisciplinary and multidisciplinary research program with his first *ELMAU Conference* "Synergetics – Cooperative Phenomena in Multicomponent Systems." (See Chapter 6d). That this was done intentionally and not by chance is reflected in the introduction of the proceedings of the conference:[709]

> „[The topics of this book] range from phase-transition-like phenomena of chemical reactions, lasers and electrical currents to biological systems, like neuron networks and membranes, to population dynamics and sociology."

The scientific topics touched upon comprised physics and chemistry, biology and informatics, as well as sociology. We have to keep in mind that Haken, being the sole organiser of the conference, personally selected and invited the speakers. *Synergetics* thus acquired the status of a structural science, searching for common structures underlying the different research fields. This became particularly clear in the conference segment that dealt with "General Structures," where the later famous Serbian-American physicist Mihajlo D. Mesarovic,[710] gave a talk on "Hierarchy of Structures." Haken's broadly conceived concept had been an ambitious venture in the 1970s, because exceeding the limits of one's own specialist field was not commonplace and in doing so one's reputation would have been at stake. Even years later, Haken remembered the difficulty of his decision:

> "People would say "it's such a nonsense what he makes. And how arrogant proclaiming to open a completely new field of science." It has been the anxiety that one was seen as notoriously arrogant and presumptuous to proclaim a new research field."[711]

Notwithstanding the fact that Hermann Haken, from the very beginning, adopted a multi-disciplinary approach, he never left the field of theoretical physics, contrary to some other colleagues.[712]

Only a short time later, his considerations concerning *Synergetics* got a boost when Haken was able to demonstrate that the mathematical methods he had developed for the laser were also able to solve other related problems. His approach worked successfully, solving the Bénard effect, the Brusselator and the

---

[709] (Hermann Haken, 1973).

[710] See footnote 652.

[711] Interview with Hermann Haken from 20.09.2010, p. 32 (Archive Haken (University Archive Stuttgart)).

[712] For instance Nobel Prize laureate Donald Glaser, who shifted to neurology after his invention of the bubble chamber in elementary particle physics.

"Lorenz equations" of Hydrodynamics as well. Finally in 1977, he summarised his work in the publication of his seminal textbook "Synergetics – An Introduction. Nonequilibrium Phase Transitions and Self-Organization in Physics, Chemistry and Biology." Not until this moment, did he actively use and promote the word *Synergetics* in preference to the term "cooperative phenomena" and thus, simultaneously focused the research activities of his university institute on this new scientific field. Five years later, he was able to expand his activities significantly, when *Synergetics* was chosen as a priority program of the *Volkswagen Foundation* and was sponsored by considerable funds. (See Chapter 7a and b). Helpful for *Synergetics* winning the status of a priority program was the 1977 Chemistry Nobel Prize awarded to Ilya Prigogine that generated great public and scientific awareness for the concepts of "dissipative (open) systems" and of self-organisation. The priority program of the *Volkswagen Foundation* progressed for ten years and supported nearly fifty scientific research groups. On the promotion of his ideas, Hermann Haken organised eleven *ELMAU Conferences* with respect to *Synergetics* and self-organisation; not least, in cooperation with the Springer publishing company, he founded the book edition "Springer Series in Synergetics" that finally grew to 77 volumes during his senior-editorship.

Haken focused his own research activities on developing more detailed mathematical concepts of phase transition in systems far from thermal equilibrium. During this work, the concepts of order parameters and of the "slaving principle," in the form of the generalised Ginzburg-Landau equations, took their final shape. During this process, Haken succeeded in a stochastic justification of the phenomenological Ginzburg-Landau theory, a result, he was very proud of:

> "The Ginzburg-Landau theory is brilliant, but it is a phenomenological theory. First one has to construct an expression for the "free" energy, then one has to develop this expression by a small parameter and finally the order parameters come about. It was phenomenological. I was able to demonstrate how the Ginzburg-Landau theory is connected with and can be constructed by microscopic (statistical) theories. We can deduce the order parameters by the determination of the unstable modes which subsequently are eliminated etc. This is done in such a way that also the spatial dependencies of the order parameters are retained. It was new for me that the Ginzburg-Landau equations "do not fall from the sky" or are constructed from thermodynamics, but result from a microscopic (statistical) theory."[713]

As previously done, the research and work performed from 1977 until 1983 was systematically structured by Haken and then published in another textbook titled "Advanced Synergetics" (1983). The mathematical formulation of the first

[713] Interview with Hermann Haken from 9.10.2012 (Archive Haken (University Archive Stuttgart)).

phase of *Synergetics* – its foundation from a microscopic approach – was nearly finished when these two textbooks appeared in 1977 and 1983.

In the following years, Hermann Haken most intensively occupied himself with research on the question, how phase transitions far from thermal equilibrium order parameters can be deduced, if only macroscopic measurements are available, i.e., no microscopic approach seemed possible. (See Chapter 8). This situation often occurs in biology or medicine, for instance when measured data of cardiac action or from brainwaves are only available. Haken found a solution, applying the solution as the "Maximum Information Entropy Principle" that the US-American physicist Edwin T. Jaynes had formulated in the 1950s. With the aid of this principle, Haken was able to demonstrate how the order parameters and the "slaved modes" could be calculated. The application of the deduced set of equations then resulted in the prediction of the measured data.[714]

> "Our approach will start from correlation functions, i.e. moments of observed variables from which we may then reconstruct the order parameters and the enslaved modes. Incidentally, we can also construct the macroscopic pattern, or in other words we may automatize the recognition of the evolving patterns which are produced in a non-equilibrium phase transition".

This method turned out to be a powerful mathematical tool for experimentalists and was so important for Hermann Haken that he called it "the second foundation of *Synergetics*."[715] In addition to the microscopic approach starting from the dynamics of the individual components of a system, a macroscopic approach from measured (aggregate) data was now possible. By using this method, Haken's doctoral student Rudolf Friedrich was able to demonstrate that a malfunction of the brain, the "petit mal epilepsy," can be mathematically described by only three order parameters.[716] The macroscopic approach was also used in the work done in automatic pattern recognition with the "synergetic computer," research activities performed by Haken and his institute in the late 1980s. The proof that the coordination of complex muscular movements is another synergetic example of self-organisation and is not steered by a fixed "motor program" (as was thought until that time) was another important medical topic that was verified by Haken in collaboration with the US-American neuro-physiologist Scott Kelso.

Thus, we can state that the phenomena of pattern recognition and pattern formation in the visual system, as well as questions of movement coordination, directed Haken's research interest into the medical and physiological fields. Thus, pattern recognition and pattern formation lead to a fourth research period starting in the 1990s. For a long time, Haken had thought of the human brain as the most complex synergetic system possible. He was the first to propose the idea that

---

714 (Hermann Haken, 1988c), p. 35.
715 (Hermann Haken, 1988c), p. 33.
716 (Friedrich & Uhl, 1992).

thoughts might be order parameters of the neural network of the brain; the synergetic approach proved to be most fruitful in psychology. (See Chapter 8e). The terms of phase transition, order parameters and the harmonic cooperation of parts leading to self-organisation was met with a positive response. This new approach allowed for a fresh perspective of well-known phenomena and made new attempts for the treatment of mental disorder possible.[717]

Looking in retrospect at the broad scope of applications in different fields that Hermann Haken used to develop *Synergetics*, his claim of *Synergetics* being a comprehensive systems theory becomes understandable. Nevertheless, he was always aware of the fact that he did not invent a new science, for him, *Synergetics* allowed for a new perspective on known phenomena. That is why, on the one hand, he emphasised similarities to other fields showing how *Synergetics* could be fruitful within this subject, on the other hand, he isolated himself in certain aspects to accent the uniqueness of the synergetic approach. Table 14 provides a survey of differences and similarities with other scientific areas that Hermann Haken himself expressed on different occasions.

Table 14 Differences and similarities of *Synergetics* with other fields of knowledge

Subject	Synergetics
Thermodynamics and information theory[718] - Valid only in thermal equilibrium - Counting occupation numbers, therefore static - Irreversible thermodynamics is only valid near the point of thermal equilibrium	**Synergetics**[718] - Is dealing with systems far from thermal equilibrium - Is also dealing with new phenomena like oscillations that do not exist in thermodynamics - Dynamic theory - Order parameters are different to "Observables" in thermodynamics
Phase transitions - Symmetry breaking, critical "Slowing Down," fluctuations	**Synergetics**[718] - Symmetry breaking, critical "Slowing Down," fluctuations are central features as well - Phase transitions <u>far from</u> thermal equilibrium show more features than <u>in</u> thermal equilibrium: Oscillations, special structures, chaos etc. - Initial conditions and boundary conditions in Synergetics play an important role in phase transitions

[717] (H. Haken & Schiepek, 2006).
[718] (Hermann Haken, 1983), p. 314 – 316.

Table 14 (*continued*)

Cybernetics	**Synergetics**
- Deals with control and stabilisation of existing systems[720]	- Deals with the emergence of new structure hierarchies.[720]
Systems theory	**Synergetics**
- Is searching for general principles[718] - Is looking for isomorphism (in the mathematical sense) between the elements of different systems[719]	- Sets the focus on systems that underwent dramatic transitions[718] - Is not looking for isomorphism of the elements but for isomorphism of the order parameters[720] - Allows for general statements on a large class of systems[718] - Explains and deals with "coherent states"[718] - Complements dynamical systems theory with new elements like "slaving principle," fluctuations, etc.[718]
Bifurcation theory	**Synergetics**
- Does not include noise and fluctuations[718] - Deals only with solutions at the different bifurcation points[718]	- Shows that fluctuations at the bifurcation point are crucial[718] - Investigates the entire stochastic dynamics of the time-dependent order parameters[718] - Gives statements on the stability of the branching and of its pattern development in time[718] - Connects phase transition theory with bifurcation theory[718] - If one neglects fluctuations and omitting time-dependent relaxant solutions, the general concepts and methods of *Synergetics* contain conventional bifurcation theory as special cases[721]
Engineering Sciences	**Synergetics**
- Electrical engineering works with linear structures and networks (cables)[718]	- Deals with different materials[718] - Deals with spatially extended structures[718]

---

719 (Albrecht, 1995).
720 (Hermann Haken, 1999b).
721 (Hermann Haken, 1988a), p. 233.

Table 14 (*continued*)

Chaos theory	**Synergetics**
- Dealing with few degrees of freedom	- Demonstrates the way complex systems are determined by few degrees of freedom (order parameters), and how these few degrees of freedom generate chaotic behaviour. "Chaos theory, in a certain sense, builds upon *Synergetics* and in a certain sense is part of *Synergetics.*"[722] - "In this respect chaos theory, quite rightly, might be seen as a special field of *Synergetics*, because in *Synergetics* the behavior of complex systems is reduced to the behavior of few degrees of freedom, subsequently investigating the behavior of these degrees of freedom"[723]
Catastrophe theory	**Synergetics**
- Requires a so-called "potential function" and is only applicable to Hamiltonian systems[724] - Is a static theory, fluctuations are not known	- In systems far from thermal equilibrium, totally different requirements and systems of equations are necessary[725] - Is a dynamic theory including fluctuations

The advantages and disadvantages of the synergistic multidisciplinary approach also had an impact in its adoption by other scientists:

> "If a scientist is familiar with self-organizing phenomena in his field he may take two positions. Either he says: the regularities identified by synergetics allow me to understand easily another phenomenon in a different field. Or he argues: There's nothing new for me, so what's all about?
> On the other hand, we have to realize that every field has its specialties, for instance in the examination of the properties of the different parts of a system. At that point *Synergetics* is only

722 (Albrecht, 1995), p. 259.
723 (Hermann Haken, 1999b).
724 (Hermann Haken, 1988b), p. 214.
725 (Hermann Haken, 1983).

> of little help. *Synergetics* is more involved in the understanding of macroscopic structures."[726]

In retrospect, Haken's search for unified principles in the sciences matched the "zeitgeist." The term "evolutionary vision," coined by the US-American economist Kenneth E. Boulding,[727] received a tremendous boost from the publication of a book by Erich Jantsch, "*The volutionary vision – Toward a unifying paradigm of physical, biological, and sociocultural evolution*," written in 1981:[728]

> „The evolutionary vision has always been the source of profound inspiration for humanity. In Eastern mysticism and philosophy, especially in Buddhism, it has remained alive over millennia. In Western thinking, it has become temporarily subdued by an emphasis on entities (things).
> [...]
> A scientific foundation of the evolutionary vision had to wait for the emergence of a new self-organization paradigm which constitutes perhaps as the crowning scientific achievement of the 1970s, already recognizable as a great decade for science in many respects."

During the 1970s, the different roads and approaches for a theory of self-organisation started to amalgamate. (See the sketch in Chapter 9). The approaches comprised general systems theory from Ludwig von Bertalanffy, cybernetics of first and second order from Norbert Wiener and Heinz von Foerster, the hyper-cycle theory from Manfred Eigen and Peter Schuster, the theory of dissipative systems from Ilya Prigogine, chaos theory and *Synergetics*. These fields of research all developed in the second half of the twentieth century. Haken's view of this fact is shown in his following statement:[729]

> „In view of the vast variety of different disciplines the possibility to develop some universal approach is certainly not self-evident. Nevertheless I think it is worthwhile to search for and further develop universal approaches. They seem to be the only way to understand or at least to describe our increasingly complex world".

---

726 (Hermann Haken, 1999a).

727 Kenneth E. Boulding (1910 - 1993), was an US-American economist teaching at *Harvard University* (from 1948). Boulding promoted the idea of an "evolutionary economy". (Boulding, 1981). See also (Fontaine, 2010).

728 (Jantsch, 1981), p. 2.

729 (Hermann Haken, 1979), p. 38.

Evaluating the significance of Hermann Haken's scientific achievements, one also has to take a look at his methodology. His way of proceeding can only be understood by taking into account that he was a mathematically educated scientist.[730] This perspective also comprises two concomitant circumstances that are self-evident for every scientist, but nevertheless should be mentioned. There is the plain fact that in every science there are facts and assumptions that are taken for granted, thus, there is no need to mention them. In the respective scientific environment, it can be expected that all other scientists in the field know these assumptions. For a mathematician there is no need to explain that in general the solution of a partial differential equation is not given analytically but has to be found numerically. On the other hand, every scientist is aware of the fact that his or her results depend on many discussions, suggestions and stimulations from co-workers and other colleagues, even his or her opponents. That is why it was natural for Hermann Haken to cite correctly and comprehensively existing literature on the subject under consideration and to quote all sources of information. It seems understandable that he was annoyed when some of his work was not cited correctly, especially from some US-American colleagues.

Hermann Haken was not a "bookman." He looked for the regular face-to-face exchange with colleagues and put emphasise on in person discussions with the members of his institute. He especially promoted his co-workers and students by enabling them to participate in international conferences and meetings.[731] Guest-professorships, invited regularly to his Stuttgart institute, were another important communication vehicle. Another important element of his communication concept was the first hand contact with other scientists. Of course, for the development of *Synergetics,* the eleven *ELMAU Conferences* were central. In addition to these meetings, Hermann Haken was an ardent congress attendee, regularly engaging in the discussions. He did not shy away from presenting his arguments to non-specialists, although sometimes, in the beginning, he was misunderstood, due to the use of different languages in different fields.

Despite these "border crossings," Haken always stayed attached to the scientific field. Political activities, administrative tasks in scientific or public institutions are only rarely to be found.

Haken's scientific methodology manifested itself very early and is distinctly summarised in Friedrich Hund's evaluation of his post-doc dissertation (habilitation):[732]

> "For me the principal value of the scientific work of Mr. Haken seems to be that he tackles the problems in question at the

---

[730] "I would feel much more comfortable, if you could see me as being a normal scientist and not as a person of public interest". (Cited from an email from Hermann Haken to the author.)

[731] This fact is accentuated by many acknowledgements in the dissertations of his students.

[732] Copy of the "second opinion concerning the habilitation work Hermann Haken" (Personal file Haken p. 63; University Archive Stuttgart).

> beginning in a very wide and general manner. Then, he takes the best obtainable general methods, uses a proper tool that allows a deep insight into the structure of the problem and only then gets the solution of the special problem. This approach preserves the connection to vividness".

This rigorous education in the mathematical approach and way of thinking led Hermann Haken to proceed systematically and step by step in his research. First, he looked for the simplest example of a problem. After solving the problem, he then incrementally proceeded to more and more complex applications. Paradigmatically, this can be seen in the development of the laser theory: he started with "simple" rate equations, then proceeded to the development of the semi-classical laser theory, where only the photons (light-field) are quantized. In the next step, he moved to the fully quantized laser theory, including second quantization. Having reached this goal, he again started with the description of one mode, continuing with two modes and thereafter many modes. Only then did he vary different forms of fluctuations and boundary conditions.[733]

Another important aspect of Hermann Haken's methodology is the ability of abstraction, something that also originated from his mathematical education. Looking from a deductive perspective, a mathematical equation is not only a tool for solving a numerical problem, but – in conclusions of analogy – also allows for the assignment to other problems. In scientific heuristic, analogies play a crucial role.[734]

A "deep" analogy is only given, "if it allows for a farther-reaching network between source and target region," as is required by the analogy structure transfer theory of Dedre Gentner.[735] In the natural sciences, for Hermann Haken, these far-reaching structural analogies were given in the case of phase transitions in open systems. That is also the reason for the importance of the laser to him. If we follow the arguments in the synopsis of his former student and colleague Robert Graham, the laser represented a complex many particle system that allowed for a rare casual glance of the microscopic world with its inherent fluctuations and, radiating a coherent laser beam, with the empirically observable macroscopic world.

---

[733] Competing scientists at the time simultaneously made different assumptions that make a historical evaluation very difficult for the non-specialist of the theoretical field. For instance, fluctuation and noise can be treated and applied to the theory in quite different forms mathematically.

[734] For the central role analogies play in the natural sciences see the comprehensive survey of (Hentschel, 2010), including an extensive bibliography.

[735] (Gentner, 1983).

This system could be mathematically (exactly) solved and manifested itself as an open system that, at laser threshold, underwent a phase transition far from thermal equilibrium. These open systems are much more frequent in nature than energetically closed systems and show a much wider spectrum of features: in addition to common symmetry breaking, "critical slowing down," hysteresis and rising fluctuations, in phase transitions in systems far from thermal equilibrium oscillations, competition between modes, self-organisation, auto-catalysis and other phenomena can be detected. For the first time, the idea of generalised thermodynamic potentials describing non-equilibrium phenomena and the "slaving" of fast modes by slow modes (creating the order parameters) has also been demonstrated in laser theory.

In addition, Hermann Haken showed that the laser was the first realistic example for the realisation of deterministic chaos, proposed by the hydrodynamic equations of Edward Lorenz. Moreover, the laser exhibits a hierarchy of phase transitions that could be found elsewhere in other systems, for instance in flow patterns of hydrodynamics. All in all, Graham summarised: [736]

> „The laser can therefore be seen at the crossroads between quantum and classical physics, between equilibrium and non-equilibrium phenomena, between phase-transitions and self-organization and between regular and chaotic dynamics. At the same time, it is a system which we understand, on the basis of the theory initiated by Haken, both on a microscopic quantum mechanical and on a classical macroscopical level. It is a solid ground for discovering general concepts of non-equilibrium physics, and has therefore rightly been called by Haken a 'trailblazer of synergetics'."

Therefore, the laser is a far-reaching and comprehensive analogy model. The more Hermann Haken in his applications of *Synergetics* departed from phenomena in the natural sciences, the more abstract concepts of the theory came to the forefront. Analogy profoundness was still given in the experiments by Kelso concerning movements of the hand (see Chapter 8b), because many attributes and relations could be found, such as phase transition, critical "slowing down," hysteresis or the coordination of subsystems (muscles).

Transferring the methodology of *Synergetics* to the more remote scientific field of psychology, the analogies watered down. Even though in Schiepek's evidence based method managing change processes in psychotherapy[737] close analogies to the original synergetic concept can be found, in other applications of synergetic psychology, one can not help thinking that only mere use of metaphors is

[736] (Robert Graham, 1987), p. 6.

[737] (H. Haken & Schiepek, 2006), especially chapter 5.5. An introduction on an elementary level is given by (Schiepek, Tominschek, Eckert, & Caine, 2007).

expressed in words like self-organisation, phase transition or the spontaneous creation of order after a "critical phase." Haken himself has been fully aware of the limitations of analogy transfer to other scientific fields: [738]

> "Needless to say that once the universal approach exists we can go from one field to another and use the results of one field to promote another field. However, we should never forget limitations of "universal approaches". It is hiqhly dangerous to apply such an approach, if it has worked in a certain domain, to other domains as a dogma. Using any universal approach you must again and again check whether the prepositions made are fulfilled by the objects to which these approaches are applied. Going to more and more abstractions where we must heavily rely on mathematics which, after all, is the Queen of science. "

Nevertheless, this precaution in transferring *Synergetics* from physics to other domains did not obstruct his conviction that "[the principles of *Synergetics*] can be applied to nearly all sciences and technical fields, even to the internal development of science itself."[739] For Hermann Haken, who once described himself as "being a tremendously curious person,"[740] this search for new fields of applications[741] has not come to an end.

---

[738] (Hermann Haken, 1979), p. 38.
[739] See the preface from Haken in (Brunner, Tschacher, & Kenklies, 2010).
[740] Hermann Haken, private communication July 2012.
[741] This is exemplified in the application of synergetics to robotics in his latest book, published in 2012. (Hermann Haken & Levi, 2012).

Erratum to: Hermann Haken and the "Stuttgart School" 1960 – 1970: Their Contribution to the Development of Laser Theory

Erratum to:
Chapter 5 in: B. Kröger, *Hermann Haken: From the Laser to Synergetics*, DOI: 10.1007/978-3-319-11689-1_5

In the original version of the chap. 5, some text were missing in Figs. 6 and 8. The erratum chapter and the book have been updated with the change.

The updated original online version for this chapter can be found at
DOI: 10.1007/978-3-319-11689-1_5

B. Kröger
Tübingen, Germany
e-mail: drberndkroeger@t-online.de

B. Kröger, *Hermann Haken: From the Laser to Synergetics*,
DOI: 10.1007/978-3-319-11689-1_11

Bibliography

Albrecht, H.: Hermann Haken im Gespräch. In: Albrecht, H. (ed.) Forschung und Technik in Deutschland nach 1945, pp. 250–259. Deutscher Kunstverlag, Deutsches Museum Bonn (1995)

Albrecht, H.: Laserforschung in Deutschland 1960 - 1970. Habilitation thesis University of Stuttgart, Stuttgart (1997)

Altner, G.: Die Welt als offenes System - eine Kontroverse um das Werk von Ilya Prigogine. Fischer alternativ, Frankfurt a. M. (1986)

Anderson, P.W.: Gregory Wannier. Physics Today 37(5), 100 (1984)

Andronov, A.A., et al.: Theory of Bifurcations of Dynamic Systems on a Plane. John Wiley, New York (1973)

Arecchi, F., Schulz-Dubois, E.: Laser Handbook. North Holland Publ., Amsterdam (1972)

Arecchi, F.T., Berné, A., Bulamacchi, P.: Higher Order Fluctuations in a Single-Mode Laser Field. Physcal Review Letters 16, 32–35 (1966)

Armstrong, J., Smith, A.: Intensity Fluctuations in a GaaS Laser. Physical Review Letters 14, 68–70 (1965)

Ashburn, E.: Laser Literature - A permuted Bibliography 1958 - 1966. Western Periodicals, North Hollywood (1967)

Ashby, R.W.: Principles of the Self-Organizing Dynamic System. Journal of General Psychology 37, 125–128 (1947)

Aubin, D.: A CulturalHistory of Catastrophes and Chaos: Around the Institut des Hautes Études Scientifiques, France 1958 - 1980. Princeton University, Princeton (1998)

Aubin, D.: Forms of Explanations in the Catastrophe Theory of René Thom: Topology, Morphogenesis, and Structuralism. In: Wise, M.N. (ed.) Growing Explanations: Historical Perspective on Recent Science, pp. 95–130. Duke University Press, Durham (2004)

Aubin, D., Delmedico, A.: Writing the History of Dynamical Systems and Chaos: Longue Durée and Revolution, Disciplines and Culture. Historia Mathematica 29, 1–67 (2002)

Balescu, R., Prigogine, I.: His Life, his Work. Advances in Chemical Physics 135, 1–81 (2007)

Bartz, O.: Der Wissenschaftsrat - Entwicklungslinien der Wissenschaftspolitik in der Bundesrepublik Deutschland 1957 - 2007. Franz Steiner, Stuttgart (2007)

Basar, E., Flohr, H., Mandell, A., Haken, H.: Synergetics of the Brain. Springer, Berlin (1983)

Basov, N., Vul, B., Popov, Y.: Quantum-Mechanical Semiconductor Generators and Amplifiers of Electromagnetic Oscillations. Zh. Eksp. Teor. Fiz. 37, 587 (1959)

Becker, N., Quarthal, F.: Quarthal, Die Universität Stuttgart nach 1945. Geschichte, Entwicklungen, Persönlichkeiten (zum 175. Juniläum der Universität). Jan Thorbecke, Stuttgart (2004)

Bénard, H.: Les tourbillons cellulaires dans une nappe liquide. Rev. Géneral des Sciences Pures et Appl. 11, 1261–1268 (1900)

Bénard, H.: Les tourbillions cellulaires dans une nappe liquide transportant de la chaleur par convection en régime permanent. Ann. Chimie Phys. 23, 62–144 (1901)

Bennett, W.R.: Gaseous Optical Masers. Appl. Opt. Supplement 1, 24–61 (1962)

Berding, C.: Die Entwicklung raumzeitlicher Strukturen in der Morphogenese. Diploma thesis, University of Stuttgart (1981)

Berding, C.: Zur theoretisch-physikalischen Behandlung von Nichtgleichgewichts-Phasenübergängen: die Entwicklung zeitlicher und räumlicher Strukturen in biologischen Systemen. Dissertation, University of Stuttgart (1985)

Bernstein, J.: Three degrees above zero: Bell Laboratories in the information age. Cambridge University Press, Cambridge (1987)

von Bertalanffy, L.: Zu einer allgemeinen Systemlehre. Biologia Generalis 19, 114–129 (1949)

von Bertalanffy, L.: An outline of General Systems Theory. British Journal for the Philosophy of Science 1, 139–164 (1950)

von Bertalanffy, L.: The Theory of Open Systems in Physics and Biology. Science 111, 23–29 (1950)

von Bertalanffy, L.: Biophysik des Fließgleichgewichts - Einführung in die Physik offener Systeme und ihre Anwendung in der Biologie. Braunschweig: F. Vieweg (1953)

von Bertalanffy, L.: The History and Status of General Systems Theory. In: Klir, G.L. (ed.) Trends in General Systems Theory, pp. 21–41. Wiley-Interscience, New York (1972)

Bertolotti, M.: Twenty-five years of the laser - The European contribution to its development. Optica Acta 32, 962–980 (1985)

Bertolotti, M.: Masers and Lasers - An historical Approach. Adam Hilger Ltd., Bristol (1983)

Bertolotti, M.: History of the Laser. Institute of Physics, Bristol (2005)

Bestehorn, M.: Musterbildung beim Bénard-Problem der Hydrodynamik. Diploma thesis, University of Stuttgart (1983)

Bestehorn, M.: Verallgemeinerte Ginzburg-Landau-Gleichungen für die Musterbildung bei Konvektions-Instabilitäten. Dissertation, University of Stuttgart (1988)

Birman, J.L., Cummins, H.Z.: Melvin Lax, 1922 - 2002 (=Biographical Memoirs 87 of the National Academy of Sciences). The National Academic Press, Washington D.C. (2005)

Bloch, F.: Über die Quantenmechanik der Elektronen in Kristallgittern. Zeitschrift für Physik 52, 555–600 (1928)

Bluma, L.: Norbert Wiener und die Entstehung der Kybernetik im Zweiten Weltkrieg: eine historische Fallstudie zur Verbindung von Wissenschaft, Technik und Gesellschaft, Bochum (2005)

Bonifacio, R.: Quantum Statistical theory of superradiation. Physical Review A4, 302–313 (1971)

Borck, C.: Hirnströme: Eine Kulturgeschichte der Elektroenzephalographie. Wallstein, Göttingen (2005)

Boulding, K.E.: Evolutionary Economics. Sage Publications, Beverly Hill (1981)

Brazier, M.: A history of the electrical activity of the brain; the first half-century. MacMillan, New York (1961)

Bromberg, J.L.: The Birth of the Laser. Physics Today (10), 26–33 (1988)

Bromberg, J.L.: The Laser in America, 1950 - 1970. The MIT Press, Cambridge, MasP. (1991)

Brown, G.: Hans Bethe and his physics. World Scientific Books, New Jersey (2006)

Brown, L., Pais, A., Pippard, B.: Twentieth Century Physics, vol. II. Institute of Physics Publishing, Bristol (1995)

Brunner, E.J., Tschacher, W., Kenklies, K. (eds.): Selbstorganisation in Wissenschaft. Edition Paideia, Jena (2010)
Brush, S.G.: Scientist as Historians. Osiris 10, 214–231 (1995)
Bushev, M.: Synergetics - Chaos, Order, Self-Organization. World Scientific, Singapore (1994)
Capra, F.: Das Tao der Physik. O.W.Barth-Verlag, Bern (1977)
Carroll, J.: Todesstrahlen? - Die Geschichte des Laser. Ullstein, Berlin (1964)
Cassidy, D.C.: Uncertainty - the life and science of Werner Heisenberg, 1st pr. edn. Freeman, New York (1992)
Cech, T.R.: RNA as an Enzyme. Scientific American 255, 76–84 (1986)
Chandrasekhar, P.: Hydrodynamic and Hydromagnetic Stability. Clarendon Press, Oxford (1961)
Cohen, L., Scully, M., Scully, R., Lamb Jr., W.E.: Biographical Memoir. National Academy of Sciences, Washington, D.C. (2009)
Conway, F., Siegelman, J.: Dark Hero of the Information Age: in search of Norbert Wiener, the father of cybernetics. Basic Books, New York (2005)
Cormack, A.M.: Early Two-Dimensional Reconstruction and Recent Topics Stemming from It. In: Lindsten, J. (ed.) Nobel Lectures, Physiology or Medicine 1971-1980, pp. 551–563. World Scientific Publ., Singapore (1992)
Corning, P.A.: Synergy and Self-Organization in the Evolution of Complex Systems. Systems Research 12, 89–121 (1995)
Crick, F.: Life itself. Its origin and nature. Simon and Schuster, New York (1981)
Darrigol, O.: From c-numbers to q-numbers, the classical analogy in the history of quantum theory. Univ. of California Press, Berkeley (1992)
Davidson, M.: Uncommon Sense The life and thought of Ludwig von Bertalanffy, Father of General Systems Theory. J. P. Tarcher, Los Angeles (1983)
Dawson Jr., J.W.: Das logische Dilemma. Leben und Werk von Kurt Gödel. Springer, Wien (2007)
De Duve, C.: Ursprung des Lebens. Spektrum Akademischer Verlag, Heidelberg (1994)
Descartes, R.: Discours de la méthode pour bien conduire sa raison, et chercher la verité dans les sciences plus La dioptrique, Les météores et La géométrie, qui sont les essais de cette Méthode. Jean Robert Armogathe. Fayard, Paris (1987)
de Witt, C., Blandin, A., Cohen-Tannoudji, C. (eds.): Quantum Optics and Electronics. Proc. Summerschool Les Houches. Gordon and Breach, New York (1964)
Ditzinger, T., Haken, H.: Oscillations in the Perception of Ambiguous Patterns - a Model Based on Synergetics. Biological Cybernetics 61, 279–287 (1989)
Dress, A., Hendrichs, H., Küppers, G.: Selbstorganisation - Die Entstehung von Ordnung in Natur und Gesellschaft. Piper, München (1986)
Ebert, T. (ed.): Phaidon: Übersetzung und Kommentar. Vandenhoeck & Ruprecht, Göttingen (2004)
Eckert, M.: Sommerfeld und die Anfänge der Festkörperphysik. Wissenschaftliches Jahrbuch des Deutschen Museums 7, 33–71 (1990)
Eckert, M.: Die Atomphysiker. Eine Geschichte der theoretischen Physik am Beispiel der Sommerfeld Schule. Vieweg, Braunschweig (1993)
Edelman, G.M., Tononi, G.: Gehirn und Geist - wie aus Materie Bewusstsein entsteht. Beck, München (2002)
Eigen, M., Schuster, P.: The Hypercycle. Naturwissenschaften 64, 541–565 (1977)
Eigen, M., Schuster, P.: The Hypercycle. Naturwissenschaften 65, 341–369 (1978)
Eigen, M., Winkler-Oswatitsch, R.: Transfer-RNS, an early Gene? Naturwissenschaften 68, 282–292 (1982)
Eigen, M., Winkler, R.: Alkali ion carriers: specificity, architecture and mechanisms. In: Proc. of the Second Int. Conference on Theoretical Physics and Biology, pp. 251–260. Editions du CRNS, Paris (1971)

Eigen, M.: Selforganization of matter and the evolution of biological macromolecules. Naturwissenschaften 58, 465–523 (1971)

Eigen, M., Schuster, P.: The Hypercycle. Springer, Berlin (1979)

Einstein, A.: Zur Quantentheorie der Strahlung. Physikalische Zeitschrift 18, 121–128 (1917)

Feigenbaum, M.: Quantitative Universality for a class of Nonlinear transformations. Journ. Stat. Physics 19, 25–52 (1978)

Fenstermacher, P.R., Swinney, H., Gollub, J.P.: Dynamical Instabilities and the transition to chaotic Taylor Vortex Flow. J. Fluid Mechanics 94, 103–128 (1979)

Field, R.J., Körös, E., Noyes, R.M.: Oscillations in chemical systems. II. Thorough analysis of temporal oscillation in the bromate-cerium-malonic acid system. J. Am. Chem. Soc. 94(25), 8649–8664 (1972)

Fischer, E.P.: Laser. Siedler, München (2010)

Fontaine, P.: Stabilizing American Society: Kenneth Boulding and the Integration of the Social Sciences, 1943-1980. Science in Context 23, 221–265 (2010)

Forbes, J.D.: Biographical account of Professor Louis Albert Necker, of Geneva, Honorary Member of the Royal Society of Edinburgh. Proc. Royal Soc. Edinburgh 5, 53–76 (1863)

Fox, J.: Optical Masers (=Proceedings of the Symposium on optical Masers). Polytec Press, New York (1964)

Frängsmyr, T. (ed.): Nobel Lectures, Physics 1981-1990. World Scientific, Singapore (1993)

Freed, C., Haus, H.A.: Measurement of Amplitude Noise in Optical Cavity Masers. Apl. Physics Letters 6, 85–87 (1965)

Friedrich, R., Fuchs, A., Haken, H.: Spatio-temporal EEG patterns. In: Haken, H., Köpchen, H.P. (eds.) Synergetics of Rhythms in Biological Systems. Springer, Berlin (1992)

Friedrich, R., Haken, H.: Static, Wave-Like, and Chaotic Thermal-Convection in Spherical Geometries. Physical Review, A 34, 2100–2120 (1986)

Friedrich, R., Haken, H.: Time-dependent and chaotic behaviour in systems with O(3)-symmetry. In: Güttinger, W., Dangelmayer, G. (eds.) The Physics of Structure Formation, pp. 334–345. Springer, Berlin (1987)

Friedrich, R., Uhl, C.: Synergetic analysis of human electroencephalograms: Petit-mal epilepsy. In: Friedrich, R., Wunderlin, A. (eds.) Evolution of Dynamical Structures in Complex Systems. Springer, Berlin (1992)

Friedrich, R., Wunderlin, A.: Evolution of Dynamical Structures in Complex Systems. Springer, Berlin (1992)

Friedrich, R.: Höhere Instabilitäten beim Taylor-Problem der Flüssigkeitsdynamik. Diploma thesis, University of Stuttgart (1982)

Fröhlich, H.: Elektronentheorie der Metalle. Springer, Berlin (1936)

Fröhlich, H.: Isotope Effect in Superconductivity. Proc. Roy. Soc. London, A 63, 778 (1950)

Fröhlich, H.: Theory of the Superconducting State I. Phys. Review 79, 845–856 (1950)

Fuller, R.B.: Synergetics - explorations in the geometry of thinking. MacMillan, New York (1975)

Gentner, D.: Structure-mapping: A theoretical framework for analogy. Cognitive Science 7, 155–170 (1983)

Gierer, A., Meinhardt, H.: Biological Pattern Formation Involving Lateral Inhibition. Lectures on Mathematics in Life Science, vol. 7, pp. 163–183 (1974)

Gierer, A., Meinhardt, H.: Theory of Biological Pattern Formation. Kybernetik 12, 30–39 (1972)

Glansdorff, P., Prigogine, I.: Thermodynamic Theory of Structure. Wiley Interscience, New York (1971)

Glauber, R.: Coherence and Quantum Detection. In: Quantum Optics (=Proc. Int. School "E.F." XLII). Academic Press, New York (1969)

Gleick, J.: Chaos - die Ordnung des Universums (Orig.: Chaos - Making a new science). Droemer Knaur, München (1988) (Engl.1987)

Gollub, J.P., Swinney, H.L.: Onset of turbulence in a rotating fluid. Physical Review Letters 35, 927–930 (1975)

Gordon, J.P., Zeiger, H.J., Townes, C.H.: Molecular Microwave Oscillator and new Hyperfine Structure in the Microwave Spectrum of NH3. Physical Review 95, 282–284 (1954)

Graham, R., Haken, H.: Laserlight - First Example of a Second-Order Transition Far Away from Thermal Equilibrium. Zeitschrift für Physik 237, 31–46 (1970)

Graham, R., Haken, H.: Fluctuations and Stability of Stationary Non-Equilibrium Systems in Detailed Balance. Zeitschrift für Physik 245, 141–153 (1971)

Graham, R., Haken, H.: Generalized Thermodynamic Potential for Markoff Systems in Detailed Balance and Far from Thermal Equilibrium. Zeitschrift für Physik 243, 289–302 (1971)

Graham, R., Haken, H., Weidlich, W.: Flux Equilibria in Quantum Systems Far Away from Thermal Equilibrium. Physics Letters, A 32, 129–130 (1970)

Graham, R.: Generalized Thermodynamic Potential for the Convection Instability. Physical Review Letters 31, 1479–1482 (1973)

Graham, R., Haken, H.: Onset of Cooperative Behavior in Nonequilibrium Steady States. In: Nicolis, G., Dewel, G., Turner, J.W. (eds.) Order and Fluctuations in Equilibrium and Non-Equilibrium Statistical Mechanics, pp. 235–288. Reidel, Dordrecht (1981)

Graham, R., Haken, H.: Contributions of Hermann Haken to our understanding of Coherence and Selforganization in Nature. In: Graham, R., Wunderlin, A. (eds.) Lasers and Synergetics, pp. 2–13. Springer, Berlin (1987)

Graham, R., Wunderlin, A.: Lasers and Synergetics (to honor the 60th birthday of Hermann Haken). Springer, Berlin (1987)

Grandin, K. (ed.): Les Prix Nobel. The Nobel Prizes 2005. Nobel Foundation, Stockholm (2006)

Grivet, P., Bloembergen, N. (eds.): Quantum electronics (=Proceedings of the third international Congress). Columbia University Press, New York (1964)

Grossmann, S., Thomae, S.: Invariant distributions and stationary correlation functions of one dimensional discrete processes. Zeitschrift für Naturforschung 33a, 1353–1363 (1977)

Gurel, O., Rössler, O.: Bifurcation theory and its applications in scientific disciplines. New York Academy of Sciences, New York (1977)

Haken, H.: Nonlinear Theory of Laser Noise and Coherence. I. Zeitschrift für Physik 181, 96–124 (1964)

Haken, H.: A Nonlinear Theory of Laser Noise and Coherence. Teil 2. Zeitschrift für Physik 182, 346–359 (1965)

Haken, H., Graham, R.: Synergetik - die Lehre vom Zusammenwirken. Umschau in Wissenschaft und Technik 1971, 191–195 (1971)

Haken, H., Haken-Krell, M.: Erfolgsgeheimnisse der Wahrnehmung. Verlagsanstalt DVA, Stuttgart (1992)

Haken, H., Haken-Krell, M.: Gehirn und Verhalten. Deutsche Verlagsanstalt, Stuttgart (1997)

Haken, H., Kelso, J.A.S., Bunz, H.: A Theoretical-Model of Phase-Transitions in Human Hand Movements. Biological Cybernetics 51, 347–356 (1985)

Haken, H., Koepchen, H.P.: Rhythms in Physiological Systems (Berlin: Springer. Springer, Berlin (1991)

Haken, H., Olbrich, H.: Analytical Treatment of Pattern Formation in Gierer-Meinhardt Model of Morphogenesis. Journal of Mathematical Biology 6, 317–331 (1978)

Haken, H., Sauermann, H.: Frequency Shifts of Laser Modes in Solid State and Gaseous Systems. Zeitschrift für Physik 176, 47–62 (1963)

Haken, H., Sauermann, H.: Nonlinear Interaction of Laser Modes. Zeitschrift für Physik 173, 261–275 (1963)

Haken, H., Schiepek, G.: Synergetik in der Psychologie. Selbstorganisation verstehen und gestalten. Hogrefe Verlag, Göttingen (2006)

Haken, H., Schottky, W.: Allgemeine optische Auswahlregeln in periodischen Kristallgittern. Zeitschrift für Physik 144, 91–107 (1956)

Haken, H., Schottky, W.: Die Behandlung des Exzitons nach der Vielelektronentheorie. Zeitschrift für Physik/Chemie NF 16, 218–244 (1958)

Haken, H., Wagner, M.: Cooperative Phenomena (Festschrift Herbert Fröhlich). Springer, Berlin (1973)

Haken, H., Weidlich, W.: Quantum Noise Operators for N-Level System. Zeitschrift für Physik 189, 1–9 (1966)

Haken, H., Wunderlin, A.: Die Selbstrukturierung der Materie. Synergetik in der unbelebten Natur. Vieweg, Braunschweig (1991)

Haken, H.: Eine modellmäßige Behandlung der Wechselwirkung zwischen einem Elektron und einem Gitteroszillator. Zeitschrift für Physik 135, 408–430 (1953)

Haken, H.: On the problem of radiationless transitions. Physica 20, 1013–1016 (1954)

Haken, H.: Application of Feynmans New Variational Procedure to the Calculation of the Ground State Energy of Excitons. Nuovo Cimento 4, 1608–1609 (1956)

Haken, H.: Berechnung der Energie des Exzitonen-Grundzustandes im polaren Kristall nach einem neuen Variationsverfahren von Feynman.1. Zeitschrift für Physik 147, 323–349 (1957)

Haken, H.: Der heutige Stand der Exzitonen-Forschung in Halbleitern. In: Schottky, W. (ed.) Halbleiterprobleme, pp. 1–48. Vieweg, Berlin (1957)

Haken, H.: Sur la Theorie des Excitons et leur Role dans les Transferts dEnergie a létat solide. Journal De Chimie Physique Et De Physico-Chimie Biologique 55, 613–620 (1958)

Haken, H.: On the Theory of Excitons in Solids. Journal of Physics and Chemistry of Solids 8, 166–171 (1959)

Haken, H.: Theory of Coherence of Laser Light. Physical Review Letters 13, 329–331 (1964)

Haken, H.: Laser Theory (= Handbuch der Physik Band XXV/2c (Flügge, S. (Hg.)). Springer, Berlin (1970)

Haken, H.: Laserlicht - ein neues Beispiel für eine Phasenumwandlung? In: Schottky, W. (ed.) Festkörperprobleme, pp. 351–365. Vieweg, Braunschweig (1970)

Haken, H.: Introduction to Synergetics. In: Haken, H. (ed.) Synergetics - Cooperative Phenomena in Multicomponent Systems (Proc. Schloß Elmau 1972). Teubner-Verlag, Stuttgart (1973)

Haken, H.: Quantentheorie des Festkörpers. Teubner Verlag, Stuttgart (1973)

Haken, H.: Stability and Fluctuations of Multimode Configurations near Convection Instability. Physics Letters A A 46, 193–194 (1973)

Haken, H.: Synergetics - Cooperative Phenomena in Multicomponent Systems (=Proc. Conf. Schloss Elmau 1972). Teubner, Stuttgart (1973)

Haken, H.: Cooperative Effects - Progress in Synergetics. North Holland, Amsterdam (1974)

Haken, H.: Analogy between higher instabilities in fluids and lasers. Physics Letters 53A, 77–78 (1975)

Haken, H.: Cooperative Phenomena in Systems Far from Thermal Equilibrium and in Nonphysical Systems. Reviews of Modern Physics 47, 67–121 (1975)

Haken, H.: Generalized Ginzburg-Landau Equations for Phase Transition-Like Phenomena in Lasers, Nonlinear Optics, Hydrodynamics and Chemical-Reactions. Zeitschrift für Physik B 21, 105–114 (1975)

Haken, H.: Higher-Order Corrections to Generalized Ginzburg-Landau Equations of Nonequilibrium Systems. Zeitschrift für Physik B 22, 69–72 (1975)

Haken, H.: Statistical Physics of a Chemical-Reaction Model. Physics Letters A A 51, 125–126 (1975)

Haken, H.: Statistical Physics of Bifurcation, Spatial Structures and Fluctuations of Chemical Reactions. Zeitschrift für Physik B20, 413–420 (1975)

Haken, H.: Synergetics - An Introduction Nonequilibrium Phase Transitions and Self-Organization in Physics. Springer, Berlin (1977)

Haken, H.: The Laser - Trailblazer of Synergetics. In: Mandel, L., Wolf, E. (eds.) Coherence and Quantum Optics (= Proc. of the 4th Rochester Conference), pp. 49–62. Plenum Press, New York (1978)

Haken, H.: Synergetics: Some recent trends and developments. Suppl. of the Report of Progress of Theoretical Physics 64, 21–34 (1978)

Haken, H.: Pattern Formation and Pattern Recognition - An Attempt at a Synthesis. In: Pattern Formation by Dynamic Systems and Pattern Recognition. Springer, Berlin (1979)

Haken, H.: Synergetics and a new approach to bifurcation theory. In: Güttinger, W., Eikemeier, H. (eds.) Structural Stability in Physics, pp. 31–40. Springer, Berlin (1979)

Haken, H.: Synergetics and Bifurcation Theory. Annals of the New York Academy of Sciences 316, 357 (1979)

Haken, H.: Licht und Materie Band 1: Elemente der Quantenoptik. BI Wissenschaftsverlag, Mannheim (1979) (2nd edn. 1989)

Haken, H.: Selforganization through increase of number of components. Supplement of the Progress of Theoretical Physics 69, 30–40 (1980)

Haken, H.: Erfolgsgeheimnisse der Natur: Synergetik - die Lehre vom Zusammenwirken. Deutsche Verlagsanstalt DVA, Stuttgart (1981)

Haken, H.: Synergetik - Eine Einführung. Springer, Berlin (1982)

Haken, H.: Advanced Synergetics - Instability Hierarchies of Self-Organizing Systems and Devices. Springer, Berlin (1983)

Haken, H.: Von der Laserphysik zur Synergetik. In: Schneider, C. (ed.) Forschung in der Bundesrepublik Deutschland. Im Auftrag der Deutschen Forschungsgemeinschaft DFG, pp. 515–518. Verlag Chemie, Weinheim (1983)

Haken, H.: Nachruf auf Professor Serge Nikitine. Physikalische Blätter 42, 11 (1986)

Haken, H.: Computational Systems - Natural and Artificial. Springer Series in Synergetics Band 38. Springer, Berlin (1987)

Haken, H.: Synergetik und ihre Anwendung auf psychosoziale Probleme. In: Stierlin, H., Simon, F.B., Schmidt, G. (eds.) Familiäre Wirklichkeiten. Der Heidelberger Kongreß, pp. 36–50. Klett Cotta, Stuttgart (1987)

Haken, H.: Entwicklungslinien der Synergetik. Teil 1. Naturwissenschaften 75, 163–172 (1988)

Haken, H.: Entwicklungslinien der Synergetik. Teil 2. Naturwissenschaften 75, 225–234 (1988)

Haken, H.: Geschichte der Synergetik. In: Niedersen, U. (ed.) Komplexität - Zeit - Methode, vol. 3, pp. 198–231 (1988)

Haken, H.: Information and Self-Organization. A Macroscopic Approach to Complex Systems. Springer, Berlin (1988)

Haken, H.: Neural and Synergetic Computers. Springer Series in Synergetics Band 42. Springer, Berlin (1988)

Haken, H.: Synergetics: an overview. Report on Progress in Physics, 515–553 (1989)

Haken, H.: Synergetischer Computer - ein nichtbiologischer Zugang. Physikalische Blätter 45, 168–171 (1989)

Haken, H.: Offene Systeme - die merkwürdige Welt des Nichtgleichgewichts. Physikalische Blätter 46, 203–208 (1990)

Haken, H.: Synergetics as a tool for the conceptualization and mathematization of Cognition and Behaviour - How far can we go? In: Haken, H., Stadler, M. (eds.) Synergetics of Cognition, pp. 2–31. Springer, Berlin (1990)

Haken, H.: Je mehr wir Grenzen ausloten, um so mehr erfahren wir vom Menschen - hoffentlich. In: Oesterreicher-Mollwo, M. (ed.) Was uns bewegt: Naturwissenschaftler sprechen über sich und ihre Welt, pp. 186–193. Beltz, Weinheim (1991)

Haken, H.: Synergetik und Naturwissenschaften (anschl. Kritiken von 33 Wissenschaftlern und Replik hierauf von H. Haken S. 658 - 675). Ethik und Naturwissenschaften 7, 587–594 (1996)

Haken, H.: Synergetik: Vergangenheit, Gegenwart, Zukunft. In: Mainzer, K. (ed.) Komplexe Systeme und nichtlineare Dynamik in Natur und Gesellschaft. Komplexitätsforschung in Deutschland auf dem Weg ins nächste Jahrhundert, pp. 30–48. Springer, Berlin (1999)

Haken, H.: Über Beziehungen zwischen der Synergetik und anderen Disziplinen. In: Eisenhardt, P. (ed.) Der Weg der Wahrheit. Aufsätze zur Einheit der Wissenschaftsgeschichte, pp. 167–173. Olms, Hildesheim (1999)

Haken, H.: Nel Senso della Sinergetica (Autobiography in Italian). Renzo Editore, Rom (2005)

Haken, H., Haken-Krell, M.: Entstehung von Biologischer Information und Ordnung. Wissenschaftliche Buchgesellschaft, Darmstadt (1989)

Haken, H., Levi, P.: Synergetic Agents: From Multi Robot-Systems to Molecular Robotics. Wiley-VCH, Weinheim (2012)

Haken, H., Wunderlin, A.: New Interpretation and size of strange attractor of the Lorenz model of turbulence. Physics Letters 133, 133–134 (1977)

Handel, K.: Historische Entwicklung der mikroskopischen Theorie der Supraleitung. Diploma thesis IGN University of Hamburg, Hamburg (1994)

Haken, H.: Research styles in particle theory and solid state theory: the historical development of the microscopic theory of superconductivity. In: The Emergence of Modern Physics, pp. 371–386. Universita degli Studi di Pavia, Pavia (1996)

Haken, H.: Anfänge der Halbleiterforschung und - entwicklung. Dargestellt an Biographien von vier deutschen Halbleiterpionieren. Dissertation RWTH Aachen, Aachen (1999)

Hankins, T.L.: In Defence of Biography: the use of biography in the history of science. History of Science 17, 1–16 (1979)

Haus, H.A.: The Measurement of G and its Signal-to-Noise Ratio. In: Glauber, R. (ed.) Quantum Optics, Academic Press, New York (1969)

Hebb, D.: The organization of behavior. Wiley, New York (1949)

Hecht, J.: Beam the race to make the laser. Oxford University Press, Oxford (2005)

Hedrich, R.: Die Entdeckung der Komplexität: Skizzen einer strukturwissenschaftlichen Revolution. Harri Deutsch, Thun - Frankfurt a. M. (1994)

Heilbron, J.L.: Applied History of Science. ISIS 78, 552–563 (1987)

Heims, S.: Constructing a Social Science for Postwar America. The Cybernetics Group, 1946–1953. MIT Press, Cambridge, Mass. (1993)

Hemminger, H.: Über Glaube und Zweifel – Das New Age in der Naturwissenschaft. In: Hemminger, H.-J. (ed.) Die Rückkehr der Zauberer: New Age – eine Kritik, pp. 115–185. Rowohlt, Reinbek (1987)

Hempstedt, R., Lax, M.: Classical Noise. VI. Noise in Self-Sustained Oscillators near Threshold. Physical Review A 161, 350–366 (1967)

Hentschel, K.: Interpretationen und Fehlinterpretationen der speziellen und der allgemeinen Relativitätstheorie durch Zeitgenossen Albert Einsteins. Birkhäuser, Basel (1990)

Hentschel, K.: Finally, Some Historical Polyphony! In: Dörries, M. (ed.) Michael Frayns Copenhagen in Debate, pp. 31–37. University of California, Berkeley (2005)

Hentschel, K.: Die Funktion von Analogien in den Naturwissenschaften, auch in Abgrenzung zu Metaphern und Modellen. Acta Historica Leopoldina 56, 13–66 (2010)

Hentschel, K. (ed.): Zur Geschichte von Forschungstechnologien: Generizität - Interstitialität - Transfer. GNT-Verlag, Diepholz (2012)

Hentschel, K., Hentschel, A.: Physics and National Socialism - An Anthology of Primary Sources. Birkhäuser, Basel (1996)

Hermann, A.: Frühgeschichte der Quantentheorie (1899-1913). Physik-Verlag (1969)

Heuser-Keßler, M.-L.: Die Produktivität der Natur - Schellings Naturphilosophie u.d. neue Paradigma d. Selbstorganisation in d. Naturwissenschaften. Duncker und Humblot (1986)

Hoddeson, L., Braun, E., Teichmann, J., Weart, S.: Out of the Crystal Maze. Chapters from the History of Solid-State Physics. Oxford University Press, Oxford (1992)

Hoddeson, L., Daitch, V.: True genius: the life and science of John Bardeen. The John Henry Press, Washington D. C. (2002)

Holton, G.J.: Thematic origins of scientific thought, Kepler to Einstein. Harvard Univ. Press, Cambridge (1973)

Hopfield, J.J.: Neural networks and physical systems with emergent collective computational abilities. Proc. Nat. Acad. Sci. 79, 2554–2558 (1982)

Hounsfield, G.N.: Computed Medical Imaging. In: Lindsten, J. (ed.) Nobel Lectures, Physiology or Medicine 1971-1980, pp. 568–586. World Scientific Publ, Singapore (1992)

Hund, F.: Geschichte der Quantentheorie, 3rd edn. Bibliographisches Institut, Mannheim (1984)

Husbands, P., Holland, O.: Warren McCulloch and the British Cyberneticians. Interdisciplinary Science Reviews 37, 237–253 (2012)

Hyland, G.: Herbert Fröhlich FRS, 1905 - 1991: A physicist ahead of his time. In: Hyland, G. (ed.) Herbert Fröhlich FRS - A physicist ahead of his time, pp. 247–344. The Universitys of Liverpool, Liverpool (2006)

Institution, The Royal, Faraday Symposium - The Physical Chemistry of Oscillatory Processes. Journal Chem. Soc., Faraday Transactions, London (1975)

Jammer, M.: The conceptual development of quantum mechanics, 2nd edn. Tomash Publ., Los Angeles (1989)

Jantsch, E.: The Evolutionary Vision - Towards a unifying paradigm of physical, biological and sociocultural evolution (AAAS Selected Symposium 61). Westview Press, Boulder (1981)

Javan, A., Bennett, W., Herriott, D.: Population Inversion and Continuous Optical Maser Oscillation in a Gas Discharge Containing a He-Ne Mixture. Physical Review Letters 6, 106–110 (1961)

Jaynes, E.T.: Information Theory and Statistical Mechanics. Physical Review 106, 620–630 (1957)

Jiu-Li, L., Broeck, C., Nicolis, G.: Stability criteria and fluctuations around nonequilibrium states. Zeitschrift für Physik B Condensed Matter 56, 165–170 (1984)

Joseph, D.: Elementary Stability and Bifurcation Theory. Springer, Berlin (1980)

Judson, H.F.: The eighth day of creation, makers of the revolution in biology. Cold Spring Harbor Laboratory Press, Cold Spring Harbor, NY (1996)

Kamal, A.: Laser Abstracts. Plenum Press, New York (1964)

Kant, I.: Kritik der Urteilskraft, Berlin (1790)

Kauffman, S.: At Home in the Universe, the search for laws of complexity. Oxford University Press, Oxford (1995)

Keizer, J., Fox, R.: Qualms Regarding the Range of Validity of the Glansdorff-Prigogine Criterion for Stability of Non-Equilibrium States. Proc. Nat. Acad. Sci. USA 71, 192–196 (1974)

Keizer, J.: Fluctuations, stability, and generalized state functions at nonequilibrium steady states. J. Chem. Physics 65, 4431–4445 (1976)

Keller, E.F.: Organisms, Machines, and Thunderstorms: A History of Self-Organization, Part One. Stud. in the Natural Sciences 38, 45–75 (2008)

Keller, E.F.: Organisms, Machines, and Thunderstorms: A History of Self-Organization, Part Two: Complexity, Emergence, and Stable Attractors. Hist. Stud. in the Natural Sciences 39, 1–31 (2009)

Kellert, S.H.: A philosophical evaluation of the chaos theory "revolution". Proceedings of the Biennial Meeting of the Philosophy of Science Association, pp. 33–49 (1992)

Kelly, P., Lax, B., Tannenwald, P. (eds.): Proceedings of the Physics of Quantum Electronics Conference. McGraw Hill, New York (1966)

Kelso, J.A.S., Schöner, G., Scholz, J.P., Haken, H.: Nonequilibrium Phase-Transitions in Coordinated Movements Involving Many Degrees of Freedom. Annals of the New York Academy of Sciences 504, 293–296 (1987)

Klenk, H.: Anwendung der verallgemeinerten Ginzburg-Landau Gleichungen auf die Theorie der Bénard-Instabllität eines Plasmas. Dissertation, University of Stuttgart (1979)

Koschmieder, E.L.: Instabilities in Fluid Dynamics. In: Haken, H. (ed.) Synergetics - A Workshop, pp. 70–79. Springer, Berlin (1977)

Kragh, H.: An Introduction to the Historiographie of Science. Cambridge University Press, Cambridge (1987)

Kragh, H.: Quantum Generations - A History of Physics in the Twentieth Century. Princeton University Press, Princeton (1994)

Krohn, W., Küppers, G.: Die Selbstorganisation der Wissenschaft. Kleine Verlag, Bielefeld (1987)

Krohn, W., Krug, H.-J., Küppers, G.: Konzepte von Chaos und Selbstorganisation in der Geschichte der Wissenschaften. Uwe Niedersen, Selbstorganisation - Jahrbuch für Komplexität in den Natur-, Sozial- und Gesiteswissenschaften Band 2. Duncker & Humblot, Berlin (1992)

Krohn, W., Küppers, G., Paslack, R.: Selbstorganisation - Zur Genese und Entwicklung einer wissenschaftlichen Revolution. In: Der Diskurs des Radikalen Konstruktivismus, pp. 441–465. Suhrkamp, Frankfurt (1987)

Kubo, R., Kamimura, H.: Dynamical Processes in Solid State Optics (1966 Tokyo Summer Lectures in Theoretical Physics). W.A. Benjamin and Syokabo, Tokyo and New York (1967)

Küppers, B.O.: Ordnung aus dem Chaos. Piper, München (1987)

Lamb, W., Schleich, W., Scully, M., Townes, C.: Laser physics: Quantum controversy in action. Reviews of Modern Physics 71, 263–273 (1999)

Lamb, W.: Quantum Mechanical Amplifiers. In: Lectures in Theoretical Physics, vol. 2, Interscience Publishers, New York (1960)

Lamb, W.E.: Theory of optical masers. Physical Review 134, A1429–A1450 (1964)

Landauer, R.: Fluctuations in Bistable Tunnel Diode Circuits. Journal of Applied Physics 33, 2209–2216 (1962)
Landauer, R.: Poor Signal to Noise Ratio in Science. In: Dynamic Patterns in Complex Systems. World Scientific, Singapore (1988)
Lax, M., Louisell, W.H.: Quantum Fokker-Planck Solution for Laser Noise. IEEE Journal of Quantum Electronics QE3, 47–58 (1967)
Lax, M.: Quantum Noise V: Phase Noise in a Homogeneously Broadenend Maser. In: Kelley, P., et al. (eds.) Proceedings of the Physics on Quantum Electronic Conference. McGraw Hill, New York (1966)
Lax, M.: Quantum Noise VII: The Rate Equations and Amplitude Noise in Lasers. IEEE Journal of Quantum Electronics 3, 37–46 (1967)
Lax, M.: Quantum Noise. X. Density-Matrix Treatment of Field and Population-Difference Fluctuations. Physical Review 157, 213–231 (1967)
Lax, M.: Fluctuations and Coherence Phenomena in classical and quantum physics. In: Statistical Physics, Phase Transition and Superfluidity, vol. 2. Gordon Breach, New York (1968)
Leiber, W.: On the impact of deterministic chaos on modern science and philosophy of science: Implications for the philosophy of technology? Society for Philosophy and Technology (1998)
Lemmerich, J.: Zur Geschichte der Entwicklung des Lasers. D.A.V.I.D, Berlin (1987)
Lewin, R.: Complexity. Life at the Edge of Chaos. Macmillan Pub., New York (1992)
Liapunov, A.: Problème général de la stabilité du mouvement (=französische Übersetzung einer 1892 auf russisch erschienenen Ausgabe). Princeton University Press, Princeton (1907); Reprint (1949)
Libchaber, A., Maurer, J.: Une Experience de Rayleigh-Benard en geometrie reduite: multiplication, accrochage et demultiplication des frequences. Journal de Physique, Colloques 41-C3(41), 51–56 (1980)
Lifson, S.: Chemical selection, diversity, teleonomy and the second law of thermodynamics - Reflections on Eigens theory of self-organization of matter. Biophysical Chemistry 26, 303–311 (1987)
Little, W.A.: The existence of persistent states in the brain. Mathematical Biosciences 19, 101–120 (1974)
Lorenz, E.: On the prevalence of aperiodicity in simple systems. In: Global Analysis. Lecture Notes in Mathematics, vol. 755. Springer, Berlin (1979)
Lorenz, E.N.: Deterministic Nonperiodic Flow, Journ. Atmosph. Sciences 20, 130–141 (1963)
Lorenz, E.N.: Irregularity: a fundamental property of the atmosphere. Tellus 36A, 98–110 (1984)
Lotka, A.J.: Elements of Physical Biology (Nachdruck als: Elements of Mathematical Biology). Williams and Wilkins (Reprint: Dover), Baltimore (Reprint: New York) (1925)
Louisell, W.H.: Quantum statistical properties of radiation. John Wiley, New York (1973)
Lovelock, J.: Gaia - a new look at life on Earth. Oxford University Press, Oxford (1979)
Madelung, O.: Schottky - Spenke - Welker. Physikalische Blätter 55(6), 54–58 (1999)
Maiman, T.: Stimulated Optical Radiation in Ruby. Nature 118, 493–494 (1960)
Maiman, T.H.: The Laser Odyssey. Laser Press, Blaine (2000)
Mandel, L., Wolf, E.: Coherence and Quantum Optics (=Proc. of the 4th Rochester Conference). Plenum, New York (1978)
Mandell, A.J.: Coming of the Middle Age. Summit Books, New York (1977)
Marage, P., Wallenborn, G. (eds.): The Solvay Councils and the Birth of modern Physics (=Science Networks - Historical Studies Vol. 22). Birkhäuser, Basel (1999)

Marois, M.: Theoretical Physics and Biology (Proc. of the first Intern. Conference, Versailles, June 26-30, 1967). North Holland, Amsterdam (1969)
Marois, M.: From Theoretical Physics to Biology. In: Proc. of the Second Intern. Conf., June 30-July 5, 1969. Editions CNRS, Paris (1971)
Marois, M.: From theoretical Physics to biology. In: Proc. of the Third Intern. Conf., June 21-26, 1971. Karger, Basel (1973)
Marois, M.: From theoretical physics to biology. In: Proc. of the Fourth Intern. Conf., May 28- June 2, 1973. North Holland Publ., Amsterdam (1976)
Marois, M.: Documents for History - Life and Human Destiny. The Institut de la Vie; Editons Rive Droite, Paris (1997)
Marois, M.: Documents pour lhistoire - Tome II Les Grandes Conférences internationales: Problèmes de vie. LInstitut de la Vie; Editions Rive Droite, Paris (1997)
Marois, M.: Institut de la Vie - Documents pour lhistoire. Editions Rive Droite, Paris (1998)
Martin, P.C.: Instabilities, Oscillations, and Chaos. Journal de Physique 37(1), 1–57 (1976)
Marx, K.: Analytische und numerische Behandlung der zweiten Instabilität beim Taylor-Problem der Flüssigkeitsdynamik. Dissertation, University of Stuttgart (1987)
Maturana, H.: Biology of Cognition (Biological Computer Laboratory Research Report BCL 9.0). University of Illinois, Urbana (1970)
Maurer, J., Libchaber, A.: A Raleigh Bénard Experiment: Helium in a small box. Journal de Physique Lettres 40, 419–423 (1979)
McCulloch, W.S., Pitts, W.: A Logical Calculus of the Ideas Immanent in Nervous Activity. Bull. Math. Biophysics 5, 115–133 (1943)
McFarlane, R., Bennett, R., Lamb, W.: Single Mode Tuning Dip in the Power Output of an He-Ne Optical Maser. Appl. Phys. Lett. 2, 189–190 (1963)
McLaughlin, J.B., Martin, P.C.: Transition to Turbulence of a Statically, Stressed Fluid. Physical Review Letters 33, 1189–1192 (1974)
Mehra, J.: The Solvay Conferences on Physics. D. Reidel, Dordrecht (1975)
Mehra, J.: Einstein, Hilbert, and the theory of gravitation, historical origins of general relativity theory. Reidel, Dordrecht (1974)
Mehra, J., Rechenberg, H.: The historical development of quantum theory, 9 vols., p. 1982. Springer (9 Bände), New York (1982)
Meinhardt, H.: The Spatial Control of Cell Differentiation by Autocatalysis and Lateral Inhibition. In: Haken, H. (ed.) Synergetics, a Workshop, pp. 214–223. Springer, Berlin (1977)
Mesarovic, M.D., Pestel, E.: Mankind at the turning Point: the 2. report to the Club of Rome. Hutchinson, London (1975)
Mesarovich, M.D., Takahara, Y.: General Systems Theory: Mathematical Foundations. Academic Press, New York (1975)
von Meyenn, K.: Die Biographie in der Physikgeschichte. In: von Meyenn, K. (ed.) Die Großen Physiker, pp. 7–15. C.H. Becksche Verlagsbuchhandlung, München (1997)
Meyer, K.P., Brändli, H.P., Dändliker, R. (eds.): Proceedings of the International Symposium on Laser-Physics and Applications (=Zs. f. angewandte Math. und Physik, 16 (1965), pp. 1–184. Birkhäuser, Basel (1965)
Meyers, R.A.: Encyclopedia of Complexity and System Science, 10 vols. Springer, Heidelberg (2009)
Miles, P. (ed.): Quantum electronics and Coherent Light (=Proc. XXXI Int. School "Enrico Fermi"). Academic Press, New York (1964)
Miller, A.I.: Albert Einsteins special theory of relativity, emergence (1905) and early interpretation (1905 - 1911). Addison-Wesley, Reading (1981)

Millman, S.(Hg.): A History of Engineering and Science in the Bell System - Physical Sciences (1925 - 1980). AT&T Bell Laboratories, o.O. (1983)
Mittelstraß, J.: Enzyklopädie Philosophie und Wissenschaftstheorie (6 Bände), 2nd edn., 6 vols. Metzler, Stuttgart (2005)
Monod, J.: Le hasard et la necessité. Ed. du Seuil, Paris (1970)
Monod, J.: Zufall und Notwendigkeit. Piper Verlag, München (1971)
Müller, A.: Eine kurze Geschichte des BCL - Heinz von Foerster und das Biological Computer Laboratory. Österreichische Zeitschrift für Geschichtswissenschaften 11, 9–30 (2000)
Müller, A., Müller, K.H. (eds.): An Unfinished Revolution? Heinz von Foerster and the Biological Computer Laboratory 1958 - 1976. Verlag Echoraum, Wien (2007)
Mutschler, H.-D.: Physik – Religion – New Age. Echter, Würzburg (1990)
Nicolaysen, R.: Der lange Weg zur VolkswagenStiftung. Eine Gründungsgeschichte im Spannungsfeld von Politik, Wirtschaft und Wissenschaft. Vandenhoeck & Ruprecht, Göttingen (2002)
Nicolis, G., Dewel, G., Turner, J.W.: Order and Fluctuations in Equilibrium and Non-Equilibrium Statistical Mechanics. John Wiley, New York (1981)
Nicolis, G., Prigogine, I.: Self-Organization in Non-Equilibrium Systems, New York (1977)
Nye, M.J.: Scientific Biography: History of Science by Another Means? Isis 97, 322–329 (2006)
Olbricht, H.: Die Bildung von raumzeitlicher Strukturen infolge von Mehrfachinstabilitäten. Diploma thesis, University of Stuttgart (1977)
Olby, R.C.: The path to the double helix. Macmillan, London (1974)
Packard, N., Crutchfield, J., Shaw, R., Farmer, J.: Geometry from a time series. Phys. Review Letters, 712–716 (1980)
Pais, A.: Raffiniert ist der Herrgott... Albert Einstein; eine wissenschaftliche Biographie. Vieweg, Braunschweig (1986)
Pal, L.: Statistical Physics - Proceedings of the International Conference. North Holland Publ., Budapest (1975)
Paslack, R.: Urgeschichte der Selbstorganisation: zur Archäologie eines wissenschaftlichen Paradigmas. Vieweg, Braunschweig (1991)
Paslack, R., Knost, P.: Zur Geschichte der Selbstorganisationsforschung - Ideengeschichtliche Einführung und Bibliographie (1940 - 1990). Kleine Verlag, Bielefeld (1990)
Peitgen, H.-O., Richter, P.H.: The Beauty of Fractals, Berlin (1986)
Perny, G.: Centenaire de la Naissance de LHumaniste alsacien Alfred Kastler (1902 - 1984): Aspects de son oeuvre. In: Annuaire (ed.) Sélestat: Société des amis de la bibliothèque de Sélestat, pp. 149–155 (2003)
Peyenson, L.: Who the guys were: Prosopography in the History of Science. History of Science 15, 155–188 (1977)
Pias, C.: Cybernetics | Kybernetik. The Macy-Conferences 1946-1953, 2 vols., pp. 1946–1953. diaphanes, Zürich (2003)
Poincaré, H.: Science et Méthode. Flammarion, Paris (1908)
Polanyi, M.: Personal knowledge - towards a post-critical philosophy. Routledge & Kegan Paul, London (1958)
Pöppel, E.: Grenzen des Bewußtseins. Über Wirklichkeit und Welterfahrung. dtv, München (1987)
Pouvreau, D.: Une Biographie non officielle de Ludwig von Bertalanffy (1901-1972) (2006), `http://www.bertalanffy.org` (accessed January 6, 2013)
Prigogine, I.: Vom Sein zum Werden. Zeit und Komplexität in den Naturwissenschaften. Piper, München (1979)

Prigogine, I., Nicolis, G.: On symmetry-breaking instabilites in dissipative systems. J. Chem. Physics 46, 3542–3550 (1967)

Prigogine, I.: Introduction to Thermodynamics of Irreversible Processes. Interscience Publishers, New York (1955)

Prigogine, I.: Non-Equilibrium statistical Mechanics. Interscience Publishers, New York (1962)

Prigogine, I.: Physique, temps et devenir. Masson, Paris (1980)

Prigogine, I.: Autobiography. In: Frängsmyr, T., Forsen, S. (eds.) Nobel Lectures, Chemistry 1971-1980. World Scientific Pub., Singapore (1993)

Prigogine, I., Rice, S. (eds.): Proc. of the Conference on Instability and dissipative Structures. Advances in Chemical Chemistry, vol. 32 (1973)

Prigogine, I., Stengers, I.: La nouvelle alliance, métamorphose de la science. Gallimard, Paris (1979)

Prigogine, I., Stengers, I.: Das Paradox der Zeit, Zeit, Chaos und Quanten. Piper, München (1993)

Prokhorov, A.: Quantum Radiation. In: Nobel Lectures, Physics 1963 - 1970, pp. 110–116. Elsevier, Amsterdam (1972)

Risken, H., Schmid, C., Weidlich, W.: Fokker-Planck equation for atoms and light mode in a laser model with quantum mechanically determined dissipation and fluctuation coefficients. Physics Letters 20, 489–491 (1966)

Prigogine, I., Stengers, I.: Fokker-Planck Equation, Distribution and Correlation Functions for Laser Noise. Zs. für Physik 194, 337–359 (1966)

Risken, H., Schmid, C., Weidlich, W.: Fokker-Planck equation, distribution and correlation functions for laser noise. Zs. für Physik 193, 37–51 (1966)

Prigogine, I., Stengers, I.: Quantum Fluctuations, Master Equation and Fokker-Planck Equation. Zs. für Physik 193, 37–51 (1966)

Risken, H.: Distribution- and Correlation-Functions for a Laser Amplitude. Zs. für Physik, 186 (1965), 85 - 98.

Röss, D.: Laser - Lichtverstärker und -Oszillatoren (incl. bibliography with 3140 numbers). Akademische Verlagsgesellschaft, Frankfurt (1966)

Rubin, E.: Visuell wahrgenommene Figuren. Studien in psychologischer Analyse (Aus dem Dänischen übersetzt nach "Synsoplevede Figurer", 1915). Gyldendalske Boghandel, Kopenhagen (1921)

Ruelle, D., Takens, F.: On the nature of turbulence. Comm. Math. Physics 20, 167–192 (1971)

Rupieper, H.-J. (ed.): Beiträge zur Geschichte der Martin-Luther-University of 1502–2002. Mitteldeutscher Verlag, Halle (2002)

Ruß, H.: Der neue Mystizismus. Östliche Mystik und moderne Naturwissenschaft im New Age - Denken. Königshausen & Neumann, Würzburg (1993)

Sargent III, M.: A Note on Semiclassical Laser Theory. Optics Communications 11, I–III (1974)

Sargent III, M., Scully, M., Lamb, W.: Laser Physics. Addison Wesley, Reading (1974)

Sattinger, D.: Topics in Stability and Bifurcation Theory. Springer, Berlin (1973)

Sauermann, H.: Dissipation und Fluktuationen in einem Zwei-Niveau-System (Dissertation). Zs. für Physik 188, 480–505 (1965)

Sauermann, H.: Quantenmechanische Behandlung des optischen Maser (Dissertation). Zs. für Physik 189, 315–334 (1965)

Schawlow, A.L.: From Maser to Laser. In: Impact of Basic Research on Technology. Plenum Press, New York (1973)

Schawlow, A., Townes, C.: Infrared and Optical Masers. Physical Review 112, 1940–1949 (1958)

Schiepek, G., Tominschek, I., Eckert, H., Caine, C.: Monitoring - Der Psyche bei der Arbeit zuschauen. Psychologie Heute, 42–47 (2007)

Schleich, W., Vollmer, H.D.: Zum Tode von Hannes Risken. Physikalische Blätter 50, 469 (1994)

Schleich, W., Walther, H., Lamb, W.: Ode to a Quantum Physicist - A Festschrift in Honor of Marlan O. Scully. Elsevier, Amsterdam (2000)

Schmidt, S.J.: Der Radikale Konstruktivismus: Ein neues Paradigma im interdisziplinären Diskurs. In: Schmidt, S.J. (ed.) Der Diskurs des Radikalen Konstruktivismus, pp. 11–87. Suhrkamp, Frankfurt a. M. (1987)

Schröder, M. (ed.): Hundert Jahre Friedrich Hund: Ein Rückblick auf das Wirken eines bedeutenden Physikers. Nachrichten der Akademie der Wissenschaften in Göttingen, Göttingen (1996)

Schrödinger, E.: What is life? Cambridge University Press, Cambridge (1944)

Schweber, S.: Nuclear Forces. The making of the physicist Hans Bethe. Harvard University Press, Harvard (2012)

Scully, M., de Giorgio, V.: Analogy between the Laser Threshold Region and a Second-Order Phase Transition. Physical Review A2, 1170–1177 (1970)

Scully, M., Lamb, W.: Quantum theory of an optical maser. Physical Review Letters 16, 853–855 (1966)

Scully, M.: The Quantum Theory of a Laser. In: Quantum Optics (Proc. Int. School “Enrico Fermi” XLII). Academic Press, New York (1969)

Seeger, A.: Sogar theoretische Physik kann praktisch sein! - Ulrich Dehlinger. In: Die University of Stuttgart nach 1945. Jan Thorbecke, Stuttgart (2004)

Seifert, J.: Das Leib-Seele-Problem in der gegenwärtigen philosophischen Diskussion. Wissenschaftliche Buchgesellschaft, Darmstadt (1979)

Serchinger, R.: Walter Schottky - Atomtheoretiker und Elektrotechniker - Sein Leben und Werk bis ins Jahr 1941. GNT Publisher, Berlin (2008)

Shortland, M., Yeo, R.: Telling lives in science - Essays on scientific biography. Cambridge University Press, Cambridge (1996)

Singer, J.: Advances in Quantum Electronics (=Proc. 2nd Quantum Electronics Conf.). Columbia University Press, New York (1961)

Skar, J.: Introduction: self-organization as an actual theme. Phil. Trans. R. Soc. London A 361, 1049–1056 (2003)

Smale, S.: Differentiable dynamical systems. Bull. Amer. Math. Soc, 747–817 (1967)

Sondheimer, E.H.: Sir Alan Herries Wilson. 2 July 1960 - 30 September 1995. In: Biographical Memoirs of Fellows of the Royal Society Bd. 45, pp. 548–562. The Royal Society, London (1999)

Spenke, E.: Elektronische Halbleiter. Springer, Berlin (1954)

Stadler, M., Kruse, P.: The Self-Organization Perspective in Cognition Research: Historical Remarks and New Experimental Approaches. In: Haken, H., Stadler, M. (eds.) Synergetics of Cognition, pp. 32–53. Springer, Heidelberg (1990)

Stegmüller, W.: Unvollständigkeit und Unentscheidbarkeit. Die metamathematischen Resultate von Goedel, Church, Kleene, Rosser und ihre erkenntnistheoretische Bedeutung. Springer, Wien (1973)

Suzuki, M.: Ryogo Kubo. Physics Today 49, 87 (1996)

Swinney, H.L., Fenstermacher, P.R., Gollup, J.P.: Transition to turbulence in a fluid flow. In: Synergetics - A Workshop. Springer, Berlin (1977)

Szöke, A., Javan, A.: Isotope Shift and Saturation Behaviour of the 1.15m Transition of Neon. Phys. Rev. Letters 10, 521–524 (1963)

Teichmann, J.: Zur Geschichte der Festkörperphysik. Farbzentrenforschung bis 1940. Boethius, Wiesbaden (1988)

Teichmann, J.: Pohl, Robert Wichard. Neue Deutsche Biographie 20, 586 (2001)

Thom, R.: Stabilité structurelle et morphogénèse. Benjamin, Reading (1972)

Tomiyasu, K.: The Laser Literature - an annotated guide. Plenum Press, New York (1968)

Townes, C.H.: Nobel Lecture 1964. In: Nobel Lectures 1963-1970. Elsevier Publishing Comp., Amsterdam (1972)

Townes, C.: Quantum electronics - A Symposium. Columbia University Press, New York (1960)

Townes, C.M.: How the Laser happened. Oxford University Press, New York (1999)

Tschacher, W.: Bericht von der Herbstakademie "Selbstorganisation und Klinische Psychologie" in Bamberg 1. - 5. 10 1990. System Familie 4, 124–126 (1991)

Turing, A.M.: The Chemical Basis of Morphogenesis. Phil. Transactions of the Royal Society B237, 37–72 (1952)

Tyson, J.J., Kagan, M.L.: Spatiotemporal Organization in Biological and Chemical Systems: Historical Review. In: Markus, M., Müller, S.C., Nicolis, G. (eds.) From Chemical to Biological Organization, pp. 14–21. Springer, Berlin (1988)

Varela, F.J., Maturana, H.R., Uribe, R.: Autopoiesis, the organization of Living Systems: Its Characterization and a Model. Biosystems 5, 187–196 (1974)

Vavilin, V., Zhabotinsky, A.M., Zaikin, A.: Effect of Ultraviolet Radiation on the Oscillating Oxidation Reaction of Malonic Acid Derivatives. Russ. J. Phys. Chem. 42, 1649 (1968)

Vidal, C., Pacault, A.: Nonlinear Phenomena in Chemical Dynamics (Bordeaux 1981). Springer, Berlin (1981)

Voigt, J.: University of Stuttgart. Phasen ihrer Geschichte. Konrad Wittwer, Stuttgart (1981)

Voigt, J. (Hg.): Festschrift zum 150 jährigen Bestehen der University of Stuttgart (Die University of Stuttgart Volume 2). DVA, Stuttgart (1979)

Volkswagen-Stiftung: Bericht - Stiftung Volkswagenwerk; 1990. Vandenhoek und Ruprecht, Göttingen (1991)

Volkswagen-Stiftung: Impulse geben - Wissen stiften - 40 Jahre VolkswagenStiftung. Vandenhoeck & Ruprecht, Göttingen (2002)

Volkswagenwerk, S.: Bericht - Stiftung Volkswagenwerk; 1980. Vandenhoek & Ruprecht, Göttingen (1981)

Volterra, V.: Lécon sur la Théorie Mathematique de la lutte pour la vie, Paris (1931)

Wachter, C.: Die Professoren und Dozenten der Friedrich-Alexander-University of Erlangen 1743-1960. Teil 3: Philosophische Fakultät, Naturwissenschaftliche Fakultät. Universitätsbibliothek, Erlangen (2009)

Wagner, W., Birnbaum, G.: Theory of Quantum Oscillators in a multimode Cavity. Journal of Applied Physics 32, 1185–1192 (1961)

Waldrop, M.: Complexity - the emerging science at the edge of chaos and complexity. Viking, London (1992)

Watson, J.D.: DNA. The secret of life, 1st edn. Knopf, New York (2003)

Weidlich, W., Haake, F.: Coherence-Properties of the Statistical Operator in a Laser Model. Zs. für Physik 185, 30–47 (1965)

Weidlich, W., Haake, F.: Master-equation for the Statistical Operator of Solid State Laser. Zs. für Physik 186, 203–221 (1965)

Weidlich, W., Risken, H., Haken, H.: Quantummechanical Solutions of Laser Masterequation. Part I. Zeitschrift für Physik 201, 396–410 (1967)

Weidlich, W., Haake, F.: Quantummechanical Solutions of the Laser Masterequation. II. Zs. für Physik 204, 223–239 (1967)

Wiener, N.: Cybernetics or control and communication in the animal and the machine. Wiley, New York (1948)

Wigner, E.P.: The probability of the existence of a self-reproducing unit. In: The Logic of Personal Knowledge. Free Press, Glencoe (1961)

Wilkins, M.: The third man of the double helix, the autobiography of Maurice Wilkins. Oxford University Press, Oxford (2003)

Wilson, A.: The Theory of electronic semi-conductors I und II. Proc. Roy. Soc. London A133, 458–491 (1931)

Winfree, A.T.: The Prehistory of the Belousov-Zhabotinsky Oscillator. Journal of Chemical Education 61, 661–663

Winkler-Oswatitsch, R.: Manfred Eigen - Scientist and Musician. Biophysical Chemistry 26, 103–115 (1987)

Winnacker, A.: Helmut Volz - Der Gründungsvater der Technischen Fakultät. In: 40 Jahre Technische Fakultät - Festschrift, Erlangen, pp. 42–46 (2006)

Wissenschaftsrat: Empfehlungen zum Ausbau der Wissenschaftlichen Einrichtungen. Teil I: Wissenschaftliche Hochschulen. J.C.B. Mohr (Paul Siebeck), Tübingen (1960)

Wittgenstein, L.: Philosophical Investigations. Blackwell, Oxford (1953)

Wunderlin, A.: Die Behandlung von elektronischen Kollektivanregungen Im Festkörper mit Hilfe von Quasiwahrscheinlichkeitsverteilungen. Diploma thesis, University of Stuttgart (1971)

Wunderlin, A.: Über statistische Methoden und ihre Anwendung auf Gleichgewichts- und Nichtgleichgewichtssysteme. Dissertation, University of Stuttgart (1975)

Yovits, M., Jacobi, G., Goldstein, G. (eds.): Self-Organizing Systems 1962. Spartan Books, Washington D.C. (1962)

Yovits, M., Cameron, S. (eds.): Self-Organizing Systems - Proceedings of an Interdisciplinary Conference. Pergamon Press, New York (1960)

Zhabotinsky, A.M.: A History of Chemical Oscillations and Waves. Chaos 1, 379–386 (1991)

Zhaboutinsky, A.M.: Periodic processes of the oxidation of malonic acid in solution. Biofizika 9, 306 (1964)

Zorell, J.: Stochastische Modelle für Nichtgleichgewichtsphasenübergänge bei chemischen Reaktionen. University of Stuttgart (1976)

Appendix 1
Bibliography of Hermann Haken's Works

1. Haken, H., Volz, H.: Zur Quantentheorie des Mehrelektronenproblems im Festkörper. Zeitschrift für Physikalische Chemie 198, 61 (1951)
2. Haken, H.: Zum Identitätsproblem bei Gruppen. Mathematische Zeitschrift 56, 335–362 (1952)
3. Haken, H.: Eine modellmässige Behandlung der Wechselwirkung zwischen einem Elektron und einem Gitteroszillator. Zeitschrift für Physik 135, 408–430 (1953)
4. Haken, H.: On the problem of radiationless transitions. Physica 20, 1013–1016 (1954)
5. Haken, H.: Eine Methode zur strengen Behandlung der Wechselwirkung zwischen einem Elektron und mehreren Gitteroszillatoren. Zeitschrift für Physik 138, 56–70 (1954)
6. Haken, H.: Zum Energieschema des Systems Elektron - schwingendes Gitter. Zeitschrift für Physik 139, 66–87 (1954)
7. Haken, H.: Über die Struktur der Lösung des Mehrelektronenproblems im Festkörper und ein Theorem von Bloch. Zeitschrift für Naturforschung 9, 228–235 (1954)
8. Haken, H.: Einige Zusatzbemerkungen zum Referat Pfirsch. In: Schottky, W. (ed.) Halbleiterprobleme I, pp. 70–75. Vieweg, Berlin (1954)
9. Haken, H.: Die Unmöglichkeit der Selbstlokalisation von Elektronen im störstellenfreien Kristallgitter. Zeitschrift für Naturforschung 10, 253–254 (1955)
10. Haken, H.: Die Bewegung elektronischer Ladungsträger in polaren Kristallen. In: Schottky, W. (ed.) Halbleiterprobleme II, pp. 1–39. Vieweg, Berlin (1955)
11. Haken, H.: Zur Quantentheorie des Mehrelektronensystems im schwingenden Gitter. Teil 1. Zeitschrift für Physik 146(5), 527–554 (1956)
12. Haken, H.: Zur Quantentheorie des Mehrelektronensystems im schwingenden Gitter. Teil 2. Zu Fröhlichs eindimensionalem Supraleitungsmodell. Zeitschrift für Physik 146, 555–570 (1956)
13. Haken, H.: Kopplung nichtrelativistischer Teilchen mit einem quantisierten Feld.Teil 1. Das Exziton im schwingenden, polaren Kristall. Nuovo Cimento 3, 1230–1253 (1956)
14. Haken, H.: Application of Feynmans New Variational Procedure to the Calculation of the Ground State Energy of Excitons. Nuovo Cimento 4, 1608–1609 (1956)
15. Haken, H.: Zu Fröhlichs eindimensionalem Modell der Supraleitung. Zeitschrift für Naturforschung 11, 96–98 (1956)

16. Haken, H.: Theorie des Excitons dans les Cristaux Polaires. Journal De Physique Et Le Radium 17, 826–828 (1956)
17. Haken, H.: Die freie Weglänge des Exzitons im polaren Kristall. Zeitschrift für Naturforschung 11, 875–876 (1956)
18. Haken, H., Schottky, W.: Allgemeine optische Auswahlregeln in periodischen Kristallgittern. Zeitschrift für Physik 144, 91–107 (1956)
19. Haken, H.: Berechnung der Energie des Exzitonen-Grundzustandes im polaren Kristall nach einem neuen Variationsverfahren von Feynman 1. Zeitschrift für Physik 147, 323–349 (1957)
20. Haken, H.: Der heutige Stand der Exzitonen-Forschung in Halbleitern. In: Schottky, W. (ed.) Halbleiterprobleme IV, pp. 1–48. Vieweg, Berlin (1957)
21. Haken, H.: Theorie des Excitons dans les cristaux polaires. Journal de Physique et le Radium 17, 826–828 (1957)
22. Haken, H.: Sur la Theorie des Excitons et leur Role dans les Transferts d'Energie a l'état solide. Journal De Chimie Physique et de Physico-Chimie Biologique 55, 613–620 (1958)
23. Haken, H.: Die Theorie des Exzitons im festen Körper. Fortschritte der Physik 6, 271–334 (1958)
24. Haken, H.: Theorie stationärer und nichtstationärer Zustände des Exzitons im polaren Kristall. In: Schön, M., Welker, H. (eds.) Halbleiter und Phosphore (= Proc. Int. Kolloquium Garmisch-Partenkirchen 1956). Vieweg, Braunschweig (1958)
25. Haken, H., Schottky, W.: Die Behandlung des Exzitons nach der Vielelektronentheorie. Zeitschrift für Physik/Chemie NF 16, 218–244 (1958)
26. Nikitine, S., et al.: Transferts d'Energie lumineuse et photosensibilisation – Discussion. Journal de Chimie Physique et de Physico-Chimie Biologique 55, 657–660 (1958)
27. Haken, H.: Über den Einfluss von Gitterschwingungen auf Energie und Lebensdauer des Exzitons. Zeitschrift für Physik 155, 223–246 (1959)
28. Haken, H.: Theory of Excitons in Solids. Uspekhi Fizicheskikh Nauk 68, 565–619 (1959)
29. Haken, H.: On the Theory of Excitons in Solids. Journal of Physics and Chemistry of Solids 8, 166–171 (1959)
30. Haken, H., Englert, F.: Mass Renormalization and effective interaction of electrons in semiconductors. Unveröffentlicht (1959)
31. Nikitine, S., Haken, H.: Sur la possibilité de provoquer une migration dirigée d'excitons. Comptes Rendus Hebdomadaires des Seances de l' Academie des Sciences 250, 697–699 (1960)
32. Haken, H.: Nonlinear interactions of laser modes. In: International Conference on Optical Pumping, Heidelberg (1962) (unpublished)
33. Haken, H.: Theory of exciton absorption in an electric field. In: International Conference on the Physics of Semiconductors, p. 462. Institute of Physics and the Physical Society of London, Exeter (1962)
34. Haken, H.: Effective particle interaction via a Bose-Field. Annalen der Physik 11(7.Folge), 123–131 (1963)
35. Haken, H., Haken, E.: Zur Theorie des Halbleiter-Lasers. Zeitschrift für Physik 176, 421–428 (1963)
36. 35a. Haken, H.: Professor Dr. E. Fues zum 70. Geburtstag am 17. Januar 1963. Annalen der Physik 466, 1–3 (1963)

37. Haken, H., Sauermann, H.: Nonlinear Interaction of Laser Modes. Zeitschrift für Physik 173, 261–275 (1963)
38. Haken, H., Sauermann, H.: Frequency Shifts of Laser Modes in Solid State and Gaseous Systems. Zeitschrift für Physik 176, 47–62 (1963)
39. Abate, E., Haken, H.: Exakte Behandlung eines Laser-Modells. Zeitschrift für Naturforschung A 19, 857 (1964)
40. Haken, H.: Nonlinear Theory of Laser Noise and Coherence. I. Zeitschrift für Physik 181, 96–124 (1964)
41. Haken, H.: Theory of Coherence of Laser Light. Physical Review Letters 13, 329–331 (1964)
42. Haken, H.: Theory on Excitons. Zeitschrift für Physikalische Chemie 43, 264 (1964)
43. Haken, H., Sauermann, H.: Theory of laser action in solid-state, gaseous and semiconductor systems. In: Miles, S.A. (ed.) Quantum Electronics and Coherent Light (Proceedings Int. Summerschool "Enrico Fermi" XXXI), pp. 111–155. Academic Press, New York (1964)
44. Haken, H.: Der heutige Stand der Theorie des Halbleiterlasers. Advances in Solid State Physics 4, 1–26 (1965)
45. 43a. Haken, H.: Der Nobelpreis 1964 für den Maser. Physikalische Blätter 21, 109–114 (1965)
46. Haken, H.: Nonlinear Theory of Laser Noise and Coherence. Zeitschrift für Angewandte Mathematik und Physik 16, 14 (1965)
47. Haken, H.: A Nonlinear Theory of Laser Noise and Coherence. Teil 2. Zeitschrift für Physik 182, 346–359 (1965)
48. Haken, H., Der Agobian, R., Pauthier, M.: Theory of Laser Cascades. Physical Review 140, A437–A447 (1965)
49. Haken, H., et al.: Nonlinear Theory of Laser Noise and Coherence. Zeitschrift für Angewandte Mathematik und Physik 16, A437 (1965) (Conference "Physics of Quantum Electronics" Puerto Rico (1965))
50. Arzt, V., et al.: Quantum Theory of Noise in Gas and Solid State Lasers with an Inhomogeneously Broadened Line. Part I. Zeitschrift für Physik 197, 207–227 (1966)
51. Haken, H.: Theory of Noise in Solid State Gas and Semiconductor Lasers. IEEE Journal of Quantum Electronics Qe 2, R19 (1966)
52. Haken, H.: Theory of Intensity and Phase Fluctuations of a Homogeneously Broadened Laser. Zeitschrift für Physik 190, 327–356 (1966)
53. Haken, H., Weidlich, W.: Quantum Noise Operators for N-Level System. Zeitschrift für Physik 189, 1–9 (1966)
54. Haken, H.: Dynamics of Nonlinear Interaction between Radiation and Matter. In: Kubo, R., Kamimura, H. (eds.) Dynamical Processes in Solid State Optics, pp. 168–194. W.A. Benjamin, Tokyo (1967)
55. Haken, H., Haug, H.: Theory of Noise in Semiconductor Laser Emission. Zeitschrift für Physik 204, 262–275 (1967)
56. Haken, H., Risken, H., Weidlich, W.: Quantum Mechanical Solutions of Laser Masterequation. Teil 3: Exact Equation for a Distribution Function of Macroscopic Variables. Zeitschrift für Physik 206, 355–368 (1967)
57. Haken, H., et al.: Theory of Laser Noise in Phase Locking Region. Zeitschrift für Physik 206, 369–393 (1967)
58. Haken, H., Weidlich, W.: A Theorem on Calculation of Multi-Time-Correlation Functions by Single-Time Density Matrix. Zeitschrift für Physik 205, 96–102 (1967)

59. Strobl, G., Haken, H.: Exact treatment of coherent and incoherent triplet exciton migration. In: Zahlan, A.B. (ed.) The Triplet State, pp. 311–314. Cambridge University Press, Cambridge (1967)
60. Weidlich, W., Risken, H., Haken, H.: Quantummechanical Solutions of Laser Masterequation. Part I. Zeitschrift für Physik 201, 396–410 (1967)
61. Weidlich, W., Risken, H., Haken, H.: Quantummechanical Solutions of Laser Masterequation. Part 2. Zeitschrift für Physik 204, 223–239 (1967)
62. Graham, R., et al.: Quantum Mechanical Correlation Functions for Electromagnetic Field and Quasi-Probability Distribution Functions. Zeitschrift für Physik 213, 21–32 (1968)
63. Graham, R., Haken, H.: Theory of Quantum Fluctuations of Parametric Oscillator. IEEE Journal of Quantum Electronics QE4, 345–346 (1968)
64. Graham, R., Haken, H.: Quantum-Fluctuations of Optical Parametric Oscillator. Part I. Zeitschrift für Physik 210, 276–302 (1968)
65. Graham, R., Haken, H.: Quantum Theory of Light Propagation in a Fluctuating Laser-Active Medium. Zeitschrift für Physik 213, 420–450 (1968)
66. Haken, H.: Exact Generalized Fokker-Planck Equation for Arbitrary Dissipative Quantum Systems. Physics Letters A A28, 286–287 (1968)
67. Haken, H.: Exact Stationary Solution of a Fokker-Planck-Equation for Multimode Laser-Action. Physics Letters A A27, 190–191 (1968)
68. Haken, H., Pauthier, M.: Nonlinear Theory of Multimode Action in Loss Modulated Lasers. IEEE Journal of Quantum Electronics QE4, 454 (1968)
69. Haken, H., Reineker, S.: Comments on the interaction of Excitons and Phonons. In: Zahlan, A.B. (ed.) Dynamical Processes in Solid State Optics, p. 185. Cambridge University Press, Cambridge (1968)
70. Graham, R., Haken, H.: Analysis of Quantum Field Statistics in Laser Media by Means of Functional Stochastic Equations. Physics Letters A A29, 530–531 (1969)
71. Haken, H.: Exact Stationary Solution of a Fokker-Planck Equation for Multimode Laser Action Including Phase Locking. Zeitschrift für Physik 219, 246–268 (1969)
72. Haken, H.: Exact Generalized Fokker-Planck Equation for Arbitrary Dissipative and Nondissipative Quantum-Systems. Zeitschrift für Physik 219, 411–433 (1969)
73. Bivas, A., et al.: Excitons Absorption and Emission of cuprous chloride during excitation by a powerful light source. Optics Communications 2, 227–230 (1970)
74. Graham, R., Haken, H.: Laserlight - First Example of a Second-Order Transition Far Away from Thermal Equilibrium. Zeitschrift für Physik 237, 31–46 (1970)
75. Graham, R., Haken, H.: Functional Fokker-Planck Treatment of Electromagnetic Field Propagation in a Thermal Medium. Zeitschrift für Physik 234, 193–206 (1970)
76. Graham, R., Haken, H.: Microscopic Reversibility, Stability and Onsager Relations in Systems Far from Thermal Equilibrium. Physics Letters A A33, 335–336 (1970)
77. Graham, R., Haken, H.: Functional Quantum Statistics of Light Propagation in a 2-Level System. Zeitschrift für Physik 235, 166–180 (1970)
78. Graham, R., Haken, H., Weidlich, W.: Flux Equilibria in Quantum Systems Far Away from Thermal Equilibrium. Physics Letters A A32, 129–130 (1970)
79. Haken, H.: Theory of Multimode Effects Including Noise in Semiconductor Lasers. IEEE Journal of Quantum Electronics QE6, 325 (1970)
80. Haken, H.: Laserlicht - ein neues Beispiel für eine Phasenumwandlung? In: Schottky, W. (ed.) Festkörperprobleme X. Vieweg, Braunschweig (1970)
81. Haken, H.: Laser Theory. In: Flügge, G. (ed.) Handbuch der Physik Band, vol. XXV/2. Springer, Berlin (1970)

82. Haken, H.: Quantum Optics. Klinische Wochenschrift (Organ der Naturforscher und Ärzte) 48, 1 (1970)
83. Haken, H.: Quantum Fluctuations in Nonlinear Optics. Opto-Electronics 2, 161–167 (1970)
84. Graham, R., Haken, H.: Generalized Thermodynamic Potential for Markoff Systems in Detailed Balance and Far from Thermal Equilibrium. Zeitschrift für Physik 243, 289–302 (1971)
85. Graham, R., Haken, H.: Fluctuations and Stability of Stationary Non-Equilibrium Systems in Detailed Balance. Zeitschrift für Physik 245, 141–153 (1971)
86. Haken, H.: Quantenoptik. Naturwissenschaften 58, 188–194 (1971)
87. Haken, H., Graham, R.: Synergetik - die Lehre vom Zusammenwirken. Umschau in Wissenschaft und Technik, 191–195 (März 1971)
88. Haken, H., Nikitine, S.: Spontaneous and Stimulated Emissions from Excitons at High Concentration. Akad. Nauk SSSR, Ser. Fiz. 37, 220 (1971) (in Russisch)
89. Haken, H., Vollmer, H.D.: Fokker-Planck Equation of a Laser with Many Modes and Multi-Level Atoms. Zeitschrift für Physik 242, 416–431 (1971)
90. Reineker, S., Haken, H.: The Coupled Coherent and Incoherent Motion of Excitons and its influence on the line shape of optical absorption. Zeitschrift für Physik 249, 253–268 (1971)
91. Haken, H.: The Theory of coherence, noise and photon statistics of laser light. In: Arecchi, A.T., Schulz-Dubois, E.O. (eds.) Laser Handbook, pp. 115–150. North Holland Publ., Amsterdam (1972)
92. Haken, H., Reineker, S.: Coupled Coherent and Incoherent Motion of Excitons and Its Influence on Line Shape of Optical-Absorption. Zeitschrift für Physik 249, 253–268 (1972)
93. Haken, H., Schenzle, A.: Giant Polaritons. Physics Letters A A41, 405–406 (1972)
94. Schenzle, A., Haken, H.: Self-induced Transparency of Excitons. Optics Communications 6, 96–97 (1972)
95. Levy, R., et al.: Shift of Exciton-Exciton Emission-Line in Cadmium Sulfide. Solid State Communications 10, 915–917 (1972)
96. Reineker, S., Haken, H.: Influence of Exciton Motion on Spin-Resonance Absorption. Zeitschrift für Physik 250, 300–323 (1972)
97. Schwarze, E., Haken, H.: Moments of Coupled Coherent and Incoherent Motion of Excitons. Physics Letters A A42, 317–318 (1972)
98. Haken, H.: Stability and Fluctuations of Multimode Configurations near Convection Instability. Physics Letters A A46, 193–194 (1973)
99. Haken, H.: Resonant Interaction between Excitons and Intense Coherent Light. Zeitschrift für Physik 262, 119–134 (1973)
100. Haken, H.: Distribution Function for Classical and Quantum Systems Far from Thermal Equilibrium. Zeitschrift für Physik 263, 267–282 (1973)
101. Haken, H.: Laser Light Emission from Interacting Chaotic Fields. Zeitschrift für Physik 265, 105–118 (1973)
102. Haken, H.: Distribution Function for Systems Far from Thermal Equilibrium. Physics Letters A A44, 303–304 (1973)
103. Haken, H.: Correlations in Classical and Quantum Systems Far from Thermal Equilibrium. Zeitschrift für Physik 265, 503–510 (1973)
104. Haken, H.: Quantentheorie des Festkörpers. Teubner Publisher, Stuttgart (1973)

105. Haken, H.: Introduction to Synergetics. In: Haken, H. (ed.) Synergetics - Cooperative Phenomena in Multicomponent Systems (Proc. Schloß Elmau 1972). Teubner-Publisher, Stuttgart (1973)
106. Haken, H.: Synergetics - Towards a New Discipline. In: Haken, H., Wagner, M. (eds.) Cooperative Phenomena (Festschrift Herbert Fröhlich), pp. 363–373. Springer, Berlin (1973)
107. Haken, H.: Cooperative Phenomena in Systems far from Thermal Equilibrium. In: Marois, M. (ed.) From Theoretical Physics to Biology (Proc. of the 3rd Int. Conference, Versailles 1971). Karger, Basel (1973)
108. Haken, H.: Synergetics - Cooperative Phenomena in Multicomponent Systems (Proc. Schloß Elmau 1972). Teubner, Stuttgart (1973)
109. Haken, H., Schenzle, A.: Giant Polaritons and Self-Induced Transparency of Frenkel-Excitons. Zeitschrift für Physik 258, 231–241 (1973)
110. Haken, H., Schwarzer, E.: Theory of Line Shape of Optical Absorption and Emission Spectra in Molecular Crystals. Optics Communications 9, 64–68 (1973)
111. Haken, H., Strobl, G.: An Exactly Solvable Model for Coherent and Incoherent Exciton Motion. Zeitschrift für Physik 262, 135–148 (1973)
112. Nikitine, S., Haken, H.: Spontaneous and Stimulated Emissions from Excitons at High Concentration. In: Williams, F.E. (ed.) Luminiscence of Cristals, Molecules and Solutions, pp. 4–19. Plenum Press (1973)
113. Wöhrstein, H.G., Haken, H.: Theory of Second-Order Mode-Locking in Semiconductor Lasers. IEEE Journal of Quantum Electronics QE9, 318–323 (1973)
114. Wöhrstein, H.G., Haken, H.: Stationary Solution of Fokker-Planck Equation near Detailed Balance. Physics Letters A A45, 231–232 (1973)
115. Wöhrstein, H.G., Haken, H.: Atom-Field Correlation, Conservation Laws and the Phase Transition of the Laser. Optics Communications 9, 123–127 (1973)
116. Haken, H.: Exact Stationary Solution of Master Equation for Systems Far from Thermal Equilibrium in Detailed Balance. Physics Letters A A46, 443–444 (1974)
117. Haken, H.: Solution of Master Equation for Quantum Systems Weakly Coupled to Reservoirs and Far from Thermal Equilibrium. Zeitschrift für Physik 266, 265–269 (1974)
118. Haken, H.: Synergetics. Basic Concepts and Mathematical Tools. In: Haken, H. (ed.) Cooperative Effects. North Holland Publ., Amsterdam (1974)
119. Haken, H., Goll, J., Schenzle, A.: Resonant Interaction between Excitons and Coherent Light. IEEE Journal of Quantum Electronics QE10, 743 (1974)
120. Haken, H., Schwarzer, E.: Theory of the Influence of Coherent or Incoherent Motion of Triplet Excitons on NMR. Chemical Physics Letters 27, 41–46 (1974)
121. Merle, J.C., Nikitine, S., Haken, H.: Temperature-Dependence of 1s Line of Yellow Series of Cu2O. Physica Status Solidi B 61, 229–239 (1974)
122. Schenzle, A., Haken, H.: Giant Polaritons and Self-induced Transparency of Excitons. In: Burstein, E., De Martini, F. (eds.) Polaritons (Proceedings of the First Taormina Research Conference 1972). Pergamon, Oxford (1974)
123. Haken, H.: Higher-Order Corrections to Generalized Ginzburg-Landau Equations of Nonequilibrium Systems. Zeitschrift für Physik B 22, 69–72 (1975)
124. Haken, H.: Cooperative Phenomena in Systems Far from Thermal Equilibrium and in Nonphysical Systems. Reviews of Modern Physics 47, 67–121 (1975)
125. Haken, H.: Statistical Physics of Bifurcation, Spatial Structures, and Fluctuations of Chemical-Reactions. Zeitschrift für Physik B 20, 413–420 (1975)

126. Haken, H.: Analogy between Higher Instabilities in Fluids and Lasers. Physics Letters A 53, 77–78 (1975)
127. Haken, H.: Critical Fluctuations in Continuous Nonequilibrium Systems. Zeitschrift für Physik B 22, 73–77 (1975)
128. Haken, H.: Generalized Ginzburg-Landau Equations for Phase Transition-Like Phenomena in Lasers, Nonlinear Optics, Hydrodynamics and Chemical-Reactions. Zeitschrift für Physik B 21, 105–114 (1975)
129. Haken, H.: Statistical Physics of a Chemical-Reaction Model. Physics Letters A A51, 125–126 (1975)
130. Haken, H.: Exzitonen hoher Dichte. Physik in Unserer Zeit 6, 78–86 (1975)
131. Haken, H., Goll, J., Schenzle, A.: Polaritons at High Light Intensities and in Bose Condensed Exciton Systems. In: Nikitine, S., Haken, H. (eds.) Excitons at High Density, pp. 285–295 (1975)
132. Haken, H., Nikitine, S.: Theory of stimulated Emissions by Excitons. In: Nikitine, S., Haken, H. (eds.) Exitons at High Density. Springer, Berlin (1975)
133. Wunderlin, A., Haken, H.: Scaling Theory for Nonequilibrium Systems. Zeitschrift für Physik B 21, 393–401 (1975)
134. Haken, H.: Generalized Onsager-Machlup Function as Solution of a Multidimensional Fokker-Planck Equation. Physics Letters A 55, 323–324 (1976)
135. Haken, H.: Higher Order Corrections to generalized Ginzburg-Landau Equations of Non-Equilibrium Systems (Erratum). Zeitschrift für Physik B 23, 388 (1976)
136. Haken, H.: Generalized Onsager-Machlup Function and Classes of Path Integral Solutions of Fokker-Planck Equation and Master Equation. Zeitschrift für Physik B 24, 321–326 (1976)
137. Haken, H.: Stimulated Emission at High Concentration of Excitons in Molecular Spectroscopy of Dense Phases. In: Grosmann, M., Elkomoss, S.G., Ringeisen, J. (eds.) Molecular Spectroscopy of Dense Phases (= XIIth European Congress on Molecular Spectroscopy Strasbourg 1975). Elsevier Publ., Amsterdam (1976)
138. Haken, H.: Statistical Theory of Self-Organizing Structures. In: Pal, L., Szépfaluszy, S. (eds.) Statistical Physics (=Proc. IUPAP Conference Budapest 1975), pp. 49–68. North Holland Publ., Amsterdam (1976)
139. Haken, H.: Synergetics. Europhysics News S7 (1976)
140. Haken, H., Ohno, H.: Theory of Self-Pulsing Lasers. Optics Communications 18, 19 (1976)
141. Haken, H., Ohno, H.: Theory of Ultrashort Laser Pulses. Optics Communications 16, 205–208 (1976)
142. Haken, H., Reineker, S.: The Coupled Coherent and Incoherent Motion of Frenkel Excitons in Molecular Crystals. In: Chalvet, O. (ed.) Localization and Delocalization in Quantum Chemistry. D. Reidel Publ., Dordrecht (1976)
143. Ohno, H., Haken, H.: Transient Ultrashort Laser-Pulses. Physics Letters A 59, 261–263 (1976)
144. Egler, W., Haken, H.: Theory of Damping and Fluctuations of Polariton Due to Its Interaction with Lattice-Vibrations. Zeitschrift für Physik B 28, 51–60 (1977)
145. Haken, H.: Die Kohärenzeigenschaften des Lichts. Acta Physica Austriaca 47, 59–81 (1977)
146. Haken, H.: Some Aspects of Synergetics. In: Haken, H. (ed.) Synergetics - A Workshop (Proc. Elmau Konferenz 1977). Springer, Berlin (1977)
147. Haken, H.: Synergetics - An Introduction Nonequilibrium Phase Transitions and Self-Organization in Physics, Chemistry and Biology. Springer, Berlin (1977)

148. Haken, H.: Synergetics. Physics Bulletin 1977, 412 (1977)
149. Haken, H., Wunderlin, A.: New Interpretation and Size of Strange Attractor of Lorenz Model of Turbulence. Physics Letters A 62, 133–134 (1977)
150. Schwarzer, E., Haken, H.: Theory of Influence of Exciton Motion on Proton Relaxation in Organic Solids. Physica Status Solidi B 84, 253–267 (1977)
151. Goll, J., Haken, H.: Self-Induced Transparency of Excitons and Dispersion Law of Steady-State Exciton-Photon Pulses. Physical Review A 18, 2241–2252 (1978)
152. Goll, J., Haken, H.: Exciton Self-Induced Transparency and Dispersion Law of Steady-State Exciton-Photon Pulses. Optics Communications 24, 1–4 (1978)
153. Haken, H.: Synergetics - Some Recent Trends and Developments. Supplement of the Progress of Theoretical Physics 64, 21–34 (1978)
154. Haken, H.: Non-Equilibrium Phase-Transitions and Bifurcation of Limit-Cycles and Multi-Periodic Flows. Zeitschrift für Physik B 29, 61–66 (1978)
155. Haken, H.: Non-Equilibrium Phase-Transitions and Bifurcation of Limit-Cycles and Multi-Periodic Flows in Continuous Media. Zeitschrift für Physik B 30, 423–428 (1978)
156. Haken, H.: Synergetik. Nachrichten aus Chemie, Technik und Laboratorium 26 (1978)
157. Haken, H.: Synergetics - a field beyond irreversible thermodynamics. In: Garrido, L. (ed.) Stochastic Processes in Nonequilibrium Systems, pp. 139–167. Springer, Berlin (1978)
158. Haken, H.: Nonequilibrium Phase Transitions and Instability Hierarchy of the Laser, an example from Synergetics. In: Pacault, A., Vidal, C. (eds.) Synergetics far from Equilibrium, pp. 22–33. Springer, Berlin (1978)
159. Haken, H.: The Laser - Trailblazer of Synergetics. In: Mandel, L., Wolf, E. (eds.) Coherence and Quantum Optics (= Proc. of the 4th Rochester Conference), pp. 49–62. Plenum Press, New York (1978)
160. Haken, H., Goll, J.: Optical Resonance of Frenkel Excitons. Solid State Communications 27, 371–373 (1978)
161. Haken, H., Ohno, H.: Onset of Ultrashort Laser-Pulses - 1st or 2nd Order Phase-Transition? Optics Communications 26, 117–118 (1978)
162. Haken, H., Olbrich, H.: Analytical Treatment of Pattern Formation in Gierer-Meinhardt Model of Morphogenesis. Journal of Mathematical Biology 6, 317–331 (1978)
163. Goll, J., Haken, H.: Theory of Exciton-Superradiance and of Exciton Free Induction Decay. Journal of Luminescence 18, 719–723 (1979)
164. Haken, H.: Synergetics - Non-Equilibrium Phase-Transitions and Self-Organization in Physics, Chemistry and Biology. Chimia 33, 334–335 (1979)
165. Haken, H.: Theorie ultrakurzer Impulse und ihrer Wechselwirkung mit kondensierter Materie. Acta Physica Austriaca, Suppl. XX, 107–116 (1979)
166. Haken, H.: Synergetics and a new approach to bifurcation theory. In: Güttinger, W., Eikemeier, H. (eds.) Structural Stability in Physics, pp. 31–40. Springer, Heidelberg (1979)
167. Haken, H.: Pattern formation and pattern recognition - an attempt at a synthesis (Elmau 1979). In: Haken, H. (ed.) Pattern Formation by Dynamic Systems and Pattern Recognition, pp. 2–13. Springer, Berlin (1979)
168. Haken, H.: Synergetics and Bifurcation Theory. Annals of the New York Academy of Sciences 316, 357 (1979)

169. Haken, H.: Licht und Materie Band 1 - Elemente der Quantenoptik. Bibliographisches Institut (BI), Mannheim (1979)
170. Haken, H., Wunderlin, A.: Laser Instabilities - an example from synergetics. In: Casati, G., Ford, J. (eds.) Stochastic Behaviour in Classical and Quantum Hamiltonian Systems, pp. 213–231. Springer, Berlin (1979)
171. Goll, J., Haken, H.: Theory of Optical Bistability of Excitons. Physica Status Solidi B 101, 489–501 (1980)
172. Haken, H.: Selforganization through Increase of Number of Components. Supplement of the Progress of Theoretical Physics 69, 30–40 (1980)
173. Haken, H.: Synergetics - Are Cooperative Phenomena Governed by Universal Principles? Naturwissenschaften 67, 121–128 (1980)
174. Haken, H.: Die Synergetik - Ordnung aus dem Chaos. Bild der Wissenschaft 1980(3), 83 (1980)
175. Haken, H.: Selbstorganisation und Evolution - Der Beitrag synergetischer Ansätze. In: Jensen, H.J. (ed.) Kooperative Systeme in Biologie und Technik, pp. 17–20. Oldenbourg, München (1980)
176. Haken, H.: Transition Phenomena in Nonlinear Systems. In: Arnold, L. (ed.) Stochastic Nonlinear Systems in Physics, Chemistry and Biology, Springer, Berlin (1980)
177. Haken, H.: Lines of Development of Synergetics. In: Haken, H. (ed.) Dynamics of Synergetic Systems (=Proc. Int. Symposium on Synergetics, Bielefeld 1979), pp. 2–19. Springer, Berlin (1980)
178. Haken, H., Klenk, H.: Ordnung durch Instabilität - ein Beispiel aus der Plasmaphysik. Acta Physica Austriaca 52, 187–191 (1980)
179. Haken, H., Wolf, H.C.: Atom-und Quantenphysik. Springer, Berlin (1980)
180. Haken, H.: Fundamental Theoretical Aspects of Luminescence. Journal of Luminescence 24/25, 21–22 (1981)
181. Haken, H.: Synergetik, Nichtgleichgewichte, Phasenübergänge und Selbstorganisation. Naturwissenschaften 68, 293–299 (1981)
182. Haken, H.: Synergetik - Die Entstehung von Ordnung aus dem Chaos. In: Festvortrag zum 150. Jahrestag der Universität Stuttgart, Universitätsarchiv Stuttgart, Unveröffentlicht (1981)
183. Haken, H.: Erfolgsgeheimnisse der Natur, Synergetik - die Lehre vom Zusammenwirken. Deutsche Verlagsanstalt DVA, Stuttgart (1981)
184. Haken, H.: Synergetics. Is Self-Organization governed by Universal Principles? In: Jantsch, E. (ed.) The Evolutionary Vision. Westview Press, Boulder (1981)
185. Haken, H.: Licht und Materie Band 2, Laser. Bibliographisches Institut BI, Mannheim (1981)
186. Haken, H., Light, I.: Waves, Photons, Atoms. North Holland, Amsterdam (1981)
187. Haken, H.: Synergetics - a study of complex systems at points of critical behavior. In: Brulin, O., Hsieh, R.K.T. (eds.) Continuum Models of Discrete Systems. North Holland, Amsterdam (1981)
188. Haken, H.: Chaos and Order in Nature. In: Haken, H. (ed.) Chaos and Order in Nature, pp. 2–12. Springer, Berlin (1981)
189. Haken, H.: Ottica Quantistica - Enciclopedia del Novecento V, pp. 19–34. I.d.E. Italiana, Ed. (1981)
190. Haken, H.: Physical Thought in the 1980s. Physics Bulletin 32, 396–397 (1981)
191. Haken, H., Mayerkress, G.: Chapman-Kolmogorov Equation and Path-Integrals for Discrete Chaos in Presence of Noise. Zeitschrift für Physik B 43, 185–187 (1981)

192. Haken, H., Mayerkress, G.: Chapman-Kolmogorov Equation for Discrete Chaos. Physics Letters A 84, 159–160 (1981)
193. Mayerkress, G., Haken, H.: The Influence of Noise on the Logistic Model. Journal of Statistical Physics 26, 149–171 (1981)
194. Mayerkress, G., Haken, H.: Intermittent Behavior of the Logistic System. Physics Letters A 82, 151–155 (1981)
195. Wunderlin, A., Haken, H.: Generalized Ginzburg-Landau Equations, Slaving Principle and Center Manifold Theorem. Zeitschrift für Physik B 44, 135–141 (1981)
196. Benk, H., Haken, H., Sixl, H.: Theory of the Interaction of Frenkel Excitons with Local and Extended Perturbations in One-Dimensional Molecular-Crystals - Application of the Greens-Function Formalism to the Energy Funnel and Cluster Problem. Journal of Chemical Physics 77, 5730–5747 (1982)
197. Berding, C., Haken, H.: Pattern-Formation in Morphogenesis - Analytical Treatment of the Gierer-Meinhardt Model on a Sphere. Journal of Mathematical Biology 14, 133–151 (1982)
198. Haken, H.: Synergetics - an Approach to Complex Dynamic-Systems. Advances in Applied Probability 14, 197 (1982)
199. Haken, H.: Mathematical Methods of Synergetics for Applications to Self-Organizing Systems. In: Ricciardi, L.M., Scott, A.C. (eds.) Biomathematics in 1980. North Holland, Amsterdam (1982)
200. Haken, H.: Synergetics. Nonequilibrium Phase Transitions and Self-Organization in Biological Systems. In: Lamprecht, I., Zotin, A.J. (eds.) Thermodynamics and Kinetics of Biological Processes. De Gruyter, Berlin (1982)
201. Haken, H.: Synergetics - An Outline in Mechanical and Thermal Behaviour of Metallic Materials. Soc. Italiana di Fisica (1982)
202. Haken, H., Mayer-Kress, G.: Transition to Chaos for Maps with positive Schwarzian Derivative. In: Haken, H. (ed.) Evolution of Order and Chaos, pp. 183–186. Springer, Berlin (1982)
203. Haken, H., Wunderlin, A.: Slaving Principle for Stochastic Differential-Equations with Additive and Multiplicative Noise and for Discrete Noisy Maps. Zeitschrift für Physik B 47, 179–187 (1982)
204. Haken, H., Wunderlin, A.: Some Exact Results on Discrete Noisy Maps. Zeitschrift für Physik B 46, 181–184 (1982)
205. Berding, C., Harbich, T., Haken, H.: A Pre-Pattern Formation Mechanism for the Spiral-Type Patterns of the Sunflower Head. Journal of Theoretical Biology 104, 53–70 (1983)
206. Bestehorn, M., Haken, H.: A Calculation of Transient Solutions Describing Roll and Hexagon Formation in the Convection Instability. Physics Letters A 99, 265–267 (1983)
207. Goll, J., Haken, H.: Saturation of Interband-Transitions in Semiconductors and the Effect of Optical Bistability. Physical Review A 28, 910–928 (1983)
208. Haken, H.: At Least One Lyapunov Exponent Vanishes If the Trajectory of an Attractor Does Not Contain a Fixed-Point. Physics Letters A 94, 71–72 (1983)
209. Haken, H.: Von der Laserphysik zur Synergetik. In: Schneider, C. (ed.) Forschung in der Bundesrepublik Deutschland (im Auftrag der Deutschen Forschungsgemeinschaft DFG), pp. 515–518. Verlag Chemie, Weinheim (1983)

210. Haken, H.: Some Applications of Basic Ideas and Models of Synergetics to sociology. In: Frehland, E. (ed.) Synergetics - From Microscopic to Macroscopic Orders, pp. 174–182. Springer, Berlin (1983)
211. Haken, H.: Synopsis and Introduction. In: Basar, E., et al. (eds.) Synergetics of the Brain, pp. 3–25. Springer, Berlin (1983)
212. Haken, H.: Advanced Synergetics - Instability Hierarchies of Self-Organizing Systems and Devices. Springer, Berlin (1983)
213. Haken, H.: Synergetik - Selbstorganisationsvorgänge in Physik, Chemie und Biologie. A. v. Humboldt Mitteilungen 43 (1983)
214. Mayerkress, G., Haken, H.: Type-Iii-Intermittency in a Smooth Perturbation of the Logistic System. Lecture Notes in Physics, vol. 179, pp. 237–238 (1983)
215. Nagashima, T., Haken, H.: Chaotic Modulation of Correlation-Functions. Physics Letters A 96, 385–388 (1983)
216. Nara, S., Haken, H.: An Approach to Pattern-Formation in Crystal-Growth. 1. Needle Growth in a Simplified Model. Journal of Crystal Growth 63, 400–406 (1983)
217. Renardy, M., Haken, H.: Bifurcation of Solutions of the Laser Equations. Physica D 8, 57–89 (1983)
218. Shimizu, H., Haken, H.: Co-Operative Dynamics in Organelles. Journal of Theoretical Biology 104, 261–273 (1983)
219. Bestehorn, M., Haken, H.: Transient Patterns of the Convection Instability - a Model-Calculation. Zeitschrift für Physik B 57, 329–333 (1984)
220. Bunz, H., Ohno, H., Haken, H.: Subcritical Period Doubling in the Duffing Equation-Type-3 Intermittency, Attractor Crisis. Zeitschrift für Physik B 56, 345–354 (1984)
221. Haken, H.: Synergetics-83 Autowaves in Biology, Chemistry, and Physics. Physica D 11, 265–266 (1984)
222. Haken, H.: Synergetics - the Spontaneous Creation of Structures in the Living and Nonliving Environment. Helvetica Physica Acta 57, 140–156 (1984)
223. Haken, H.: Synergetics. Physica B & C 127, 26–36 (1984)
224. Haken, H.: Synergetics - Theory of Nonequilibrium Phase Transition and Formation of Spatio-Temporal Patterns. In: Grünewald, H. (ed.) Chemistry for the Future. Pergamon Press, Oxford (1984)
225. Haken, H.: Indeterminismus, Wahl und Freiheit - wie sind diese Begriffe im Bereich des Anorganischen zu verstehen? In: Koslowski, S., Kreuzer, S., Löw, R. (eds.) Evolution und Freiheit. Zum Spannungsfeld von Naturgeschichte und Mensch (Civitas Resultate Band 5), pp. 13–23. Hirzel, Stuttgart (1984)
226. Haken, H.: Spatial and temporal patterns formed by systems far from equilibrium. In: Vidal, C. (ed.) Non Equilibrium Dynamics in Chemical Systems, p. 7. Springer, Berlin (1984)
227. Haken, H.: Synergetik - eine stochastische Theorie von Selbstorganisationsvorgängen. In: Repges, R. (ed.) Strukturen und Prozesse. Neue Ansätze in der Biometrie, p. 1. Springer, Berlin (1984)
228. Haken, H.: Termodinamica irreversibile e sinergetica. In: Enciclopedia del Novecento. Enciclopedia Trecchani. Ed., Roma (1984)
229. Haken, H.: Towards a Dynamic Information Theory. In: Lamprecht, I., Zotin, A.I. (eds.) Thermodynamics and Regulation of Biological Processes, p. 93. Walter de Gruyter, Berlin (1984)

230. Mayerkress, G., Haken, H.: Attractors of Convex-Maps with Positive Schwarzian Derivative in the Presence of Noise. Physica D 10, 329–339 (1984)
231. Wang, Z.C., Haken, H.: Theory of 2-Photon Lasers. Part 3. 2-Photon Laser with Injected Signal. Zeitschrift für Physik B 56, 83–90 (1984)
232. Wang, Z.C., Haken, H.: Theory of 2-Photon Lasers. Part 1. Semiclassical Theory. Zeitschrift für Physik B 55, 361–370 (1984)
233. Wang, Z.C., Haken, H.: Theory of 2-Photon Lasers. Part 2. Fokker-Planck Equation Treatment. Zeitschrift für Physik B 56, 77–82 (1984)
234. Daido, H., Haken, H.: Cliff and Its Inversion as Rapid Decay of Fully-Developed Chaos. Physics Letters A 111, 211–216 (1985)
235. Haken, H.: Order in Chaos. Computer Methods in Applied Mechanics and Engineering 52, 635–652 (1985)
236. Haken, H.: Some Basic Concepts of Synergetics with Respect to Pattern-Formation - Morphogenesis of Behavior. Berichte der Bunsen-Gesellschaft 89, 565–571 (1985)
237. Haken, H.: Information, Information Gain, and Efficiency of Self-Organizing Systems Close to Instability Points. Zeitschrift für Physik B 61, 329–334 (1985)
238. Haken, H.: Pattern-Formation and Chaos in Synergetic Systems. Physica Scripta T9, 111–118 (1985)
239. Haken, H.: Application of the Maximum Information Entropy Principle to Self-Organizing Systems. Zeitschrift für Physik B 61, 335–338 (1985)
240. Haken, H.: Towards a Quantum Synergetics - Pattern-Formation in Quantum-Systems Far from Thermal-Equilibrium. Physica Scripta 32, 274–276 (1985)
241. Haken, H.: Synergetics -Self-Organization in Physics, in Nonlinear dynamics of transcritical flows. In: Jordan, H.L., Oertel, H., Robert, K. (eds.) Proceedings of a DFVLR International Colloquium, Bonn, Germany, March 26, 1984. Lecture Notes in Engineering. Springer, Berlin (1985)
242. Haken, H.: Synergetik - Selbstorganisationsvorgänge in Physik, Chemie und Biologie. Naturwissenschaftliche Rundschau 38, 171–218 (1985)
243. Haken, H.: Light II - Laser Light Dynamics. North Holland, Amsterdam (1985)
244. Haken, H.: Synergetics - an Interdisziplinary Approach to Phenomena of Self-Organization. Geoforum 16, 205–211 (1985)
245. Haken, H.: The Adiabatic Elimination Principle in Dynamical Theories in Optical Instabilities. In: Boyd, Raymer, Narducci (eds.) Instabilities and Dynamics of Lasers and Nonlinear Optical Systems. Cambridge University Press, Cambridge (1985)
246. Haken, H.: Operational Approaches to Complex Systems. In: Haken, H. (ed.) Complex Systems - Operational Approaches in Neurobiology, Physics and Computers, pp. 1–15. Springer, Berlin (1985)
247. Haken, H., Bestehorn, M.: Pattern Formation and Transients in the Convection Instability. In: Haken, H. (ed.) Complex Systems - Operational Approaches in Neurobiology, Physics and Computers, pp. 300–303. Springer, Berlin (1985)
248. Haken, H., Friedrich, R.: Convection in Spherical Geometries. In: Haken, H. (ed.) Complex Systems - Operational Approaches in Neurobiology, Physics and Computers, pp. 304–310. Springer, Berlin (1985)
249. Haken, H., Kelso, J.A.S., Bunz, H.: A Theoretical-Model of Phase-Transitions in Human Hand Movements. Biological Cybernetics 51, 347–356 (1985)
250. Haubs, G., Haken, H.: Quantities Describing Local Properties of Chaotic Attractors. Zeitschrift für Physik B 59, 459–468 (1985)
251. Schnaufer, B., Haken, H.: A Theoretical Derivation of Cellular Structures of Flames. Zeitschrift für Physik B 59, 349–356 (1985)

252. Zhang, J.Y., Haken, H.: Self-Pulsing Instabilities in the Generation of Higher Harmonics. Zeitschrift für Physik B 58, 337–343 (1985)
253. Zhang, J.Y., Haken, H., Ohno, H.: Self-Pulsing Instability in Inhomogeneously Broadened Traveling-Wave Lasers. Journal of the Optical Society of America B-Optical Physics 2, 141–147 (1985)
254. Friedrich, R., Haken, H.: Static, Wave-Like, and Chaotic Thermal-Convection in Spherical Geometries. Physical Review A 34, 2100–2120 (1986)
255. Haken, H.: The Maximum-Entropy Principle for Nonequilibrium Phase-Transitions - Determination of Order Parameters, Slaved Modes, and Emerging Patterns. Zeitschrift für Physik B 63, 487–491 (1986)
256. Haken, H.: Information and Information Gain Close to Nonequilibrium Phase-Transitions - Numerical Results. Zeitschrift für Physik B 62, 255–259 (1986)
257. Haken, H.: Excitons and the Electronic Polarization in Semiconductors. Festkörperprobleme-Advances in Solid State Phyics 26, 55–66 (1986)
258. Haken, H.: A New Access to Path-Integrals and Fokker Planck Equations Via the Maximum Caliber Principle. Zeitschrift für Physik B 63, 505–510 (1986)
259. Haken, H.: Morphogenesis of Behaviour, an example of concepts of synergetics. In: Diner, S. (ed.) Dynamical Systems, A Renewal of Mechanisms (Festschrift George, David Birkhoff), p. 133. World Scientific, Singapore (1986)
260. Haken, H.: Physik und Synergetik, die Vielfalt der Phänomene und die Einheit des Denkens. In: Maier-Leibnitz, H. (ed.) Zeugen des Wissens. Aus Anlass des Jubiläums 100 Jahre Automobil 1886 - 1986, pp. 157–201. Hase&Koehler, Mainz (1986)
261. Haken, H.: How can we implant semantics into information theory? In: Folberth, O., Hackl, C. (eds.) Der Informationsbegriff in Technik und Wissenschaft, p. 127. Oldenbourg, München (1986)
262. Haken, H.: Nachruf auf Professor Serge Nikitine. Physikalische Blätter 42 (1986)
263. Haken, H., Bunz, H.: Quantitative Theory of Changes in Oscillatory Hand Movements - Application of Methods of Synergetics. In: Rensing, An der Heiden, Mackey (eds.) Temporal Disorder in Human Oscillatory Systems, pp. 102–109. Springer, Berlin (1986)
264. Haken, H., Wunderlin, A.: Synergetik, Prozesse der Selbstorganisation in der belebten und unbelebten Natur. In: Dress, A., Hendrichs, Küppers (eds.) Selbstorganisation. Die Entstehung von Ordnung in Natur und Gesellschaft, München (1986)
265. Haken, H., Wunderlin, A.: Selforganization Processes in Complex Systems. In: Proceedings of the Third International Conference on Systems Research, Informatics and Cybernetics, Baden-Baden (1986)
266. Schöner, G., Haken, H.: The Slaving Principle for Stratonovich Stochastic Differential-Equations. Zeitschrift für Physik B 63, 493–504 (1986)
267. Schöner, G., Haken, H., Kelso, J.A.S.: A Stochastic-Theory of Phase-Transitions in Human Hand Movement. Biological Cybernetics 53, 247–257 (1986)
268. Friedrich, R., Haken, H.: Time-dependent and chaotic behaviour in systems with O(3)-symmetry. In: Güttinger, W., Dangelmayer, G. (eds.) The Physics of Structure Formation, pp. 334–345. Springer, Berlin (1987)
269. Fu, H., Haken, H.: Semiclassical Dye-Laser Equations and the Unidirectional Single-Frequency Operation. Physical Review A 36, 4802–4816 (1987)
270. Haken, H.: Model of a Chemical Parallel Computer for Pattern-Recognition and Associative Memory. Journal de Chimie Physique 84, 1289–1294 (1987)

271. Haken, H.: Self-Organization and Information. Physica Scripta 35, 247–254 (1987)
272. Haken, H.: Synergetic Information versus Shannon-Information in Self-Organizing Systems. Zeitschrift für Physik B 65, 503–504 (1987)
273. Haken, H.: Synergetic Phenomena in Nature and Technology. Fresenius Zeitschrift für Analytische Chemie 327, 60 (1987)
274. Haken, H.: Information Compression in Biological-Systems. Biological Cybernetics 56, 11–17 (1987)
275. Haken, H.: Thermodynamics - Synergetics – Life. Journal of Non-Equilibrium Thermodynamics 12, 1–10 (1987)
276. Haken, H.: Sind synergetische Systeme unsterblich? In: Kamper, D., Wulf, C. (eds.) Die Sterbende Zeit, pp. 169–174. Luchterland, Darmstadt (1987)
277. Haken, H.: Die Selbstorganisation der Information in biologischen Systemen aus Sicht der Synergetik. In: Küppers, B.-O. (ed.) Ordnung aus dem Chaos, pp. 127–156. Piper, München (1987)
278. Haken, H.: Synergetik und ihre Anwendung auf psychosoziale Probleme. In: Stierlin, H., Simon, F.B., Schmidt, G. (eds.) Familiäre Wirklichkeiten. Der Heidelberger Kongreß, pp. 36–50. Klett Cotta, Stuttgart (1987)
279. Haken, H.: Synergetische Phänomene in Natur und Technik. Zeitschrift für Analytische Chemie 327, 60 (1987)
280. Haken, H.: Morphogenese des Verhaltens, Ein Beispiel der Anwendung der Konzepte der Synergetik. Physiologie Aktuell 3, 113–122 (1987)
281. Haken, H.: Synergetics, An Approach to Self-Organization. In: Yates, F.E. (ed.) Self-Organizing Systems - The Emergence of Order. Plenum, New York (1987)
282. Haken, H.: Synergetic Computers for Pattern Recognition and Associative Memory. In: Haken, H. (ed.) Computational Systems - Natural and Artificial, pp. 2–23. Springer, Berlin (1987)
283. Haken, H.: Synergetics, from pattern formation to pattern recognition. In: Velarde, M. (ed.) Synergetics, Order and Chaos, pp. 3–20. World Scientific, Singapore (1987)
284. Haken, H., Wunderlin, A.: Synergetics and its paradigm of Selforganization in Biological Systems. In: Whiting, H.T., Meijer, O.G., van Wieringen, S.C. (eds.) The Natural - Physical Approach to Movement Control. University Press, Amsterdam (1987)
285. Haken, H., Wunderlin, A.: Selforganization in Biological Systems. In: Möller, D.S. (ed.) Advances in System Analysis. Vieweg, Braunschweig (1987)
286. Hong, F., Haken, H.: A Band-Model for Dye-Lasers and the Low Threshold of the Second Instability. Optics Communications 64, 454–456 (1987)
287. Kelso, J.A.S., et al.: Nonequilibrium Phase-Transitions in Coordinated Movements Involving Many Degrees of Freedom. Annals of the New York Academy of Sciences 504, 293–296 (1987)
288. Kelso, J.A.S., et al.: Phase-Locked Modes, Phase-Transitions and Component Oscillators in Biological Motion. Physica Scripta 35, 79–87 (1987)
289. Mayerkress, G., Haken, H.: An Explicit Construction of a Class of Suspensions and Autonomous Differential-Equations for Diffeomorphisms in the Plane. Communications in Mathematical Physics 111, 63–74 (1987)
290. Obermayer, K., Mahler, G., Haken, H.: Multistable Quantum-Systems - Information-Processing at Microscopic Levels. Physical Review Letters 58, 1792–1795 (1987)
291. Peplowski, S., Haken, H.: Effects of Detuning on Hopf-Bifurcation at Double Eigenvalues in Laser Systems. Physics Letters A 120, 138–140 (1987)

292. Schöner, G., Haken, H.: A Systematic Elimination Procedure for Ito Stochastic Differential-Equations and the Adiabatic Approximation. Zeitschrift für Physik B 68, 89–103 (1987)
293. Bestehorn, M., Friedrich, R., Haken, H.: The Oscillatory Instability of a Spatially Homogeneous State in Large Aspect Ratio Systems of Fluid-Dynamics. Zeitschrift für Physik B 72, 265–275 (1988)
294. Cheng, W.Z., Haken, H.: Quantum-Theory of the 2-Photon Laser. Zeitschrift für Physik B 71, 253–259 (1988)
295. Friedrich, R., Haken, H.: Exact Stationary Probability-Distribution for a Simple-Model of a Nonequilibrium Phase-Transition. Zeitschrift für Physik B 71, 515–517 (1988)
296. Fu, H., Haken, H.: Multichromatic Operations in Cw Dye-Lasers. Physical Review Letters 60, 2614–2617 (1988)
297. Fu, H., Haken, H.: Semiclassical Theory of Dye-Lasers - the Single-Frequency and Multifrequency Steady-States of Operation. Journal of the Optical Society of America B-Optical Physics 5, 899–908 (1988)
298. Fuchs, A., Haken, H.: Pattern-Recognition and Associative Memory as Dynamical Processes in a Synergetic System. Part 2. Decomposition of Complex Scenes, Simultaneous Invariance with Respect to Translation, Rotation, and Scaling. Biological Cybernetics 60, 107–109 (1988)
299. Fuchs, A., Haken, H.: Nonequilibrium Phase-Transitions in Pattern-Recognition and Associative Memory - Numerical Results. Zeitschrift für Physik B 71, 519–520 (1988)
300. Fuchs, A., Haken, H.: Pattern-Recognition and Associative Memory as Dynamical Processes in a Synergetic System. Part 1. Translational Invariance, Selective Attention, and Decomposition of Scenes. Biological Cybernetics 60, 17–22 (1988)
301. Fuchs, A., Haken, H.: Erratum, Pattern-Recognition and Associative Memory as Dynamical Processes in a Synergetic System. Biological Cybernetics 60, 476 (1988)
302. Haken, H.: Synergetics. IEEE Circuits and Devices Magazine 4, 3–7 (1988)
303. Haken, H.: Synergetics and Computers. Journal of Computational and Applied Mathematics 22, 197–202 (1988)
304. Haken, H.: Entwicklungslinien der Synergetik. Teil 1. Naturwissenschaften 75, 163–172 (1988)
305. Haken, H.: Entwicklungslinien der Synergetik. Teil 2. Naturwissenschaften 75, 225–234 (1988)
306. Haken, H.: Learning in Synergetic Systems for Pattern-Recognition and Associative Action. Zeitschrift für Physik B 71, 521–526 (1988)
307. Haken, H.: Nonequilibrium Phase-Transitions in Pattern-Recognition and Associative Memory. Zeitschrift für Physik B 70, 121–123 (1988)
308. Haken, H.: Geschichte der Synergetik. In: Niedersen, U. (ed.) Komplexität - Zeit - Methode, vol. 3, pp. 198–231 (1988)
309. Haken, H.: Der Computer erkennt mein Gesicht. Bild der Wissenschaft 1988(8) (1988)
310. Haken, H.: Information and Self-Organization. A Macroscopic Approach to Complex Systems. Springer, Berlin (1988)
311. Haken, H.: Synergetics - From Physics to Biology. In: Lundqvist, S., March, N.H., Tosi, M.S. (eds.) Order and Chaos in Nonlinear Physical Systems. Plenum, New York (1988)

312. Haken, H.: Sinergetica, Concetti di Base e Strumenti Matematici. In: La vita. Forme i numeri Biologica. Ancona Transeuropa, Ancona (1988)
313. Haken, H.: Synergetics in pattern recognition and associative action. In: Haken, H. (ed.) Neural and Synergetic Computers, p. 2 (1988)
314. Haken, H.: Pattern Formation, Thermodynamics or Kinetics? In: Markus, M., Müller, S.C., Nicolis, G. (eds.) From Chemical to Biological Organization, pp. 6–13. Springer, Berlin (1988)
315. Haken, H.: Morphogenesis of Behaviour and Information Compression in Biological Systems. In: Lamprecht, I., Zotin, A.I. (eds.) Thermodynamics and Pattern Formation in Biology. Walter de Gruyter, Berlin (1988)
316. Haken, H.: Synergetics, Its Microscopic and Macroscopic Foundation. In: Caglioti, G., Haken, H. (eds.) Synergetics and Dynamic Instabilities (= Proc. Int. School of Physics Enrico Fermi, Course IC), pp. 1–19. Soc. Italiana die Fisica, Bologna (1988)
317. Haken, H.: Neural and Synergetic Computers. Springer, Berlin (1988)
318. Haken, H., Weimer, W.: Synergetics, a systematic approach to instabilities, fluctuations and pattern formation and the application to the baroclinic instability of the atmosphere. In: Scheidegger, A.E. (ed.) Dynamic Systems Approach to Natural Hazards (= Zeitschrift für Geomorphologie Suppl. 67), p. 103. Gebrüder Bornträger, Berlin (1988)
319. Haken, H., Wunderlin, A.: Synergetics - Processes of Self-Organization in Complex Systems. In: Carvallo, M.E. (ed.) Nature, Cognition and Systems, pp. 279–290. Kluwer Academic Publishers, Boston (1988)
320. Marx, K., Haken, H.: The Generalized Ginzburg-Landau Equations of Wavy Vortex Flow. Europhysics Letters 5, 315–320 (1988)
321. Obermayer, K., Mahler, G., Haken, H.: Multistable Quantum-Systems - Information-Processing at Microscopic Levels – Reply. Physical Review Letters 60, 658–658 (1988)
322. Peplowski, S., Haken, H.: Bifurcation with 2 Parameters in Two-Dimensional Complex-Space - Applications to Laser Systems. Physica D 30, 135–150 (1988)
323. Wunderlin, A., Haken, H.: The Slaving Principle of Synergetics - An Outline. In: Lundqvist, S., Tosi, M.S., March, N.H. (eds.) Order and Chaos in Nonlinear Systems. Plenum Press, New York (1988)
324. Banzhaf, W., Haken, H.: A new dynamical approach to the traveling salesman problem. Physics Letters A A136 (1989)
325. Bestehorn, M., et al.: Synergetics applied pattern formation and pattern recognition. In: Plath, S. (ed.) Optimal Structures in Heterogeneous Reaction Systems. Springer, Berlin (1989)
326. Bestehorn, M., Friedrich, R., Haken, H.: Traveling Waves in Nonequilibrium Systems. Physica D 37, 295–299 (1989)
327. Bestehorn, M., Friedrich, R., Haken, H.: Modulated Traveling Waves in Nonequilibrium Systems - the Blinking State. Zeitschrift für Physik B 77, 151–155 (1989)
328. Bestehorn, M., Friedrich, R., Haken, H.: Two-Dimensional Traveling Wave Patterns in Nonequilibrium Systems. Zeitschrift für Physik B 75, 265–274 (1989)
329. Ditzinger, T., Haken, H.: Oscillations in the Perception of Ambiguous Patterns - a Model Based on Synergetics. Biological Cybernetics 61, 279–287 (1989)
330. Friedrich, R., Haken, H.: A short course on synergetics. In: Proto, A. (ed.) Nonlinear Phenomena in Complex Systems, pp. 103–150. Elsevier, Amsterdam (1989)

331. Fu, H., Haken, H., Wunderlin, A.: The Convergence of the Slaving Principle in a Simplified Model. Zeitschrift für Physik B 76, 127–135 (1989)
332. Gang, H., Haken, H.: Polynomial Expansion of the Potential of Fokker-Planck Equations with a Noninvertible Diffusion Matrix. Physical Review A 40, 5966–5978 (1989)
333. Gang, H., Haken, H.: Multimode Instability Criterion in Optical Bistable Systems. Physical Review A 40, 1899–1907 (1989)
334. Gang, H., Haken, H.: Application of the Slaving Principle to Quenching Problems. Zeitschrift für Physik B 76, 537–545 (1989)
335. Haken, H.: Synergetics - an Overview. Reports on Progress in Physics 52, 515–553 (1989)
336. Haken, H.: Realization of the Logical Operation Xor by a Synergetic Computer. Progress of Theoretical Physics Supplement 99, 399–403 (1989)
337. Haken, H.: Synergetik, Vom Chaos zur Ordnung und weiter ins Chaos. In: Gerok, W. (ed.) Ordnung und Chaos in der unbelebten und belebten Natur (=115. Versammlung der Gesellschaft deutscher Naturforscher und Ärzte, Freiburg 1988), pp. 65–76. Hirzel, Stuttgart (1989)
338. Haken, H.: Synergetischer Computer - ein nichtbiologischer Zugang. Physikalische Blätter 45, 168–171 (1989)
339. Haken, H.: Synergetik - Nichtgleichgewichts-Phasenübergänge und Selbstorganisation in Physik, Chemie, Biologie. Nova Acta Leopoldina NF 60, 75–89 (1989)
340. Haken, H.: Synergetik - eine interdisziplinäre Theorie der Selbstorganisation. In: Weingartner, S., Schurz, G. (eds.) Akten des 13. Internationalen Wittgenstein-Symposiums, Kirchberg, Österreich, August 14-August 21, pp. 231–242. Hölder-Pichler-Tempsky, Wien (1989)
341. Haken, H., Haas, R., Banzhaf, W.: A New Learning Algorithm for Synergetic Computers. Biological Cybernetics 62, 107–111 (1989)
342. Haken, H., Haken-Krell, M.: Entstehung von Biologischer Information und Ordnung. Wissenschaftliche Buchgesellschaft, Darmstadt (1989)
343. Haken, H., Wunderlin, A.: A macroscopic approach to Synergetics. In: Christiansen, S., Parmentier, R. (eds.) Structure, Coherence and Chaos in Dynamical Systems. Manchester University Press, Manchester (1989)
344. Marx, K., Haken, H.: Numerical Derivation of the Generalized Ginzburg-Landau Equations of Wavy Vortex Flow. Zeitschrift für Physik B 75, 393–411 (1989)
345. Ning, C.Z., Haken, H.: Generalized Ginzburg-Landau Equation for Self-Pulsing Instability in a 2-Photon Laser. Zeitschrift für Physik B 77, 163–174 (1989)
346. Ning, C.Z., Haken, H.: Instability in Degenerate 2-Photon Running Wave Laser. Zeitschrift für Physik B 77, 157–162 (1989)
347. Weimer, W., Haken, H.: Chaotic Behavior and Subcritical Formation of Flow Patterns of Baroclinic Waves for Finite Dissipation. Journal of the Atmospheric Sciences 46, 1207–1218 (1989)
348. Banzhaf, W., Haken, H.: An Energy Function for Specialization. Physica D 42, 257–264 (1990)
349. Banzhaf, W., Haken, H.: Learning in a Competitive Network. Neural Networks 3, 423–435 (1990)
350. Bestehorn, M., Haken, H.: Traveling Waves and Pulses in a 2-Dimensional Large-Aspect-Ratio System. Physical Review A 42, 7195–7203 (1990)

351. Bestehorn, M., Haken, H.: Synergetics applied to Pattern Formation in Large - Aspect - Ratio Systems. In: Meinköhn, D. (ed.) Dissipative Structures in Transport Processes and Combustion, pp. 110–143. Springer, Berlin (1990)
352. Ditzinger, T., Haken, H.: The Impact of Fluctuations on the Recognition of Ambiguous Patterns. Biological Cybernetics 63, 453–456 (1990)
353. Gang, H., Haken, H.: Steepest-Descent Approximation of Stationary Probability-Distribution of Systems Driven by Weak Colored Noise. Physical Review A 41, 7078–7081 (1990)
354. Gang, H., Ning, C.Z., Haken, H.: Codimension-2 Bifurcations in Single-Mode Optical Bistable Systems. Physical Review A 41, 2702–2711 (1990)
355. Haken, H.: Are Synergetic Sytems Immortal? In: Robert, L. (ed.) The Theoretical Basis of Ageing (= 7. Wiener Symposium on Experimental Gerontology 1988). Facultas Universitätsverlag, Wien (1990)
356. Haken, H.: Über das Verhältnis der Synergetik zur Thermodynamik, Kybernetik und Informationstheorie. In: Niedersen, U., Pohlmann, L. (eds.) Selbstorganisation und Determination (= Selbstorganisation. Jahrbuch für Komplexität u.a.), p. 19. Duncker & Humblot, Berlin (1990)
357. Haken, H.: Synergetics, Theory of Self-Organization in Complex Systems. Methods of Operation Research. In: Fuchssteiner, B. (ed.) XIII Symposium on Operation Research, p. 39. Anton Hain, Frankfurt am Main (1990)
358. Haken, H.: Synergetic Computers - a new concept. In: Kindermann, J. (ed.) Distributed Adaptive Neural Information Processing, p. 39. Oldenbourg Publisher, München (1990)
359. Haken, H.: Synergetics as a tool for the conceptualization and mathematization of Cognition and Behaviour - How far can we go? In: Haken, H., Stadler, M. (eds.) Synergetics of Cognition, pp. 2–31. Springer, Heidelberg (1990)
360. Haken, H.: Realization of the Logical Operation XOR by a Synergetic Computer. Progress of Theoretical Physics Supplement 99, 399–403 (1990)
361. Haken, H.: Synergetik und die Einheit der Wissenschaft. In: Saltzer, W.G. (ed.) Zur Einheit der Naturwissenschaften in Geschichte und Gegenwart, p. 61. Wissenschaftliche Buchgesellschaft, Darmstadt (1990)
362. Haken, H.: Synergetic Computers - an alternative to neurocomputers. In: Garrido, L. (ed.) Proceedings of the XIth Sitges Conference on Statistical Mechanics of Neural Networks, Sitges, Barcelona, Spain, June 3-7, vol. VI, p. 477. Springer, Berlin (1990)
363. Haken, H.: From Laser Physics to Pattern Recognition, in Between science and technology. In: Sarlemijn, A., Kroes, S. (eds.) Proceedings of the International Conference Between Science and Technology, Eindhoven University of Technology, The Netherlands, June 29-30, pp. 155–178. North-Holland, Amsterdam (1990)
364. Haken, H.: Recognition of Patterns and Movement Patterns by a Synergetic Computer. In: Eckmiller, R. (ed.) Parallel Processing in Neural Networks and Computers, pp. 451–458. North-Holland, Amsterdam (1990)
365. Haken, H.: Synergetics as a Theory of Analogous Behaviour of Systems. In: van Haeringen, W. (ed.) Proceedings of the International Symposium on Analogy in Optics and Microelectronics, pp. 35–47 (1990)
366. Haken, H.: Offene Systeme - die merkwürdige Welt des Nichtgleichgewichts. Physikalische Blätter 46, 203–208 (1990)
367. Haken, H., Bestehorn, M., Friedrich, R.: Pattern formation in convective instabilities. International Journal of Modern Physics B4, 365–400 (1990)

368. Haken, H., et al.: Dynamic Pattern-Recognition of Coordinated Biological Motion. Neural Networks 3, 395–401 (1990)
369. Haken, H., Wunderlin, A.: Le Chaos Deterministe. La Recherche 21, 1248–1255 (1990)
370. Haken, H., Wunderlin, A.: Die Anwendung der Synergetik auf Musterbildung und Mustererkennung. In: Kratky, K., Wallner, F. (eds.) Grundprinzipien der Selbstorganisation, pp. 18–30. Wissenschaftliche Buchgesellschaft, Darmstadt (1990)
371. Haken, H., Wunderlin, A.: Die Entstehung der Ordnung aus dem Chaos. In: Fischer, E.S., Mainzer, K. (eds.) Die Frage nach dem Leben. Zum 100. Geburtstag des Physikers Erwin Schrödinger, p. 149. Piper, München (1990)
372. Haken, H., Wunderlin, A.: La inspiracio cientifica los fondamentod de la sinergetica. In: Haken, H., Losa, J.M. (eds.) Sobre la Imaginaci6n Científica, Una Convocatoria de Jorge Wagensberg. Tusquets, Barcelona (1990)
373. Hong, F., Haken, H.: Self-Pulsing in a Band Model for Dye-Lasers. Physical Review A 42, 4151–4163 (1990)
374. Hu, G., Haken, H.: Potential of the Fokker-Planck Equation at Degenerate Hopf-Bifurcation Points. Physical Review A 41, 2231–2234 (1990)
375. Ning, C.Z., Haken, H.: Detuned Lasers and the Complex Lorenz Equations - Subcritical and Supercritical Hopf Bifurcations. Physical Review A 41, 3826–3837 (1990)
376. Hu, G., Ning, C.Z., Haken, H.: Distribution of Subcritical Hopf Bifurcations and Regular and Chaotic Attractors in Optical Bistable Systems. Physical Review A 41, 3975–3984 (1990)
377. Ning, C.Z., Haken, H.: Multistabilities and Anomalous Switching in the Lorenz-Haken Model. Physical Review A 41, 6577–6580 (1990)
378. Ning, C.Z., Haken, H.: Quasi-Periodicity Involving Twin Oscillations in the Complex Lorenz Equations Describing a Detuned Laser. Zeitschrift für Physik B 81, 457–461 (1990)
379. Bestehorn, M., Haken, H.: Associative Memory of a Dynamic System - the Example of the Convection Instability. Zeitschrift für Physik B 82, 305–308 (1991)
380. Bestehorn, M., Haken, H.: Stationary and Travelling Pulses of the 2D Complex Ginzburg-Landau Equation. Europhysics Letters 15, 473–478 (1991)
381. Friedrich, R., Fuchs, A., Haken, H.: Modelling of Spatio-Temporal EEG patterns. In: Dvorak, I., Holden, A.V. (eds.) Mathematical Approaches to Brain Functioning Diagnostics, pp. 45–61. Manchester University Press, Manchester (1991)
382. Fu, H., Haken, H.: Multifrequency Operations in a Short-Cavity Standing-Wave Laser. Physical Review A 43, 2446–2454 (1991)
383. Haken, H.: Schöner-Haken Systematic Adiabatic Approximation for Stochastic Differential-Equations – Comment. Zeitschrift für Physik B 82, 465–465 (1991)
384. Haken, H.: Information-Flow in Synergetic Computers. Annalen der Physik 48, 97–102 (1991)
385. Haken, H.: Konzepte und Modellvorstellungen der Synergetik zum Gedächtnis. In: Schmidt, S. (ed.) Gedächtnis - Probleme und Perspektiven der interdisziplinären Gedächtnisforschung, pp. 190–205. Suhrkamp, Frankfurt am Main (1991)
386. Haken, H.: Synergetics - can it help physiology? In: Haken, H., Koepchen, H.S. (eds.) Rhythms in Physiological Systems, pp. 21–34. Springer, Berlin (1991)
387. Haken, H.: Synergetik. In: Möller, D.S. (ed.) Analyse dynamischer Systeme in Medizin, Biologie und Ökologie. Springer, Berlin (1991)

388. Haken, H.: Synergetik im Management. In: Balck, H., Kreibich, R. (eds.) Evolutionäre Wege in die Zukunft, pp. 65–91. Beltz, Weinheim (1991)
389. Haken, H.: Synergetik, Ordnung und Chaos. Selbstorganisierte Veränderung von natürlichen Systemen. Natur- und Ganzheitsmedizin 4, 134–140 (1991)
390. Haken, H.: From Physics to Synergetics. In: Suzuki, M., Kubo, R. (eds.) Trends in the Physical Sciences, pp. 203–212. Springer, Berlin (1991)
391. Haken, H.: Synergetic Computers and Cognition. Springer, Berlin (1991)
392. Haken, H.: Synergetik. Von der Mustererkennung zur Musterbildung. Wissenschaft und Fortschritt 41, 244–252 (1991)
393. Haken, H.: Je mehr wir Grenzen ausloten, um so mehr erfahren wir vom Menschen - hoffentlich. In: Oesterreicher-Mollwo, M. (ed.) Was uns bewegt, Naturwissenschaftler sprechen über sich und ihre Welt, pp. 186–193. Beltz, Weinheim (1991)
394. Haken, H., Friedrich, R., Fuchs, A.: Synergetic Analysis of Spatio-Temporal EEG Patterns. In: Holden, A.V. (ed.) Nonlinear Wave Processes in Excitable Media, pp. 23–37. Plenum Press, New York (1991)
395. Haken, H., Koepchen, H.S.: Rhythms in Physiological Systems. Springer, Berlin (1991)
396. Haken, H., Wunderlin, A.: On a macroscopic approach to synergetic systems. In: Natural Structures. Principles, Strategies and Models in Architechture and Nature. U.S. SFB230, Ed. Universität Stuttgart, Stuttgart (1991)
397. Haken, H., Wunderlin, A.: Die Selbstrukturierung der Materie. Synergetik in der unbelebten Natur. Braunschweig, Vieweg (1991)
398. Haken, H., Wunderlin, A.: Application of Synergetics to Pattern Information and Pattern Recognition. In: Babloyantz, A. (ed.) Selforganization, Emerging Properties and Learning, pp. 21–30. Plenum Press, New York (1991)
399. Ning, C.Z., Haken, H.: Phase Anholonomy in Dissipative Optical-Systems with Periodic Oscillations. Physical Review A 43, 6410–6413 (1991)
400. Ning, C.Z., Haken, H.: Phase Anholonomy and Quasiperiodicity in Optical Systems with Intensity Pulsations. In: Vilaseca, R., Corbalan, R. (eds.) Nonlinear Dynamics and Quantum Phenomena in Optical Systems, pp. 185–189. Springer, Berlin (1991)
401. Wunderlin, A., Haken, H.: Zur modernen Chaostheorie. DSV 134 (1991)
402. Bestehorn, M., et al.: Spiral Patterns in Thermal-Convection. Zeitschrift für Physik B 88, 93–94 (1992)
403. Borland, L., Haken, H.: Unbiased Determination of Forces Causing Observed Processes - the Case of Additive and Weak Multiplicative Noise. Zeitschrift für Physik B 88, 95–103 (1992)
404. Borland, L., Haken, H.: Unbiased Estimate of Forces from Measured Correlation-Functions, Including the Case of Strong Multiplicative Noise. Annalen der Physik 1(6), 452–459 (1992)
405. Borland, L., Haken, H.: Learning the Dynamics of Two-Dimensional Stochastic Markov Processes. Open Systems and Information Dynamics 1, 311–326 (1992)
406. Fantz, M., et al.: Pattern-Formation in Rotating Benard Convection. Physica D 61, 147–154 (1992)
407. Friedrich, R., Haken, H.: Nonequilibrium Phase-Transition in a System with Chaotic Dynamics - the ABCDE Model. Physics Letters A 164, 299–304 (1992)
408. Haken, H.: Some Applications of Synergetics to the Study of Sociotechnical Systems. Journal of Scientific & Industrial Research 51, 147–150 (1992)

409. Haken, H.: Synergetik, Von der Musterbildung zur Mustererkennung. Nova Acta Leopoldina NF 67, 53–71 (1992)
410. Haken, H.: Synergetics in Psychology. In: Tschacher, W., Schiepek, G., Brunner, E. (eds.) Self-Organization and Clinical Psychology, pp. 32–54. Springer, Berlin (1992)
411. Haken, H.: Von der Ordnung zum Chaos oder Chaos ist nicht gleich Chaos (=Bensberger Protokolle 69). In: Haken, H. (ed.) Auf den Spuren des Tohuwabohu. Zu neueren Ergebnissen der Chaosforschung, pp. 9–41. Thomas Morus Akademie, Bensberg (1992)
412. Haken, H.: Synergetics and Cognitive Maps. Geoforum 23, 111–130 (1992)
413. Haken, H.: Die Synergetik als Vision eines holistischen Weltbildes (Festvortrag Salzburg 1992 der Academia Scientia et Artium Europaea), Unveröffentlicht (1992)
414. Haken, H., Haken-Krell, M.: Erfolgsgeheimnisse der Wahrnehmung. Deutsche Verlagsanstalt DVA, Stuttgart (1992)
415. Haken, H., Kelso, J.A.S., Fuchs, A.: Phase Transitions in the Human Brain, Spatial Mode Dynamics. International Journal of Bifurcation and Chaos 2, 917–939 (1992)
416. Haken, H., Wolf, H.C.: Molekülphysik und Quantenchemie. Springer, Berlin (1992)
417. Haken, H., Wunderlin, A.: Synergetics and its paradigm of self-organization in biological systems. In: Whiting, A.T., Meijer, O.G. (eds.) The Natural Physical Approach to Movement Control, pp. 37–56. Free University Press, Amsterdam (1992)
418. Hu, G., Haken, H., Ning, C.Z.: A Study of Stochastic Resonance without Adiabatic Approximation. Physics Letters A 172, 21–28 (1992)
419. Ning, C.Z., Haken, H.: An Invariance Property of the Geometrical Phase and Its Consequence in Detuned Lasers. Zeitschrift für Physik B 89, 261–262 (1992)
420. Ning, C.Z., Haken, H.: Elimination of Variables in Simple Laser Equations. Applied Physics B 55, 117–120 (1992)
421. Ning, C.Z., Haken, H.: Geometrical Phase and Amplitude Accumulations in Dissipative Systems with Cyclic Attractors. Physical Review Letters 68, 2109–2112 (1992)
422. Ning, C.Z., Haken, H.: The geometric phase in nonlinear dissipative systems. Physics Letters A B6, 1541–1568 (1992)
423. Weimer, W., Haken, H.: Generalized Ginzburg-Landau Equations for 4 Unstable Baroclinic Waves. Journal of the Atmospheric Sciences 49, 453–461 (1992)
424. Bestehorn, M., et al.: Hexagonal and Spiral Patterns of Thermal-Convection. Physics Letters A 174, 48–52 (1993)
425. Borland, L., Haken, H.: On the Constraints necessary for Macroscopic Prediction of Time-dependent Stochastic Processes. Reports on Mathematical Physics 33, 35–42 (1993)
426. Borland, L., Haken, H.: Learning Networks for Process Identification and associative action. In: Mira, J., Cabestany, J., Prieto, A.G. (eds.) IWANN 1993. LNCS, vol. 686, pp. 688–693. Springer, Heidelberg (1993)
427. Dykman, M.I., et al.: Linear-Response Theory in Stochastic Resonance. Physics Letters A 180, 332–336 (1993)
428. Gang, H., et al.: Stochastic Resonance without External Periodic Force. Physical Review Letters 71, 807–810 (1993)
429. Gang, H., Haken, H., Ning, C.Z.: Nonlinear-Response Effects in Stochastic Resonance. Physical Review E 47, 2321–2325 (1993)
430. Haken, H.: Basic Concepts of Synergetics. Applied Physics A 57, 111–115 (1993)
431. Haken, H.: Information-Theory and Molecular-Biology. Nature 362, 509–509 (1993)

432. Haken, H.: Are Synergetic Systems (including Brains) Machines? In: Haken, H. (ed.) The Machine as Metaphor and Tool, pp. 123–138. Springer, Berlin (1993)
433. Haken, H.: Pattern Formation and Pattern Recognition. In: Harrison, R.G. (ed.) Nonlinear Dynamics and Spatial Complexity in Optical Systems, p. 21. Institute of Physics Publishing, Bristol (1993)
434. Haken, H.: Synergetics as a Strategy to cope with Complex Systems. In: Haken, H., Mikhailov, A. (eds.) Interdisciplinary Approaches to Nonlinear Complex Systems, pp. 5–11. Springer, Berlin (1993)
435. Haken, H.: Synergetik, eine Zauberformel für das Management? In: Rehm, W. (ed.) Synergetik, Selbstorganisation als Erfolgsrezept für Unternehmen, p. 15. Expert-Verlag, Ehingen b. Stuttgart (1993)
436. Haken, H.: An Algorithm for The Recognition of Deformes Patterns including Hand Writing. Journal of the Mathematical and Physical Sciences 25 (1993)
437. Neufeld, M., Friedrich, R., Haken, H.: Order-Parameter Equation and Model Equation for High Prandtl Number - Rayleigh-Benard Convection in a Rotating Large Aspect Ratio System. Zeitschrift für Physik B 92, 243–256 (1993)
438. Uhl, C., Friedrich, R., Haken, H.: Reconstruction of Spatiotemporal Signals of Complex-Systems. Zeitschrift für Physik B 92, 211–219 (1993)
439. Wischert, W., et al.: An Introduction to Synergetics. In: Greppin, H., Bonzon, M., Agostini, R.D. (eds.) Some Physicochemical and Mathematical Tools for the Understanding of Living Systems. University of Geneva, Geneva (1993)
440. Bestehorn, M., et al.: Spiral-Pattern Formation in Rayleigh-Benard Convection – Comment. Physical Review E 50, 625–626 (1994)
441. Biktashev, V., Krinsky, V., Haken, H.: A Wave Approach to Pattern-Recognition (with Application to Optical Character-Recognition). International Journal of Bifurcation and Chaos 4, 193–207 (1994)
442. Daffertshofer, A., Haken, H.: A New Approach to Recognition of Deformed Patterns. Pattern Recognition 27, 1697–1705 (1994)
443. Daffertshofer, A., Haken, H.: Synergetic Computers for Pattern Recognition. In: Jamiolkowski, A. (ed.) Proceedings of the XXVI Symposium on Mathematical Physics. Nicolaus Copernicus University Press, Torun (1994)
444. Haken, H.: A Brain Model for Vision in Terms of Synergetics. Journal of Theoretical Biology 171, 75–85 (1994)
445. Haken, H.: Synergetics - from Pattern-Formation to Pattern-Analysis and Pattern-Recognition. International Journal of Bifurcation and Chaos 4, 1069–1083 (1994)
446. Haken, H.: From Cybernetics to Synergetics. Cybernetica 37, 273–290 (1994)
447. Haken, H.: Can Synergetics Serve as a bridge between the Natural and Social Sciences. In: Mishra, R.K., Maaß, D., Zwierlein, E. (eds.) On Self-Organization. Springer, Berlin (1994)
448. Haken, H.: Kunstwerke rufen Instabilitäten hervor. In: Rötzer, F. (ed.) Vom Chaos zur Endophysik. Wissenschaftler im Gespräch, pp. 52–57. Boer Publisher (1994)
449. Haken, H.: Strukturentstehung und Gestalterkennung in den neuen Selbstorganisationstheorien. In: Heuser-Kessler, L., Jacobs, W. (eds.) Schelling und die Selbstorganisation (= Selbstorganisation 5), pp. 11–26. Duncker & Humblot, Berlin (1994)
450. Haken, H.: Synergetik, Vom Zusammenwirken in der Natur. Zum Naturbegriff der Gegenwart. In: Kongressdokumentation zum Projekt "Natur im Kopf". Kulturamt der Landeshauptstadt Stuttgart. Frommann-Holzboog, Stuttgart (1994)

451. Haken, H., et al.: Fragestellungen der modernen Chaostheorie. Photon 2, 8–12 (1994)
452. Haken, H., Schleich, W., Vollmer, H.D.: Risken,Hannes. Physics Today 47, 118 (1994)
453. Haken, H., Wagensberg, J.: Eine Einführung in die Synergetik. In: Mainzer, K., Schirrmacher, W. (eds.) Quanten, Chaos und Dämonen. Erkenntnistheoretische Aspekte der Modernen Physik. BI Wissenschaftsverlag, Mannheim (1994)
454. Jirsa, V.K., et al.: A Theoretical-Model of Phase-Transitions in the Human Brain. Biological Cybernetics 71, 27–35 (1994)
455. Reimann, D., Haken, H.: Stereo Vision by Self-Organization. Biological Cybernetics 71, 17–26 (1994)
456. Wischert, W., Wunderlin, A., Haken, H.: Prinzipien der Synergetik. In: Bien, G., Gil, T., Wilke, J. (eds.) Natur im Umbruch, p. 195. Fromman Publisher, Stuttgart (1994)
457. Daffertshofer, A., Haken, H.: Adaptive hierarchical structures. In: Sandoval, F., Mira, J. (eds.) IWANN 1995. LNCS, vol. 930, pp. 76–84. Springer, Heidelberg (1995)
458. Daffertshofer, A., Haken, H.: Dynamical Construction of Pattern Classes. Neural Network World 3, 255–270 (1995)
459. Gang, H., Ning, C.Z., Haken, H.: Inverse Problem and Singularity of the Integration Kernel. Physics Letters A 205, 130–136 (1995)
460. Haken, H.: Irreversibility and Self-Organization. In: Zwilling, R. (ed.) Natural Sciences and Human Thought, p. 125. Springer, Berlin (1995)
461. Haken, H.: Some basic Concepts of Synergetics with Respect to Multistability. In: Kruse, S., Stadler, M. (eds.) Perception, Phase Transitions and Formation of Meaning, Ambiguity in Mind and Nature. Springer, Berlin (1995)
462. Haken, H.: Synergetics, From Pattern Formation to Pattern Analysis and Pattern Recognition. International Journal of Bifurcation and Chaos 4, 1069–1083 (1995)
463. Haken, H.: Synergetics, From Pattern Formation to Pattern Analysis and Pattern Recognition. In: Perez Villar, V. (ed.) New Trends in Pattern Formation in Active Nonlinear Media. World Scientific, Singepore (1995)
464. Haken, H.: Ordnung aus dem Chaos. In: Gorge, V., Moser, R. (eds.) Begegnungen mit dem Chaos. Berner Universitätsschriften, Bern (1995)
465. Haken, H.: Synergetische Prinzipien bei der Ordnungsbildung. Gestalt Theory 17, 196–204 (1995)
466. Haken, H.: Laws and Chaos. In: Weinert, F. (ed.) Laws of Nature - Essays on the Philosophical, Scientific and Historical Dimension. De Gruyter, Berlin (1995)
467. Haken, H.: Synergetische Computer - ein neues Netzwerkprinzip. In: Eisenberg, W. (ed.) Synergie - Syntropie - Nichtlineare Systeme, Dynamik und Synergetik. Verlag im Wissenschaftszentrum, Leipzig (1995)
468. Haken, H.: An Application of Synergetics. Decision Making as Pattern Recognition. Zeitschrift für Wissenschaftsforschung 9/10 (1995)
469. Haken, H., et al.: Fragestellungen und Resultate der modernen Chaosforschung. Mannheimer Berichte 44, 19 (1995)
470. Haken, H., Wunderlin, A., Yigitbasi, S.: An Introduction to Synergetics. Open Systems and Information Dynamics 3, 97–130 (1995)
471. Jirsa, V.K., Friedrich, R., Haken, H.: Reconstruction of the spatio-temporal dynamics of a human magnetoencephalogram. Physica D 89, 100–122 (1995)

472. Portugali, J., Haken, H.: A Synergetic Approach to the Self-Organization of Cities and Settlements. Environment and Planning B, Planning and Design 22, 35–46 (1995)
473. Reimann, D., et al.: Vergence Eye-Movement Control and Multivalent Perception of Autostereograms. Biological Cybernetics 73, 123–128 (1995)
474. Reimann, D., Haken, H.: An Approach to the Solution of Correspondence Problems by Synergetic Computers. Neural Network World 3, 299–315 (1995)
475. Uhl, C., Friedrich, R., Haken, H.: Analysis of Spatiotemporal Signals of Complex-Systems. Physical Review E 51, 3890–3900 (1995)
476. Yigitbasi, S., et al.: Zur Modellierung von Ökosystemen unter Anwendung der Methoden der Synergetik. In: Gnauck, H., Frischmuth, A., Kraft, A. (eds.) Ökosysteme, Modellierung Modellierung und Simulation. Blottner, Taunusstein (1995)
477. Fuchs, A., et al.: Extending the HKB model of coordinated movement to oscillators with different eigenfrequencies. Biological Cybernetics 74, 21–30 (1996)
478. Gang, H., Daffertshofer, A., Haken, H.: Diffusion of periodically forced Brownian particles moving in space-periodic potentials. Physical Review Letters 76, 4874–4877 (1996)
479. Gang, H., Haken, H., Fagen, X.: Stochastic resonance with sensitive frequency dependence in globally coupled continuous systems. Physical Review Letters 77, 1925–1928 (1996)
480. Gang, H., Ning, C.Z., Haken, H.: Inverse problem with a dilated kernel containing different singularities. Physical Review E 54, 2384–2391 (1996)
481. Haken, H.: Slaving principle revisited. Physica D 97, 95–103 (1996)
482. Haken, H.: Noise in the brain, A physical network model. International Journal of Neural Systems 7, 551–557 (1996)
483. Haken, H.: Principles of Brain Functioning. A Synergetic Approach to Brain Activity, Behaviour and Cognition. Springer, Berlin (1996)
484. Haken, H.: Future Trends in Synergetics. In: Parisi, J. (ed.) Nonlinear Physics in Complex Systems, pp. 179–193. Springer, Berlin (1996)
485. Haken, H.: Synergetik und Naturwissenschaften (anschl. Kritiken von 33 Wissenschaftlern und Replik hierauf von H. Haken 658 - 675). Ethik und Naturwissenschaften 7, 587–594 (1996)
486. Haken, H.: Erfolgsgeheimnisse der visuellen Wahrnehmung. In: Bolz, N., Rueffer, U. (eds.) Das Große Stille Bild. Fink, München (1996)
487. Haken, H.: Chaos und Ordnung, Zur Selbstorganisation komplexer Systeme in Physik, Biologie und Soziologie. In: Janssen, H.-S. (ed.) Synergetik und Systeme im Sport, p. 26. Hofmann, Schorndorf (1996)
488. Haken, H.: Synergetik, Selbstorganisation in den Natur- und Geisteswissenschaften. In: Ernst-Blickle Preis 1995 (=Festreden anläßlich der Verleihung an Dr. F. Anistis). SEW-Eurodrive-Stiftung, Bruchsal (1996)
489. Haken, H., Pelster, A.: Über die Rolle der Symmetriebrechung bei der Selbstorganisation. In: Hahn, W., Weibel, S. (eds.) Evolutionäre Systemtheorie - Selbstorganisation und Dynamische Systeme (Symposium 1993 in Frankfurt/Main), pp. 121–132. Wissenschaftliche Verlagsgesellschaft, Stuttgart (1996)
490. Haken, H., et al.: A model for phase transitions in human hand movements during multifrequency tapping. Physica D 90, 179–196 (1996)
491. Haken, H., et al.: A model for phase transitions in human hand movements during multifrequency tapping. Physica D 92, 260 (1996)

492. Haken, H., Portugali, J.: Synergetics. Inter-representation Networks and Cognitive Maps. In: Portugali, J. (ed.) The Construction of Cognitve Maps, pp. 45–67. Kluwer Academic, Dordrecht (1996)
493. Haken, H., Wunderlin, A., Yigitbasi, S.: On the foundations of Synergetics. In: Weingartner, S., Schurz, G. (eds.) Law and Prediction in the Light of Chaos Research, pp. 243–279. Springer, Berlin (1996)
494. Jirsa, V.K., Haken, H.: Field theory of electromagnetic brain activity. Physical Review Letters 77, 960–963 (1996)
495. Jirsa, V.K., Haken, H.: Derivation of a field equation of brain activity. Journal of Biological Physics 22, 101–112 (1996)
496. Tass, S., Haken, H.: Synchronization in networks of limit cycle oscillators. Zeitschrift für Physik B 100, 303–320 (1996)
497. Tass, S., Haken, H.: Synchronized oscillations in the visual cortex - A synergetic model. Biological Cybernetics 74, 31–39 (1996)
498. Ditzinger, T., et al.: A synergetic model for the verbal transformation effect. Biological Cybernetics 77, 31–40 (1997)
499. Haken, H.: Visions of synergetics. International Journal of Bifurcation and Chaos 7, 1927–1951 (1997)
500. Haken, H.: Visions of synergetics. Journal of the Franklin Institute-Engineering and Applied Mathematics 334B, 759–792 (1997)
501. Haken, H.: Information und Bedeutung aus Sicht der Synergetik. In: Saltzer, W.G. (ed.) Die Erfindung des Universums? -Neue Überlegungen zur Philosophischen Kosmologie, pp. 168–177. Insel-Publisher, Frankfurt am Main (1997)
502. Haken, H.: Synergetics and Cybernetics. In: Encyclopedia of Applied Physics Bd., vol. 20. VCH Publishers, Weinheim (1997)
503. Haken, H.: Synergetics of the Brain. In: Arhem, S., Liljenströn, H. (eds.) Matter Matters? On the Material Basis of Cognitive Activity of Mind, pp. 145–176. Springer, Berlin (1997)
504. Haken, H.: Discrete Dynamics of Complex Systems. Discrete Dynamics in Nature and Society 1, 1–8 (1997)
505. Haken, H., Kein, I.Q.: für Computer. Bild der Wissenschaft 1997(10), 70–74 (1997)
506. Haken, H.: Synergetik, Chaostheorie und Komplexitätswissenschaft (Interview mit Florian Rötzer, Teleopolis (Artikel-) (1997), `http://www.heise.de/tp/r4/artikel/2/2109/2.html`
507. Haken, H., Haken-Krell, M.: Gehirn und Verhalten. Deutsche Verlagsanstalt, Stuttgart (1997)
508. Haken, H., Hönlinger, R., Vanger, S.: Anwendung der Synergetik bei der Erkennung von Emotionen im Gesichtsausdruck. In: Schiepek, G., Tschacher, W. (eds.) Selbstorganisation in Psychologie und Psychiatrie, pp. 85–101. Vieweg, Braunschweig (1997)
509. Jirsa, V.K., Haken, H.: A derivation of a macroscopic field theory of the brain from the quasi-microscopic neural dynamics. Physica D 99, 503–526 (1997)
510. Friedrich, R., et al.: Analyzing Spatio-Temporal patterns of complex systems. In: Kantz, H., Mayer-Kress, G. (eds.) Nonlinear Analysis of Physiological Data, pp. 101–116. Springer, Berlin (1998)
511. Haken, H.: Can we apply Synergetics to the human sciences? In: Altmann, G., Koch, W. (eds.) Systems - New Paradigms for the Human Sciences. Walter de Gruyter, Berlin (1998)

512. Haken, H.: Decision making and optimization in regional planning. In: Beckmann, J. (ed.) Knowledge and Networks in a Dynamic Economy, pp. 25–40. Springer, Berlin (1998)
513. Bonifacio, R., et al.: Coupled maps and scaling in high order harmonic generation. Laser Physics 9, 395–397 (1999)
514. Frank, T.D., et al.: Impacts of noise on a field theoretical model of the human brain. Physica D 127, 233–249 (1999)
515. Grigorieva, E.V., Haken, H., Kaschenko, S.A.: Theory of quasiperiodicity in model of lasers with delayed optoelectronic feedback. Optics Communications 165, 279–292 (1999)
516. Grigorieva, E.V., et al.: Travelling wave dynamics in a nonlinear interferometer with spatial field transformer in feedback. Physica D 125, 123–141 (1999)
517. Haken, H.: Visions of Synergetics. In: Huertas, J. (ed.) Visions of Nonlinear Science in the 21st Century. World Scientific, Singapore (1999)
518. Haken, H.: Über Beziehungen zwischen der Synergetik und anderen Disziplinen. In: Eisenhardt, S. (ed.) Der Weg der Wahrheit. Aufsätze zur Einheit der Wissenschaftsgeschichte, pp. 167–173. Olms, Hildesheim (1999)
519. Haken, H.: Synergetics and some applications to Psychology. In: Tschacher, W., Dauwalder, J.-S. (eds.) Dynamics, Synergetics, Autonomous Agents, pp. 3–12. World Scientific, Singapore (1999)
520. Haken, H.: Synergetik, Vergangenheit, Gegenwart, Zukunft. In: Mainzer, K. (ed.) Komplexe Systeme und Nichtlineare Dynamik in Natur und Gesellschaft. Komplexitätsforschung in Deutschland auf dem Weg ins nächste Jahrhundert, pp. 30–48. Springer, Berlin (1999)
521. Haken, H.: Cooperativity in Brain and Neural Function. In: Adelman, G., Smith, B. (eds.) Elsevier's Encyclopedia of Neuroscience, pp. 468–476. Elsevier Science, Amsterdam (1999)
522. Haken, H., Schanz, M., Starke, J.: Treatment of combinatorial optimization problems using selection equations with cost terms. Part I. Two-dimensional assignment problems. Physica D 134, 227–241 (1999)
523. Knyazewa, H., Haken, H.: Synergetics of Human Creativity. In: Tschacher, W., Dauwalder, J.-S. (eds.) Dynamics, Synergetics, Autonomous Agents, pp. 64–82. World Scientific, Singapore (1999)
524. Knyazewa, H., Haken, H.: Synergetik, zwischen Reduktionismus und Holismus. Philosophia Naturalis 37, 21–44 (1999)
525. Starke, J., Schanz, M., Haken, H.: Treatment of combinatorial optimization problems using selection equations with cost terms. Part II. NP-hard three-dimensional assignment problems. Physica D 134, 242–252 (1999)
526. Bestehorn, M., et al.: Order parameters for class-B lasers with a long time delayed feedback. Physica D 145, 110–129 (2000)
527. Frank, T.D., et al.: Towards a comprehensive theory of brain activity, Coupled oscillator systems under external forces. Physica D 144, 62–86 (2000)
528. Haken, H.: Effect of delay on phase locking in a pulse coupled neural network. European Physical Journal B 18, 545–550 (2000)
529. Haken, H.: Quasi-discrete dynamics of a neural net, The lighthouse model. Discrete Dynamics in Nature and Society 4, 187–200 (2000)
530. Haken, H.: Phase locking in the lighthouse model of a neural net with several delay times. Progress of Theoretical Physics Supplement (139), 96–111 (2000)

531. Haken, H.: Phase locking and noise in the lighthouse model of a neural net with delay. In: Broomhead, D. (ed.) Stochastic and Chaotic Dynamics in the Lakes, p. 69. American Institute of Physics, New York (2000)
532. Haken, H.: Phase locking in the lighthouse model of a neural net with several delay times. In: Robnik, M. (ed.) Let's Face Chaos Through Nonlinear Dynamics, pp. 96–111. Institute of Pure and Applied Physics, Kyoto (2000)
533. Haken, H.: Associative Memory of a Pulse Coupled Neural Network with Delays, The Lighthouse Model. In: Helbing, D. (ed.) Traffic and Granular Flow 1999. Social Traffic and Granular Dynamics, pp. 173–180. Springer, Heidelberg (2000)
534. Haken, H., Knyazeva, H.: Arbitrariness in Nature, Synergetics and the Evolutionary Laws of Prohibition. Journal for General Philosophy of Science 31, 57–73 (2000)
535. Stefanovska, A., et al.: Reversible transitions between synchronization states of the cardiorespiratory system. Physical Review Letters 85, 4831–4834 (2000)
536. Daffertshofer, A., Portugali, J., Haken, H.: Self-Organized Settlements. Environment and Planning B, Planning and Design 28, 89–102 (2001)
537. Frank, T.D., et al.: H-theorem for a mean field model describing coupled oscillator systems under external forces. Physica D 150, 219–236 (2001)
538. Haken, H.: Delay, noise and phase locking in pulse coupled neural networks. Biosystems 63, 15–20 (2001)
539. Stefanovska, A., et al.: The cardiovascular system as coupled oscillators? Physiological Measurement 22, 535–550 (2001)
540. Uhl, C., Friedrich, R., Haken, H.: A Synergetic Approach for the Analysis of Spatio-Temporal Signals. Nonlinear Phenomena in Complex Systems 4, 250–263 (2001)
541. Haken, H.: Phase-locking in a general class of integrate and fire models. International Journal of Bifurcation and Chaos 12, 2619–2623 (2002)
542. Haken, H.: Heisenberg's equations in laser theory. A historical overview. Fortschritte der Physik-Progress of Physics 50, 642–645 (2002)
543. Haken, H.: Brain Dynamics. Synchronization and Activity Patterns in Pulse-Coupled Neural Nets with Delay and Noise. Springer, Berlin (2002)
544. Haken, H.: Sind synergetische Systeme unsterblich? In: Wulf, C., Kamper, D. (eds.) Logik und Leidenschaft. Erträge Historischer Anthropologie, pp. 952–955. Dietrich Reimer Publisher, Berlin (2002)
545. Haken, H.: Die Selbstorganisationsgesellschaft. In: Bertelsmann-Stiftung (ed.) Was kommt nach der Informationsgesellschaft? 11 Antworten, pp. 152–173. Bertelsmann-Stiftung, Gütersloh (2002)
546. Haken, H.: Intelligent Behaviour - A Synergetic View. In: Tschacher, W., Dauwalder, J.-S. (eds.) The Dynamical Systems Approach to Cognition, pp. 3–16. World Scientific, Singapore (2003)
547. Haken, H.: Cooperative Phenomena. In: Arbib, M. (ed.) The Handbook of Brain Theory and Neural Networks (2. Auflage). MIT Press, Cambridge (2003)
548. Haken, H.: Synergetik der Gehirnfunktionen. In: Schiepek, G. (ed.) Neurobiologie der Psychotherapie, pp. 80–102. Schattauer, Stuttgart (2003)
549. Portugali, J., Haken, H.: The face of the city is its information. Journal of Environmental Psychology 23, 382–405 (2003)
550. Haken, H.: Quantum fluctuations of elementary excitations in discrete media. Discrete Dynamics in Nature and Society, 169–177 (2004)
551. Haken, H.: Noise and correlated transport in ion channels. Fluctuation and Noise Letters 4, L171–L178 (2004)
552. Haken, H.: Future trends in synergetics. Solid State Phenomena 97-98, 3–9 (2004)

553. Haken, H.: Ist der Mensch ein dynamisches System? In: van Schlippe, A., Kriz, W.C. (eds.) Personenzentrierung und Systemtheorie, pp. 68–77. Vandenhoek & Ruprecht, Göttingen (2004)
554. Haken, H.: Fluctuation in quantum devices. Condensed Matter Physics 7, 527–537 (2004)
555. Haken, H.: Laser Physics and the Brain, Are there Analogies? In: Becker, W., Fedorov, M.V. (eds.) Universality and Diversity in Science (=Festschrift in Honor of Naseem K. Rahmans 60th Birthday). World Scientific, New Jersey (2004)
556. Haken, H.: Nonlinearity and Beauty, A Synergetic Approach. Un approccio sinergetico alla nonlinearita e al bello. Accademia di Science e Lettere. Incontro di Studio 26, 85–102 (2004)
557. Haken, H.: Synergetik. In: Lexikon der Biologie. Spektrum Akademischer Verlag, Weinheim (2004)
558. Haken, H.: Synergetics. Introduction and Advanced Topics. Springer, Berlin (2004)
559. Haken, H.: Die Selbstorganisation Komplexer Systeme - Ergebnisse aus der Chaostheorie. Picus Publisher, Wien (2004)
560. Haken, H., Stadler, M.A.: Gestalt Phenomena. In: Scott, A.C. (hrsg.) Handbook of Nonlinear Science. Fitzroy Dearborn, London (2004)
561. Hansch, D., Haken, H.: Wie die Psyche sich selbst in Ordnung bringt. Psychologie Heute 7, 36–41 (2004)
562. Hansch, D., Haken, H.: Zur theoretischen Fundierung einer integrativen und salutogenetisch orientierten Psychosomatik. Gestalt Theory 26, 7–34 (2004)
563. Haken, H.: Synchronization and pattern recognition in a pulse-coupled neural net. Physica D 205, 1–6 (2005)
564. Haken, H.: Nel Senso della Sinergetica (Autobiografie in italienischer Sprache). Renzo Editore, Rom (2005)
565. Haken, H.: Von der Laser-Metaphorik zum Selbstorganisationskonzept im Management. In: Krieg, W., Galler, K., Stadelmann, S. (eds.) Richtiges und Gutes Management, vom System zur Praxis. Haupt Publisher, Bern (2005)
566. Haken, H.: Mesoscopic Levels in Science - Some Comments. In: Liljenström, H., Svedin, U. (eds.) Micro, Meso, Macro, Adressing Complex Systems Couplings, pp. 19–24. World Scientific, Singapore (2005)
567. Haken, H.: Synergetics, from Physics to Economics. In: Dopfer, K. (ed.) The Evolutionary Foundations of Economics, pp. 70–85. Cambridge University Press, Cambridge (2005)
568. Haken, H., et al.: Synergetics of Perception and Conciousness. Gestalt Theory 27, 8–28 (2005)
569. Portugali, J., Haken, H.: A synergetic interpretation of cue-dependent prospective memory. Cognitive Processing 6, 87–97 (2005)
570. Haken, H.: Pattern recognition and synchronization in pulse-coupled neural networks. Nonlinear Dynamics 44, 269–276 (2006)
571. Haken, H.: Synergetics of brain function. International Journal of Psychophysiology 60, 110–124 (2006)
572. Haken, H.: Beyond Attractor Neural Networks for Pattern Recognition. Nonlinear Phenomena in Complex Systems 9, 163–172 (2006)
573. Haken, H.: A Coherent Walk in Solid State Physics. In: Hyland, G., Rowland, S. (eds.) Herbert Fröhlich, FRS - A Physicist Ahead of his Time, pp. 1–6. University of Liverpool, Liverpool (2006)
574. Haken, H.: Some thoughts on Modelling of Brain Function (2006)

575. Haken, H.: Recognition of Natural and Artificial Environments by Computers, Commonalities and Differences. In: Portugali, J. (ed.) Complex and Artificial Environments, pp. 31–48. Springer, Berlin (2006)
576. Haken, H.: The interdepence between Shannonian and Semantic information (2006)
577. Haken, H.: Synergetics on its way to the Life Sciences. In: Nowak, M. (ed.) Complexus Mundi. Emergent Patterns in Nature, pp. 155–170. World Scientific, Singapore (2006)
578. Haken, H., Schiepek, G.: Synergetik in der Psychologie. In: Selbstorganisation Verstehen und Gestalten. Hogrefe Publisher, Göttingen (2006)
579. Schiepek, G., Haken, H.: Handeln und Entscheiden in komplexen Systemen. In: Daisnion-ji, Leibniz-Gemeinschaft (eds.) 2. Symposium zur Gründung einer Deutsch-Japanischen Akademie für integrative Wissenschaft, pp. 93–118. J.H. Röll Publisher, Dettelbach am Main (2006)
580. Haken, H.: Towards a unifying model of neural net activity in the visual cortex. Cognitive Neurodynamics 1, 15–25 (2007)
581. Haken, H.: Das Gehirn als Prüfstein - Vinzenz Schönfelder im Gespräch mit Hermann Haken. Gehirn & Geist 2007(10), 60–62 (2007)
582. Haken, H.: Bemerkungen zum Verhältnis zwischen Komplexitätsforschung und Synergetik. In: Leiber, T. (ed.) Dynamisches Denken und Handeln. Philosophie und Wissenschaft in einer komplexen Welt, pp. 27–30. Hirzel, Stuttgart (2007)
583. Haken, H.: Intentionality in non-equilibrium systems? The functional aspects of self-organized pattern formation. New Ideas in Psychology 25, 1–15 (2007)
584. Haken, H.: Schönheit aus einem Haufen Erde. In: Sandhoff, K., Donner, W. (eds.) Vom Urknall zum Bewußtsein - Selbstorganisation der Materie (=124. Versammlung der GNDÄ 2006), pp. 47–56. Thieme Publisher, Stuttgart (2007)
585. Haken, H.: Der menschliche Wille, eine Perspektive der Synergetik. In: Petzold, H., Sieper, J. (eds.) Der Wille, die Neurobiologie und die Psychotherapie. Aisthesis Publisher, Bielefeld (2008)
586. Haken, H.: Selbstorganisation in physikalischen Systemen. In: Breuninger, R. (ed.) Selbstorganisation. Humboldt Studienzentrum Universität Ulm, Ulm (2008)
587. Haken, H.: Synergetics, Basic Concepts. In: Meyers, R.A. (ed.) Encyclopedia of Complexity and Systems Science. Springer, Berlin (2009)
588. Haken, H., Levi, S.: Synergetic Agents, From Multi Robot-Systems to Molecular Robotics. Wiley-VCH, Weinheim (2012)

Appendix 2
List of Hermann Haken's Students and Their Diploma Thesis and Dissertations

1. Ankele, Lucas, 'Modellierung von Ordnungs-Unordnungsübergängen in selbstorganisierten Systemen mit dem Ordnungsparameter-Konzept ' (Diploma thesis, University Stuttgart, 1983).
2. Arzt, Volker, '(Title could not be identified)' (Diploma thesis, University Stuttgart, 1970).
3. Beckert, Stephan, 'Modell eines Lasers zur Mustererkennung ' (Diploma thesis, 1988).
4. Beckert, Stephan, 'Modenselektion beim Laser zur Realisierung des Synergetischen Computers ' (Dissertation, University Stuttgart, 1994).
5. Benk, Hartmut, 'Wechselwirkung von Kohärenten Frenkel-Exzitonen mit Störstellen in organischen Molekülkristallen ' (Diploma thesis, University Stuttgart, 1975).
6. Benk, Hartmut, 'Theorie zur Wechselwirkung von Frenkel-Exzitonen mit Störstellen in organischen Molekülkristallen, Kinetik der Exzitonenenergieübertragung ' (Dissertation, University Stuttgart, 1982).
7. Berding, Christoph, 'Die Entwicklung raumzeitlicher Strukturen in der Morphogenese ' (Diploma thesis, University Stuttgart, 1981).
8. Berding, Christoph, 'Zur theoretisch-physikalischen Behandlung von Nichtgleichgewichts-Phasenübergängen: die Entwicklung zeitlicher und räumlicher Strukturen in biologischen Systemen ' (Dissertation, University Stuttgart, 1985).
9. Bestehorn, Michael, 'Musterbildung beim Bénard-Problem der Hydrodynamik ' (Diploma thesis, University Stuttgart, 1983).
10. Bestehorn, Michael, 'Verallgemeinerte Ginzburg-Landau-Gleichungen für die Musterbildung bei Konvektions-Instabilitäten ' (Dissertation, University Stuttgart, 1988).
11. Beutelschieß, Jürgen, 'Theorie einer Hierarchie von Plasmainstabilitäten ' (Diploma thesis, University Stuttgart, 1983).
12. Beutelschieß, Jürgen, 'Eine nichtlineare Theorie zur Entstehung von laufenden Schichten in einer Gasentladung ' (Dissertation, University Stuttgart, 1987).
13. Beyer, Jens, 'Die Faradayinstabilität der Hydrodynamik ' (Diploma thesis, University Stuttgart, 1993).
14. Borland, Lisa Marina, 'Ein Verfahren zur Bestimmung der Dynamik stochastischer Prozesse ' (Dissertation, University Stuttgart, 1993).

15. Bunz, Herbert, 'Zur Dynamik nichtlinearer angetriebener Oszillatoren ' (Dissertation, University Stuttgart, 1987).
16. Daffertshofer, Andreas, 'Erkennung deformierter Muster mittels einer Potentialdynamik ' (Diploma thesis, University Stuttgart, 1992).
17. Daffertshofer, Andreas, 'Nichtgleichgewichtsphasenübergänge in der menschlichen Motorik und Erweiterungen des synergetischen Computers ' (Dissertation, University Stuttgart, 1996).
18. Ditzinger, Thomas, 'Oszillationen bei der Erkennung ambivalenter Muster ' (Diploma thesis, University Stuttgart, 1989).
19. Ditzinger, Thomas, 'Multistabilität in der Wahrnehmung mit einem Synergetischen Computer ' (Dissertation, University Stuttgart, 1993).
20. Egler, Wolfgang, 'Theorie der Superradiance mit zwei Photonen ' (Diploma thesis, University Stuttgart, 1972).
21. Egler, Wolfgang, 'Theorie der Dämpfung und Fluktuationen des Polaritons aufgrund seiner Wechselwirkung mit Gitterschwingungen ' (Dissertation, University Stuttgart, 1976).
22. Erhardt, Manfred, 'Computersimulation für lernfähige biomechanische Modelle ' (Diploma thesis, University Stuttgart, 1992).
23. Fantz, Marc, 'Strukturbildung beim rotierenden Bénard Problem der Flüssigkeitsdynamik ' (Diploma thesis, University Stuttgart, 1991).
24. Fischer, Eckart, 'Bestimmung einer Fokker-Planck-Gleichung aus experimentell gegebenen Zeitserien ' (Diploma thesis, University Stuttgart, 1993).
25. Fischer, Eckart, 'Ein synergetisches Modell der Datenspeicherung in Molekülschichten auf Halbleiterschichten ' (Dissertation, University Stuttgart, 1997).
26. Fischer, Karsten, 'Gitterdynamik und anharmonische Wechselwirkung in Silberchlorid-Kristallen ' (Dissertation, University Stuttgart, 1974).
27. Forster, Dieter, 'Theorie der Photon-Photon-Streuung in Kristallen ' (Diploma thesis, TH Stuttgart, 1964).
28. Frank, Till, 'Analyse von MEG-Daten ' (Diploma thesis, University Stuttgart, 1996).
29. Friedmann, Alexander, 'Zur Theorie der Fluktuationen beim elektronischen Transport im Festkörper ' (Dissertation, University Stuttgart, 1970).
30. Friedrich, Rudolf, 'Höhere Instabilitäten beim Taylor-Problem der Flüssigkeitsdynamik ' (Diploma thesis, University Stuttgart, 1982).
31. Friedrich, Rudolf, 'Stationäre, wellenartige und chaotische Konvektion in Geometrien mit Kugelsymmetrie ' (Dissertation, University Stuttgart, 1987).
32. Fuchs, Armin, 'Synergetische Systeme zur Mustererkennung und zur phänomenologischen Modellierung raum-zeitlich aufgelöst gemessener EEG's ' (Dissertation, University Stuttgart, 1990).
33. Geffers, Horst, 'Theorie erzwungener und selbsterregter Laserimpulse ' (Dissertation, University Stuttgart, 1974).
34. Goll, Joachim, 'Exzitonen-Materie ' (Diploma thesis, University Stuttgart, 1972).
35. Goll, Joachim, 'Theorie der Wannier-Exzitonen hoher Dichte und ihrer kohärenten Wechselwirkung mit dem Lichtfeld ' (Dissertation1976).
36. Graham, Robert, 'Die Quantenfluktuationen des parametrischen Oszillators' (Diploma thesis, TH Stuttgart, 1967).
37. Graham, Robert, 'Quantentheorie der Lichtausbreitung in laser-aktiven fluktuierenden Medien' (Dissertation, University Stuttgart, 1969).

38. Grauer, Thomas, 'Zur Entstehung räumlicher und zeitlicher Ordnung auf Fest-Flüssig-Grenzflächen. Eine Anwendung der Methode der verallgemeinerten Ginzburg-Landau Gleichungen ' (Dissertation, University Stuttgart, 1987).
39. Grob, Karl, 'Theorie des stimulierten Raman-Effektes. Zs. für Physik 184 (1965). 395-432 ' (Dissertation, TH Stuttgart, 1965).
40. Haas, Richard, 'Lernvorgänge bei synergetischen Computern ' (Diploma thesis, University Stuttgart, 1989).
41. Haas, Richard, 'Bewegungserkennung und Bewegungsanalyse mit dem synergetischen Computer ' (Dissertation, University Stuttgart, 1995).
42. Hanisch, Dietmar, 'Lasertätigkeit von Exzitonen ' (Diploma thesis, University Stuttgart, 1974).
43. Hanisch, Gerhard, 'Zur Theorie des Exzitons in den Alkalihalogeniden ' (Dissertation, TH Stuttgart, 1966).
44. Haubs, Georg, 'Methoden zur Beschreibung dynamischer Systeme und ihre Anwendung auf gekoppelte Josephson-Kontakte ' (Dissertation, University Stuttgart, 1986).
45. Haug, Hartmut, 'Zur Theorie der Linienform von Exzitonenabsorptionsspektren ' (Diploma thesis, TH Stuttgart, 1963).
46. Haug, Hartmut, 'Multimode-Eigenschaften verschiedener Haibleiterlasermodelle ' (Dissertation, TH Stuttgart, 1966).
47. Hauger, Wolfgang, 'Anwendung der verallgemeinerten Ginzburg-Landau Gleichung auf die Theorie der Lasertätigkeit von Atomen und Exzitonen ' (Dissertation, University Stuttgart, 1977).
48. Helm, Meinhardt, 'Dynamische Kopplung der Amplitudenfluktuationen von Laserschwingungen ' (Diploma thesis, TH Stuttgart, 1966).
49. Hofelich, Friedmar, 'Die Bewegung eines Exzitons entlang eines Polymers unter dem Einfluß der Gitterschwingung ' (Dissertation, TH Stuttgart, 1966).
50. Hölle, Bernd, 'Erkennung dreidimensionaler Körper aus einer ebenen Projektion mittels einer Potentialdynamik ' (Dissertation, University Stuttgart, 1991).
51. Hong, Fu, 'Instabilität, Selbstpulsation und multichromatische Lasertätigkeit im Bandmodell für Farbstofflaser ' (Dissertation, University Stuttgart, 1989).
52. Hönlinger, Robert, 'Erkennung von deformierten oder verrauschten Mustern als Dynamik in einem synergetischen System ' (Diploma thesis, University Stuttgart, 1989).
53. Hönlinger, Robert, 'Anwendung des synergetischen Computers auf die Erkennung mimischer Ausdrücke ' (Dissertation, University Stuttgart, 1999).
54. Huber, Armin, 'Analyse nichtlinearer Modelle von Produktionsabläufen mit Methoden der Synergetik ' (Diploma thesis, University Stuttgart, 1996).
55. Hübner, Roland, 'Exzitonenspektrum in CU2O mit und ohne äußere Felder ' (Diploma thesis, TH Stuttgart, 1964).
56. Hübner, Roland, 'Eine Momentenmethode zur Berechnung der Gitterrelaxation in der Umgebung elektronischer Störstellen in Grund- und angeregtem Zustand mit Hilfe von Green'schen Funktionen ' (Dissertation, University Stuttgart, 1969).
57. Hutt, Axel, 'Untersuchung einer neuronalen Feldgleichung mittels eines MEG-Experiments am Menschen ' (Diploma thesis, University Stuttgart, 1997).
58. Hutt, Axel, 'Methoden zur Untersuchung der Dynamik raumzeitlicher Signale' (Dissertation, University Stuttgart, 2001).
59. Jirsa, Viktor, 'Ein theoretisches Modell für Phasenübergänge im menschlichen Gehirn ' (Diploma thesis, University Stuttgart, 1993).

60. Jirsa, Viktor, 'Modellierung und Rekonstruktion raumzeitlicher Dynamik im Gehirn ' (Dissertation, University Stuttgart, 1996).
61. Klenk, Herbert, 'Exzitonen hoher Dichte als weit vom thermischen Gleichgewicht entferntes System ' (Diploma thesis, University Stuttgart, 1975).
62. Klenk, Herbert, 'Anwendung der verallgemeinerten Ginzburg-Landau Gleichungen auf die Theorie der Bénard-Instabilität eines Plasmas ' (Dissertation, University Stuttgart, 1979).
63. Kuchelmeister, Anton, 'Quantentheorie der Lasertätigkeit von Exzitonen ' (Diploma thesis, University Stuttgart, 1974).
64. Kuchelmeister, Anton, 'Theorie der spontanen und induzierten Emission von Exzitonen hoher Dichte ' (Dissertation, University Stuttgart, 1980).
65. Kühne, Reinhart, 'Quantentheorie der spontanen Emission für den optischen parametrischen Oszillator ' (Diploma thesis, University Stuttgart, 1970).
66. Kühne, Reinhart, 'Mikroskopische Begründung der zeitabhängigen Ginzburg-Landau-Gleichung mit fluktuierenden Kräften' (Dissertation, University Stuttgart, 1974).
67. Lang, Manfred, 'Der Zusammenhang zwischen Exziton und Plasmon in Isolatoren über einen gemeinsamen Grundzustand ' (Diploma thesis, TH Stuttgart, 1964).
68. Lang, Manfred, 'Zur Wechselwirkung eines quantisierten Strahlungsfeldes mit einem Nichtmetall ' (Dissertation, University Stuttgart, 1968).
69. Lassag, Johannes, 'Zur Theorie kurzer Laserpulse in Drei-Niveau Ringlasern ' (Dissertation, University Stuttgart, 1986).
70. Leutz, Rudolf Konrad, ' Wechselwirkung von Frenkel-Exzitonen mit Gitterschwingungen ' (Diploma thesis, University Stuttgart, 1973).
71. Leutz, Rudolf Konrad, 'Der Einfluss von Mehrkörper-Potentialen und Polarisationseffekten auf Bildungsenergie und Bildungsvolumen atomarer Fehlstellen in lonenkristallen mit NaCI-Struktur ' (Dissertation, University Stuttgart, 1977).
72. Lorenz, Wolfgang, 'Nichtgleichgewichts-Phasenübergänge bei Bewegungs-Koordinationen ' (Diploma thesis, University Stuttgart, 1987).
73. Lorenz, Wolfgang, 'Der Blutkreislauf als synergetisches System: numerische Datenanalyse ' (Dissertation, University Stuttgart, 1994).
74. Marx, Klaus, 'Numerische Lösung der Instabilitäts-Hierarchie des Taylorproblems der Hydrodynamik ' (Diploma thesis, University Stuttgart, 1982).
75. Marx, Klaus, 'Analytische und numerische Behandlung der zweiten Instabilität beim Taylor-Problem der Flüssigkeitsdynamik ' (Dissertation, University Stuttgart, 1987).
76. Mayer-Kress, Gottfried, 'Zur Persistenz von Chaos und Ordnung in nichtlinearen dynamischen Systemen ' (Dissertation, University Stuttgart, 1983).
77. Meuth, Hermann, 'Die Wechselwirkung von Exzitonen und Plasmonen ' (Diploma thesis, University Stuttgart, 1975).
78. Ning, Cun-Zheng, 'Versklavungsprinzip und Normalformtheorie: Die Anwendung auf Instabilitäten beim Zwei-Photonen-Laser ' (Dissertation, University Stuttgart, 1991).
79. Nuoffer, Michael, 'Einwirkung von Rauschen auf Entstehung und Form von ultrakurzen Laserimpulsen ' (Diploma thesis, University Stuttgart, 1980).
80. Ohno, Herbert, 'Mastergleichung von Systemen ohne detailliertes Gleichgewicht ' (Diploma thesis, University Stuttgart, 1975).
81. Ohno, Herbert, 'Anwendung der verallgemeinerten Ginzburg-Landau-Gleichung auf die Theorie ultrakurzer Laserpulse ' (Dissertation, University Stuttgart, 1980).

82. Olbricht, Herbert, 'Die Bildung von raumzeitlicher Strukturen infolge von Mehrfachinstabilitäten' (Diploma thesis, University Stuttgart, 1977).
83. Ossig, Martin, 'Frequenz-Kopplung zwischen oszillierenden Neuronen ' (Diploma thesis, University Stuttgart, 1991).
84. Ossig, Martin, 'Kopplung synergetischer Computer und Modellierung neuronaler Synchronisationseffekte ' (Dissertation, University Stuttgart, 1996).
85. Pluschke, Werner, 'Bifurkationsphänomene bei gekoppelten Lasermoden ' (Diploma thesis, University Stuttgart, 1983).
86. Pluschke, Werner, 'Quasiperiodische Lösungen eines reversiblen Systems, die aus einem Fixpunkt abzweigen: der differenzierbare Fall ' (Dissertation, University Stuttgart, 1989).
87. Pohl, Dieter, 'Einige Untersuchungen über die Ausstrahlungseigenschaften von Festkörpern ' (Diploma thesis, TH Stuttgart, 1964).
88. Reidl, Jürgen, 'Analytische Behandlung nichtlinearer Wellengleichungen magnetischer Gehirnaktivität ' (Diploma thesis, University Stuttgart, 1996).
89. Reimann, Dirk, 'Theorie eines synergetischen Computers zur Tiefenwahrnehmung ' (Diploma thesis, University Stuttgart, 1991).
90. Reimann, Dirk, 'Anwendung der Synergetik auf Korrespondenzprobleme wie die Stereoskopie ' (Dissertation, University Stuttgart, 1995).
91. Reineker, Peter, 'Zur Theorie der Laserkaskaden in Gas- und Festkörperlasern ' (Diploma thesis, TH Stuttgart, 1966).
92. Renz, Wolfgang, 'Theorie der Strukturbildung von Spiralnebeln ' (Diploma thesis, University Stuttgart, 1978).
93. Rössler, Joachim, 'Statistische Behandlung der Nichtgleichgewichts-Supraleitung ' (Diploma thesis, University Stuttgart, 1977).
94. Rudershausen, R., 'Wellenausbreitung in einem laseraktiven Medium ' (Diploma thesis, University Stuttgart, 1969).
95. Sauermann, Herwig, 'Zur Theorie des Lasers ' (Diploma thesis, TH Stuttgart, 1964).
96. Sauermann, Herwig, 'Theorie der Dissipation und Fluktuationen in einem Zwei-Niveau Maser und ihre Anwendung auf den optischen Maser' (Dissertation, TH Stuttgart, 1965).
97. Schanz, Michael, 'Anwendung der Theorie des deterministischen Chaos auf die Analyse komplexer Systeme. Insbesondere physiologischer ' (Diploma thesis, University Stuttgart, 1989).
98. Schanz, Michael, 'Zur Analytik und Numerik zeitlich verzögerter synergetischer Systeme ' (Dissertation, University Stuttgart, 1997).
99. Schenzle, Axel, 'Theorie der Lichtausbreitung in parametrischen Oszillatoren ' (Diploma thesis, University Stuttgart, 1970).
100. Schenzle, Axel, 'Verallgemeinerte Langevin-Gleichungen und ihre Anwendung auf Festkörper-Probleme ' (Dissertation, University Stuttgart, 1974).
101. Schindel, Martin, 'Information und Informationsgewinn bei Nichtgleichgewichtsphasenübergängen ' (Diploma thesis, University Stuttgart, 1987).
102. Schindel, Martin, 'Theorie eines Halbleitersystems zur Realisierung der Ordnungsparameterdynamik eines Synergetischen Computers ' (Dissertation, University Stuttgart, 1993).
103. Schmid, Christhard, '(Title could not be identified)' (Diploma thesis, TH Stuttgart, 1965).

104. Schmid, Christhard, 'Kohärenz- und Sättigungsverhalten bei der Erzeugung einer zweiten Harmonischen in nichtlinearen Kristallen ' (Dissertation, University Stuttgart, 1969).
105. Schnaufer, Bernd, 'Anwendung der Methoden der Synergetik auf die Musterbildung in Flammen ' (Diploma thesis, University Stuttgart, 1984).
106. Schöner, Gregor, 'Das Versklavungsprinzip für stochastische Differentialgleichungen und Anwendungen der Stochastik auf synergetische Systeme ' (Dissertation, University Stuttgart, 1985).
107. Schulz, Claus-Dieter, 'Theorie eines Lasersystems zur Mustererkennung als optische Realisierung eines synergetischen Computers ' (Dissertation, University Stuttgart, 1992).
108. Schumm, Felix, 'Photonenemission rekombinierender Exzitonen bei gleichzeitiger Kollision und Anregung weiterer Exzitonen ' (Diploma thesis, University Stuttgart, 1972).
109. Schuppert, Andreas, 'Bifurkationen solitärer Wellen in nichtlinearen dispersiven Systemen ' (Diploma thesis, University Stuttgart, 1984).
110. Schwarzer, Elmar, 'Wechselwirkung von inkohärenten Exzitonen mit Störstellen und ihre mögliche Bedeutung als Schalter in der Biologie' (Diploma thesis, University Stuttgart, 1971).
111. Schwarzer, Elmar, 'Die kohärente und inkohärente Exzitonenbewegung und ihr Einfluss auf die optische Spektroskopie und die Protonenrelaxation ' (Dissertation, University Stuttgart, 1974).
112. Strobl, Gerd-Rüdiger, 'Kohärente und inkohärente Bewegung von Excitonen in Molekülkristallen ' (Diploma thesis, TH Stuttgart, 1966).
113. Tass, Peter, 'Kollektivoszillationen im visuellen Cortex der Katze: ein synergetisches Modell ' (Diploma thesis, University Stuttgart, 1991).
114. Tass, Peter, 'Synchronisierte Oszillationen im visuellen Cortex: ein synergetisches Modell ' (Dissertation, University Stuttgart, 1993).
115. Tautenhahn, Peter, 'Zur sensibilisierten Fluoreszenz im Isolatorkristall ' (Diploma thesis, TH Stuttgart, 1964).
116. Uhl, Christian, 'Analyse von EEGen mit Hilfe der Methoden der Synergetik ' (Diploma thesis, University Stuttgart, 1991).
117. Uhl, Christian, 'Analyse raumzeitlicher Daten strukturbildender Systeme ' (Dissertation, University Stuttgart, 1995).
118. Veil, Lutz-Bodo, 'Lokale Schwingungstypen im kubisch gestörten Debye-Kristall ' (Diploma thesis, TH Stuttgart, 1967).
119. Vollmer, Hans-Dieter, 'Optische Absorption und Energiedissipation von Fremdatomen in Kristallen ' (Diploma thesis, TH Stuttgart, 1967).
120. Vollmer, Hans-Dieter, 'Quantenstatistische Behandlung des Laserrauschens durch Quasiwahrscheinlichkeitsverteilungen mit Anwendung auf den Festkörperlaser ' (Dissertation, University Stuttgart, 1970).
121. Weberruß, Volker, 'Die Erkennung von Mustern bei Nichtgleichgewichts-Phasenübergängen mit Hilfe des Maximum-Informations-Entropie Prinzips ' (Diploma thesis, University Stuttgart, 1987).
122. Weberruß, Volker, 'Eine neue Methode zur Berechnung der Lagrangeschen Parameter des Maximum-Entropie-Prinzips, eine Anwendung auf den Laser und der Übergang zu Feymannschen Wegintegralen ' (Dissertation, University Stuttgart, 1992).

123. Weimer, Wolfgang, 'Theorie der Musterbildung in der kosmischen Massenverteilung ' (Diploma thesis, University Stuttgart, 1984).
124. Weimer, Wolfgang, ' Verallgemeinerte Ginzburg-Landau-Gleichungen für die Dynamik der atmosphärischen Zyklogenese ' (Dissertation, University Stuttgart, 1989).
125. Will, Thilo, 'Deformationsdynamiken zur Bildverarbeitung ' (Dissertation, University Stuttgart, 1992).
126. Wöhrstein, Hans-Georg, 'Theorie des Selbstpulsens in Halbleiterlasern ' (Diploma thesis, University Stuttgart, 1970).
127. Wöhrstein, Hans-Georg, 'Die Fokker-Planck Gleichung in der Nichtgleichgewichtsstatistik: Lösungsmethoden und Anwendungen in der Quantenoptik und Populationsdynamik ' (Dissertation, University Stuttgart, 1974).
128. Wunderlin, Arne, 'Die Behandlung von elektronischen Kollektivanregungen Im Festkörper mit Hilfe von Quasiwahrscheinlichkeitsverteilungen ' (Diploma thesis, University Stuttgart, 1971).
129. Wunderlin, Arne, 'Über statistische Methoden und ihre Anwendung auf Gleichgewichts- und Nichtgleichgewichtssysteme ' (Dissertation, University Stuttgart, 1975).
130. Zeile, Karl, 'Ein verallgemeinertes Variationsprinzip zur Berechnung Feynmanscher Wegintegrale mit spezieller Anwendung auf das Polaronproblem ' (Diploma thesis, TH Stuttgart, 1967).
131. Zeile, Karl, 'Resolventenentwicklungen zeitabhängiger Lösungen in Spin-Bosonen-Systemen ' (Dissertation, University Stuttgart, 1990).
132. Zorell, Johannes, 'Stochastische Modelle für Nichtgleichgewichtsphasenübergänge bei chemischen Reaktionen ' (Diploma thesis, University Stuttgart, 1976).
133. Zorell, Johannes, 'Chaotische und periodische Pulslösungen der halbklassischen Lasergleichungen ' (Dissertation, University Stuttgart, 1981).

Appendix 3
Springer Series in Synergetics: (Senior Editor: Hermann Haken)

1. Haken, H.: Synergetics - An Introduction, Haken, H. (ed.). Springer Series in Synergetics, vol. 1. Springer, Berlin (1977)
2. Haken, H.: Synergetics - A workshop, Haken, H. (ed.). Springer Series in Synergetics, vol. 2. Springer, Berlin (1977)
3. Pacault, A., Vidal, C.: Synergetics - Far from Equilibrium, Haken, H. (ed.). Springer Series in Synergetics, vol. 3. Springer, Berlin (1979)
4. Güttinger, W., Eickemeier, H.: Structural Stability in Physics, Haken, H. (ed.). Springer Series in Synergetics, vol. 4. Springer, Berlin (1979)
5. Haken, H.: Pattern Formation by Dynamic Systems and Pattern Recognition, Haken, H. (ed.). Springer Series in Synergetics, vol. 5. Springer, Berlin (1979)
6. Haken, H.: Dynamics of Synergetic Systems, Haken, H. (ed.). Springer Series in Synergetics, vol. 6. Springer, Berlin (1979)
7. Blumenfeld, L.A.: Problems of Biological Physics, Haken, H. (ed.). Springer Series in Synergetics, vol. 7. Springer, Berlin (1981)
8. Arnold, L., Lefever, R.: Stochastic Nonlinear Systems in Physics, Chemistry and Biology, Haken, H. (ed.). Springer Series in Synergetics, vol. 8. Springer, Berlin (1981)
9. Della Dora, J., Demongeot, J., Lacolle, B.: Numerical Methods in the Study of Critical Phenomena, Haken, H. (ed.). Springer Series in Synergetics, vol. 9. Springer, Berlin (1981)
10. Klimontovich, Y.L.: The Kinetic Theory of Electromagnetic Processes, Haken, H. (ed.). Springer Series in Synergetics, vol. 10. Springer, Berlin (1983)
11. Haken, H.: Chaos and Order in Nature, Haken, H. (ed.). Springer Series in Synergetics, vol. 11. Springer, Berlin (1981)
12. Vidal, C., Pacault, A.: Nonlinear Phenomena in Chemical Dynamics, Haken, H. (ed.). Springer Series in Synergetics, vol. 12. Springer, Berlin (1981)
13. Gardiner, C.W.: Handbook of Stochastic Methods, Haken, H. (ed.). Springer Series in Synergetics, vol. 13. Springer, Berlin (1983)
14. Weidlich, W., Haag, G.: Concepts and Models of a Quantitative Sociology, Haken, H. (ed.). Springer Series in Synergetics, vol. 14. Springer, Berlin (1983)

15. Horsthemke, W., Lefever, R.: Noise-Induced Transitions - Theory and Applications in Physics, Chemistry, and Biology, Haken, H. (ed.). Springer Series in Synergetics, vol. 15. Springer, Berlin (1984)
16. Blumenfeld, L.A.: Physics of Bioenergetic Processes, Haken, H. (ed.). Springer Series in Synergetics, vol. 16. Springer, Berlin (1983)
17. Haken, H.: Evolution of Order and Chaos in Physics, Chemistry and Biology, Haken, H. (ed.). Springer Series in Synergetics, vol. 17. Springer, Berlin (1982)
18. Risken, H.: The Fokker-Planck Equation, Haken, H. (ed.). Springer Series in Synergetics, vol. 18. Springer, Berlin (1984)
19. Kuramoto, Y.: Chemical Oscillations, Waves and Turbulence, Haken, H. (ed.). Springer Series in Synergetics, vol. 19. Springer, Berlin (1984)
20. Haken, H.: Advanced Synergetics, Haken, H. (ed.). Springer Series in Synergetics, vol. 20. Springer, Berlin (1983)
21. Schuster, P.: Stochastic Phenomena and Chaotic Behaviour in Complex Systems, Haken, H. (ed.). Springer Series in Synergetics, vol. 21. Springer, Berlin (1984)
22. Frehland, E.: Synergetics - From Microscopic to Macroscopic Order, Haken, H. (ed.). Springer Series in Synergetics, vol. 22. Springer, Berlin (1984)
23. Basar, E., Flohr, H., Haken, H., Mandell, A.J.: Synergetics of the Brain, Haken, H. (ed.). Springer Series in Synergetics, vol. 23. Springer, Berlin (1983)
24. Kuramoto, Y.: Chaos and Statistical Methods, Haken, H. (ed.). Springer Series in Synergetics, vol. 24. Springer, Berlin (1984)
25. Nicolis, J.S.: Dynamics of Hierarchical Systems - an evolutionary approach, Haken, H. (ed.). Springer Series in Synergetics, vol. 25. Springer, Berlin (1986)
26. Ulrich, H., Probst, G.J.B.: Self-Organization and Management of Social Systems, Haken, H. (ed.). Springer Series in Synergetics, vol. 26. Springer, Berlin (1984)
27. Vidal, C.: Non-Equilibrium Dynamics in Chemical Systems, Haken, H. (ed.). Springer Series in Synergetics, vol. 27. Springer, Berlin (1984)
28. Krinsky, V.I.: Self-Organization - Autowaves and Structures far from Equilibrium, Haken, H. (ed.). Springer Series in Synergetics, vol. 28. Springer, Berlin (1984)
29. Rensing, L., Jaeger, N.I.: Temporal Order, Haken, H. (ed.). Springer Series in Synergetics, vol. 29. Springer, Berlin (1985)
30. Takeno, S.: Dynamical Problems in Soliton Systems, Haken, H. (ed.). Springer Series in Synergetics, vol. 30. Springer, Berlin (1984)
31. Haken, H.: Complex Systems - Operational Approaches, Haken, H. (ed.). Springer Series in Synergetics, vol. 31. Springer, Berlin (1985)
32. Mayer-Kress, G.: Dimensions and Entropies in Chaotic Systems, Haken, H. (ed.). Springer Series in Synergetics, vol. 32. Springer, Berlin (1986)
33. Ebeling, W., Ulbricht, H.: Selforganization by Nonlinear Irreversible Processes, Haken, H. (ed.). Springer Series in Synergetics, vol. 33. Springer, Berlin (1986)
34. Arecchi, F.T., Harrison, R.G.: Instabilities and Chaos in Quantum Optics, Haken, H. (ed.). Springer Series in Synergetics, vol. 34. Springer, Berlin (1987)
35. Schöll, E.: Nonequilibrium Phase Transitions in Semiconductors, Haken, H. (ed.). Springer Series in Synergetics, vol. 35. Springer, Berlin (1987)
36. Rensing, L., der Heiden, U., Mackey, M.C.: Temporal Disorder in Human Oscillatory Systems, Haken, H. (ed.). Springer Series in Synergetics, vol. 36. Springer, Berlin (1987)
37. Güttinger, W., Dangelmayr, G.: The Physics of Structure Formation - Theory and Simulation, Haken, H. (ed.). Springer Series in Synergetics, vol. 37. Springer, Berlin (1987)

38. Haken, H.: Computational Systems - Natural and Artificial, Haken, H. (ed.). Springer Series in Synergetics, vol. 38. Springer, Berlin (1987)
39. Markus, M., Müller, S.C., Nicolis, G.: From Chemical to Biological Organization, Haken, H. (ed.). Springer Series in Synergetics, vol. 39. Springer, Berlin (1988)
40. Haken, H.: Information and Self-Organization, Haken, H. (ed.). Springer Series in Synergetics, vol. 40. Springer, Berlin (1988)
41. Wesfreid, J.E., Brnd, H.R., Manneville, Albinet, G., Boccara, N.: Propagation in Systems far from Equilibrium, Haken, H. (ed.). Springer Series in Synergetics, vol. 41. Springer, Berlin (1987)
42. Haken, H.: Neural and Synergetic Computers, Haken, H. (ed.). Springer Series in Synergetics, vol. 42. Springer, Berlin (1988)
43. Takayama, H.: Cooperative Dynamics in Complex Physical Systems, Haken, H. (ed.). Springer Series in Synergetics, vol. 43. Springer, Berlin (1989)
44. Plath, P.J.: Optimal Structures in Heterogeneous Systems, Haken, H. (ed.). Springer Series in Synergetics, vol. 44. Springer, Berlin (1989)
45. Haken, H., Stadler, M.: Synergetics of Cognition, Haken, H. (ed.). Springer Series in Synergetics, vol. 45. Springer, Berlin (1989)
46. Atlan, H., Cohen, I.R.: Theory of Immune Networks, Haken, H. (ed.). Springer Series in Synergetics, vol. 46. Springer, Berlin (1989)
47. Jumarie, G.: Relative Information Theories and Applications, Haken, H. (ed.). Springer Series in Synergetics, vol. 47. Springer, Berlin (1990)
48. Meinköhn, D.: Dissipative Structures in Transport Processes and Combustion, Haken, H. (ed.). Springer Series in Synergetics, vol. 48. Springer, Berlin (1990)
49. Krüger, J.: Neuronal Cooperativity, Haken, H. (ed.). Springer Series in Synergetics, vol. 49. Springer, Berlin (1991)
50. Haken, H.: Synergetic Computers and Cognition - a top down approach to neural nets, Haken, H. (ed.). Springer Series in Synergetics, vol. 50. Springer, Berlin (1991)
51. Mikhailov, A.S.: Foundations of Synergetics I, Haken, H. (ed.). Springer Series in Synergetics, vol. 51. Springer, Berlin (1990)
52. Haken, H.: Foundations of Synergetics II, Haken, H. (ed.). Springer Series in Synergetics, vol. 52. Springer, Berlin (1991)
53. Zhang, W.B.: Synergetic Economics, Haken, H. (ed.). Springer Series in Synergetics, vol. 53. Springer, Berlin (1991)
54. Haake, F.: Quantum Signature of Chaos, Haken, H. (ed.). Springer Series in Synergetics, vol. 54. Springer, Berlin (1991)
55. Haken, H., Koepchen, H.: Rhythms in Physiological Systems, Haken, H. (ed.). Springer Series in Synergetics, vol. 55. Springer, Berlin (1991)
56. Gardiner, C.W.: Quantum Noise, Haken, H. (ed.). Springer Series in Synergetics, vol. 56. Springer, Berlin (1991)
57. Stratonovich, R.: Nonlinear Nonequilibrium Thermodynamics I, Haken, H. (ed.). Springer Series in Synergetics, vol. 57. Springer, Berlin (1992)
58. Tschacher, W., Schiepek, G., Brunner, E.J.: Self-Organization and Clinical Psychology, Haken, H. (ed.). Springer Series in Synergetics, vol. 58. Springer, Berlin (1992)
59. Stratonovich, R.: Nonlinear Nonequilibrium Thermodynamics II, Haken, H. (ed.). Springer Series in Synergetics, vol. 59. Springer, Berlin (1994)
60. Kratsov, Y.A.: Limits of Predictability, Haken, H. (ed.). Springer Series in Synergetics, vol. 60. Springer, Berlin (1996)

61. Mishra, R.K., Maaß, D., Zwierlein, E.: On Self-Organization, Haken, H. (ed.). Springer Series in Synergetics, vol. 61. Springer, Berlin (1994)
62. Haken, H., Mikhailov, A.S.: Interdisciplinary Approaches to Nonlinear Complex Systems, Haken, H. (ed.). Springer Series in Synergetics, vol. 62. Springer, Berlin (1993)
63. Atmanspacher, H., Dalenoort, G.J.: Inside versus Outside, Haken, H. (ed.). Springer Series in Synergetics, vol. 63. Springer, Berlin (1994)
64. Kruse, Stadler, M.: Ambiguity in Mind and Nature - Multistable Cognitive Phenomena, Haken, H. (ed.). Springer Series in Synergetics, vol. 64. Springer, Berlin (1995)
65. Mosekilde, E., Mouritsen, O.: Modelling the Dynamics of Biological Systems, Haken, H. (ed.). Springer Series in Synergetics, vol. 65. Springer, Berlin (1995)
66. Vorontsov, M.A., Miller, W.B.: Self-Organization in Optical Systems and Applications in Information Technology, Haken, H. (ed.). Springer Series in Synergetics, vol. 66. Springer, Berlin (1995)
67. Haken, H.: Principles of Brain Functioning - A Synergetic Approach to Brain Activity, Behaviour and Cognition, Haken, H. (ed.). Springer Series in Synergtics, vol. 67. Springer, Berlin (1996)
68. Grabec, I., Sachse, W.: Synergetics of Measurement, Prediction and Control, Haken, H. (ed.). Springer Series in Synergetics, vol. 68. Springer, Berlin (1997)
69. Kratsov, Y.A., Kadtke, J.B.: Predictability of Complex Dynamical Systems, Haken, H. (ed.). Springer Series in Synergetics, vol. 69. Springer, Berlin (1996)
70. Xu, J.-J.: Interfacial Wave Theory of Pattern Formation, Haken, H. (ed.). Springer Series in Synergetics, vol. 70. Springer, Berlin (1998)
71. Awrejcewicz, J., Andrianov, I.V., Manevitch, L.I.: Asymptotic Approaches in Nonlinear Dynamics, Haken, H. (ed.). Springer Series in Synergetics, vol. 71. Springer, Berlin (1998)
72. Basar, E.: Brain Function and Oscillation I, Haken, H. (ed.). Springer Series in Synergetics, vol. 72. Springer, Berlin (1998)
73. Basar, E.: Brain Function and Oscillation II, Haken, H. (ed.). Springer Series in Synergetics, vol. 73. Springer, Berlin (1999)
74. Grasman, J., van Herwaarden, O.A.: Asymptotic Methods for the Fokker-Planck Equation and the Exit Problem in Applications, Haken, H. (ed.). Springer Series in Synergetics, vol. 74. Springer, Berlin (1999)
75. Uhl, C.: Analysis of Neurophysiological Brain Functioning, Haken, H. (ed.). Springer Series in Synergetics, vol. 75. Springer, Berlin (1999)
76. Tass, P.: Phase Resetting in Medicine and Biology, Haken, H. (ed.). Springer Series in Synergetics, vol. 76. Springer, Berlin (1999)
77. Portugali, J.: Self-Organization and the City, Haken, H. (ed.). Springer Series in Synergetics, vol. 77. Springer, Berlin (2000)

Appendix 4
Participants that Attended the ELMAU Conferences as Well as the Versailles Conferences

Name	Country	Versailles - Conferences	Elmau - Conferences
Adey, W. Ross	USA	1988	1983
Anderson, James A.	USA	1975, 1981	1989
Babloyantz, Agnessa	B	1973, 1979, 1981, 1988	1979, 1985
Bergé, Paul	F	1988	1981
Bienenstock, Elie L.	USA	1979	1983
Caianiello, Eduardo R.	I	1969, 1973	1988
Careri, Giorgio	I	1967, 1969, 1971, 1973,1975, 1977, 1979, 1981, 1984, 1988	1982
Cowan, Jack D.	USA	1967, 1971, 1973, 1975, 1977, 1979, 1981, 1984	1977, 1979
Eccles, John C.	GB	1973, 1975, 1977, 1979, 1981, 1984, 1988	1985

Eigen, Manfred	D	1967, 1969, 1971, 1973, 1979	1982
Frauenfelder, Hans	D oder CH	1977, 1979, 1981	1989
Fröhlich, Herbert	GB	1967, 1969, 1971, 1973, 1975, 1977, 1979, 1981, 1984, 1988	1977
Gierer, Alfred	D	1969, 1984	1979, 1982
Glaser, Donald	USA	1969, 1971, 1977, 1979, 1981, 1988	1983
Glass, Leon	CAN	1975	1979
Hepp, Klaus	CH	1971, 1973, 1979	1983, 1985
Julesz, Béla	USA	1971, 1981	1972, 1989
Kohonen, Teuvo	FIN	1979, 1981, 1984	1979, 1983
Name	Country	Versailles - Conferences	Elmau - Conferences
Kubo, Ryoko	JPN	1967, 1969, 1971, 1979	1972
Lefever, René	B	1969, 1973, 1975, 1979	1972, 1979
Levin, Simon A.	USA	1979	1979
Libchaber, Albert J.	USA	1988	1981
Malsburg, Christoph von	D	1975, 1979, 1981, 1984, 1988	1983
Matsubara, Takeo	JPN	1971, 1973	1972

Meinhardt, Hans	D	1975	1977, 1979
Montroll, Elliot W.	USA	1973	1972
Ortoleva, Peter	USA	1979	1979
Pomeau, Yves	F	1979	1981, 1983
Reichardt, Werner	D	1967, 1981, 1988	1972, 1979, 1985, 1988
Rössler, Otto	D	1979	1977, 1979
Schuster, Peter	AU	1979, 1981	1979,1983,1985
Shimizu, H.	JPN	1981	1982,1985,
Singer, Wolf	D	1984, 1988	1983, 1989
Tomita, Kazuhisa	JPN	1979	1972,1979
Velarde, Manuel G.	ESP	1973	1982, 1987
Wagner, Max	D	1969, 1971, 1973, 1975, 1977, 1979, 1981, 1984	1972
Wilson, Hugh R.	USA	1975, 1979, 1981	1972

Appendix 5
List of Honours Bestowed on Hermann Haken

1976 Max Born Prize of the British Institute of Physics and the German Physical Society.
1981 Albert A. Michelson Medal of the Franklin Institute, Philadelphia.
1982 Honorary Doctorate University of Essen (Germany), Dr. h.c.
1982 Member of the German Academy of Natural Scientists Leopoldina (Deutsche Akademie der Naturforscher Leopoldina).
1982 Corresponding Member of the Bavarian Academy of Sciences.
1984 Member of the Order Pour Le Mérite for the Sciences and the Arts.
1985 Honorary Member of the Polish Synergetics Society.
1986 Honorary Professor of the Shanghai Institute of Mechanical Engineering.
1986 Honorary Member of the Shanghai Association of Systems Engineering.
1986 Honorary Professor of the Northwestern University Xian.
1986 Honorary Member of the Chinese Society for Systems Engineering.
1986 Großes Verdienstkreuz mit Stern of the Order of Merit of the Federal Republic of Germany.
1987 Honorary Doctorate of the Universidad Nacional de Educacion a Distancia Madrid Dr. h.c.
1988 Member of the Scientific Society of Braunschweig.
1990 Max Planck Medal of the German Physical Society.
1990 Foreign Member of the Scientific Academy of the German Democratic Republic.
1990 Full Member of the Scientific Academy of Heidelberg.
1991 Full Member of the Academia Scientiarum et Artium Europaea (Salzburg).
1991 Member of the Academia Europaea, London.
1991 Speaker ("Obmann") of the Physics Section (Theoretical Discipline) of the German Academy of Natural Scientists Leopoldina.
1992 Golden Badge of Honour of the Town of Sindelfingen.
1992 Honorary Doctorate of the Florida Atlantic University, Boca Raton (USA) Dr. h.c.
1992 Winner of the Honda Prize.
1993 Arthur-Burkhardt-Prize.
1994 Honorary Doctorate of the Universität Regensburg Dr. h.c.

1994 Lorenz-Oken Medal of the German Society of Natural Researchers and Physicians (Gesellschaft Deutscher Naturforscher und Ärzte).
1996 Honorary Chairman of the German Society of Complex Systems and Nonlinear Dynamics.
1997 Honorary Doctorate of the Technical University München Dr. phil. e.h.
1997 Yunnan University (China): Honorary President of the Haken Synergetic Institute.
1997 International Biographical Centre, Cambridge (GB): Medal "2000 Outstanding People of the 20th Century".
2002 Honorary Senator of the University of Maribor, Slowenia.
2005 Prize given for outstanding merits on the development of medicine and psychology. Donau University Krems, Austria.

Acknowledgement

This work could not have been written without the help of many other people. I take the opportunity to thank Klaus Hentschel who aroused my interest to the subject. It is a special pleasure to thank Hermann Haken for his commitment to the project. He gave me the right to inspect his personal archives and answered all my questions in several interview sessions. I am also grateful to co-workers, friends and students of Hermann Haken who volunteered for interviews: Manfred Eigen and Ruthild Winkler-Oswatitsch, Rudolf Friedrich, Robert Graham, Friedrich Haake, August Nitschke, Herbert Ohno and Wolfgang Weidlich.

Ulrike Bischler was so kind and let me inspect the few „surviving" records from the Volkswagenwerk Foundation concerning its former priority program Synergetics. Norbert Becker advised me in cataloguing the Hermann Haken Archive and made possible the inspection of Haken's personal files from the University of Stuttgart. Axel Pelster invited me to the symposium in honour of Hermann Haken's 85th birthday, where I had the chance to meet many of his former students and colleagues.

Even in times of the internet: without the help and efforts of the staff of the University libraries of Stuttgart and Tuebingen and of the Wuerttembergische Landesbibliothek Stuttgart many resources would not have been accessible. It is a pleasure to thank all of them.

Last, not least, I would like to thank my wife and my daughter for their continuous support.

Name Index

MIX
Papier aus verantwortungsvollen Quellen
Paper from responsible sources
FSC® C105338

FSC
www.fsc.org

If you have any concerns about our products,
you can contact us on
ProductSafety@springernature.com

In case Publisher is established outside the EU,
the EU authorized representative is:
Springer Nature Customer Service Center GmbH
Europaplatz 3, 69115 Heidelberg, Germany

Printed by Libri Plureos GmbH
in Hamburg, Germany